Key Technologies on New Energy Vehicles

Key Technologies on New Energy Vehicles publishes the latest developments in new energy vehicles - quickly, informally and with high quality. The intent is to cover all the main branches of new energy vehicles, both theoretical and applied, including but not limited:

- Control Technology of Driving System
- Hybrid Electric Vehicle Coupling Technology
- Cross Disciplinary design optimization technology
- Single and Group Battery Technology
- Energy Management Technology
- Lightweight Technology
- New Energy Materials and Device
- Internet of Things (IoT)
- Cloud Computing
- 3D Printing
- Virtual Reality Technologies

Within the scopes of the series are monographs, professional books or graduate textbooks, edited volumes, and reference works purposely devoted to support education in related areas at the graduate and post-graduate levels.

To submit a proposal or request further information, please contact:
Dr. Mengchu Huang, Senior Editor, Applied Sciences
Email: mengchu.huang@springer.com
Tel: +86-21-2422 5094

More information about this series at https://link.springer.com/bookseries/16377

Junqiu Li

Modeling and Simulation of Lithium-ion Power Battery Thermal Management

机械工业出版社
CHINA MACHINE PRESS

Springer

Junqiu Li
School of Mechanical Engineering
Beijing Institute of Technology
Beijing, China

ISSN 2662-2920 ISSN 2662-2939 (electronic)
Key Technologies on New Energy Vehicles
ISBN 978-981-19-0843-9 ISBN 978-981-19-0844-6 (eBook)
https://doi.org/10.1007/978-981-19-0844-6

Jointly published with China Machine Press
The print edition is not for sale in China (Mainland). Customers from China (Mainland) please order the print book from: China Machine Press.
ISBN of the Co-Publisher's edition: 978-7-111-67843-4

Translation from the Chinese Simplified language edition: "锂离子动力蓄电池热管理技术" by Junqiu Li, © China Machine Press 2021. Published by China Machine Press. All Rights Reserved.

This Springer imprint is published by the registered company Springer Nature Singapore Pte Ltd.
The registered company address is: 152 Beach Road, #21-01/04 Gateway East, Singapore 189721, Singapore

Preface

Nowadays, the world is still relying heavily on fossil fuels such as oil and coal to meet its energy needs. Under the pressure of increasingly serious environmental problems, the change in energy structure has become an inevitable trend. The development of energy-efficient and new energy vehicles is an important part of this trend and has become a major consensus on the long-term development of automobiles. Under the active guidance of national policies, China's new energy vehicle market has developed rapidly, with the market penetration rate growing from 0.3% in 2011 to over 4% in 2018. According to the goal proposed in the New Energy Vehicle Industry Development Plan (2021–2035), the sales volume of new energy vehicles will account for about 25% by 2025.

Power battery is an important source of energy for new energy vehicles, and its efficient and stable operation is the key to ensure the performance of new energy vehicles. With the advantages of high energy density, high power density and long cycle life, lithium-ion batteries are currently the first choice for automotive power batteries. However, since the suitable working temperature range of lithium-ion batteries is relatively narrow (usually 10–30 °C), while the temperature range of new energy vehicles in application scenarios is much wider, the requirements for the adaptability of power batteries to high- and low-temperature environments are also more demanding with the widespread popularity of new energy vehicles. To solve the conflict between battery temperature characteristics and application requirements, battery thermal management requires tasks such as heat dissipation, heating and temperature difference control.

Excessive temperature will accelerate the occurrence of battery side reactions and accelerate battery aging, which will seriously affect the service life of the battery. The trend of large cells leads to a decrease in the ratio of cell surface area to volume, which makes it difficult to dissipate the heat inside the battery. When the heat dissipation conditions are poor, the accumulation of heat will cause the battery temperature to rise sharply, increasing the risk of thermal run-away. In addition, as the demand for fast charging continues to grow, high rate charging has become a trend, which undoubtedly places higher demands on the heat dissipation efficiency of the battery system. Therefore, how to achieve efficient and uniform heat dissipation and avoid the high-temperature operation of the battery has always been the focus of battery thermal management research.

Under low-temperature conditions, the capacity and charge/discharge power of lithium-ion batteries will be greatly reduced, which greatly limits their application in alpine regions. To solve the limitations on the use of lithium-ion batteries at low temperature, heating and insulation of batteries are also crucial. Similarly, how to achieve fast, uniform, safe and non-damaging battery heating has become a hot spot in current thermal management research.

This book discusses the principles and methods related to the thermal management of lithium-ion batteries on the basis of the author's research practice and in combination with the research progress both in China and foreign countries. It summarizes the research status of battery thermal management, analyzes and introduces the temperature characteristics and electro-thermal coupling modeling methods of lithium-ion batteries and focuses on the modeling and simulation analysis of air cooling, liquid cooling, PTC external heating, wide wire metal film external heating and sinusoidal AC internal heating of the batteries.

The authors of this book try to dedicate the up-to-date research and development results in the field of power battery thermal management at home and abroad, as well as the research results and experience of the National Engineering Laboratory for Electric Vehicles of Beijing Institute of Technology to promoting the research on the power battery thermal management of new energy vehicles. However, this book does not cover all the knowledge related to thermal management of power battery systems due to the limited ability of the authors, and we hope that the readers will actively offer criticism and propose corrective comments on the revision and improvement of this book.

The authors would like to express their sincere gratitude to the industry colleagues who have given their support to the publication and distribution of this book!

Beijing, China Junqiu Li

Contents

Chapter 1
Current Research on Power Battery Thermal Management

1.1 New-Energy Vehicles and Power Batteries

Since the twenty-first century, with the rapid social and technological progress, issues of energy and environment have become increasingly prominent. With the continuous increase of car parc, vehicle emissions and corresponding energy consumption have gradually become part of the environment and energy issues that cannot be ignored. Battery electric vehicles are considered among the most suitable and promising vehicles for the future society because of their advantages of zero emission, zero pollution and low noise.

In order to promote the development of new energy vehicles, different countries have launched relevant policies. The United States already invested US $2.5 billion to support the development of the electric vehicle industry in 2009, and planned to achieve the goal of 1 million electric vehicles on the road by 2015 (Brown et al. 2010). In September 2018, the car parc of new energy vehicles (including plug-in hybrid and battery electric vehicles) in the United States exceeded 1 million. In addition, the US government promotes the sales of new energy vehicles by reducing the purchase tax of new energy vehicles and encouraging the government at all levels to purchase new energy vehicles. At the same time, it further promotes the development of the industry of new energy vehicles through extensive construction of new-energy infrastructure and other projects. As early as 2006, Japan promulgated a new national energy strategy, which reduced its dependence on oil by developing new energy vehicles, and launched a number of incentives and preferential policies to promote the development of new energy vehicles. In Europe, Germany has implemented preferential subsidy policy for alternative fuels; France has invested 400 million euros in clean energy vehicle projects; the British government has implemented a car ownership tax system to fund low-carbon vehicle projects (Su 2019).

The Chinese government has launched a number of supporting policies to accelerate the development of new energy vehicles, so as to narrow the gap with Germany, Japan, the United States and other traditional powers of the automotive industry. As

J. Li, *Modeling and Simulation of Lithium-ion Power Battery Thermal Management*,
Key Technologies on New Energy Vehicles,
https://doi.org/10.1007/978-981-19-0844-6_1

early as 2009, China's Ministry of Science and Technology and Ministry of Finance jointly launched the "Ten Cities, Thousand Vehicles" demonstration project of electric vehicles, and launched more than 100 hybrid buses in 13 cities including Beijing and Shanghai for the demonstration and promotion of new energy vehicles. After that, the government introduced the purchase subsidy policy for new energy vehicles to promote the application of new energy vehicles and increase their market share. In the 12th Five-Year Plan period, the Chinese government proposed to invest RMB 100 billion in the research and construction of the industry chain of new energy vehicles in the next ten years (Su 2019). Battery electric vehicles are made the development focus in the technological road map of key areas of *Made in China 2025* (Qin and Chen 2015). On March 21, 2020, it was determined at an executive meeting of the State Council to extend the purchase subsidy and exemption from purchase tax for new energy vehicles for two years, which, as the most significant and favorable policy for new energy vehicles in that year, was helpful for maintaining the good development momentum of the industry and stimulating automobile consumption. In the same year, the Ministry of Industry and Information Technology revised the *Provisions on the Admission Administration of New Energy Vehicle Manufacturing Enterprises and Products*, further liberalizing the access threshold, stimulating market vitality and promoting the high-quality development of China's new energy vehicle industry. The *Development Plan for the New Energy Vehicle Industry (2021–2035)* proposes that by 2025, the competitiveness of China's new energy vehicle market will be significantly improved, and major breakthroughs will be made in key technologies such as power batteries, drive motors and on-board operating systems, with the sales volume of new energy vehicles accounting for about 25% of the total in the market (Chen 2020). *Technology Roadmap for Energy Saving and New Energy Vehicles 2.0* once again emphasizes that China's automobile industry needs to adhere to the development strategy of battery electric vehicles, and puts forward the goal that new energy vehicles will gradually become mainstream products and the automotive industry will realize the electric transformation. The plan also states that by 2035, all the traditional energy-powered passenger vehicles in China should be hybrid, and the annual sales volume of new energy vehicles will reach more than 50% of the market's total; The car parc of fuel cell vehicles will reach about 1 million, and commercial vehicles will realize the transformation towards hydrogen power (China Society of Automotive Engineers 2020). New energy vehicles have become the future development direction of the vehicle industry.

As important energy storage components in new energy vehicles, lithium-ion batteries have been favored by the market because of their high specific power, voltage and energy density, long cycle life, zero pollution, zero memory effect and low self-discharge. Zhao et al. (2000), An and Qi (2006), Wu et al. (2011) and Rui et al. (2013). They have a wide application prospect in fields such as mobile phones, aerospace, military equipment and electric vehicles. Lithium-ion batteries have gradually replaced other batteries as major power batteries. In the next 5 to 10 years, the electric vehicle industry will surpass the consumer electronics industry and become the largest application field of lithium-ion batteries (Dai et al. 2005).

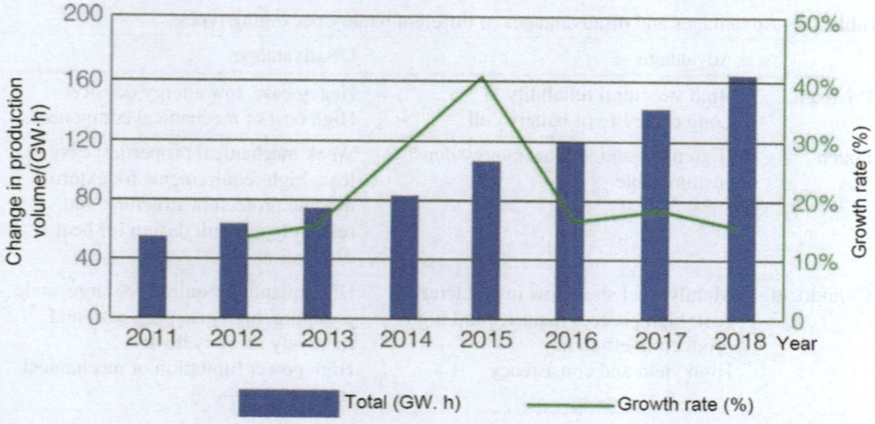

Fig. 1.1 Changes and growth rate of global lithium-ion battery production from 2011 to 2018

By the end of 2018, the global sales volume of electric vehicles exceeded 5.5 million, with more than 53% sold in China. With the development of new energy vehicles, the capacity of power batteries is being rapidly expanded. Since 2011, the global production of lithium-ion batteries has entered a period of rapid growth. Figure 1.1 shows the changes and growth rate of global lithium-ion battery production from 2011 to 2018 (Intelligence Research Group 2018).

Worldwide, the R&D and production of lithium-ion power batteries are mainly concentrated in China, Japan, South Korea and the United States. The United States, Japan and South Korea are the world's leaders in the basic R&D of lithium-ion batteries. However, China, with the highest production capacity and the vastest market of lithium-ion power batteries, has rapidly developed in the comprehensive strength of R&D and production in recent years, and has gradually narrowed the gap with industrial leaders of the world.

Meanwhile, with the continuous investment in advanced technology research by backbone enterprises in China's battery industry, the level of power battery products has also been greatly improved. In 2019, the R&D investment of Contemporary Amperex Technology Co., Limited (CATL) reached RMB 2.99 billion, up 50.3% year on year, which was nearly 3% higher than that in 2018, accounting for 6.5% of the operation revenue of the company in 2019. Gotion High-Tech invested RMB 590 million in R&D in 2019, an increase of 19.2% year-on-year, accounting for 11.9% of the company's operation revenue of that year. In 2019, BTR, a leading cathode enterprise, spent RMB 240 million on R&D, up 29.9% year-on-year, accounting for 5.4% of the company's operation revenue of that year (Yu 2020).

In terms of battery system, the lithium iron phosphate and the ternary material system are still dominant at present. Ternary power batteries have the advantages of high energy density and good low temperature performance, while LFP batteries are more advantageous in safety and cycle life.

Table 1.1 Advantages and disadvantages of different battery packaging types

	Advantage	Disadvantage
Prismatic	High structural reliability Long cycle life of battery cell	Heavy case, low energy density High cost of mechanical components
Pouch	High mass and volume energy density, customizable	Weak mechanical properties, easy to leak, high requirements for external module protection structure, and relatively difficult design for heat dissipation
Cylindrical	Mainly steel shell, low manufacturing cost, high process maturity and high production efficiency High yield and consistency	High quantity required for large-scale grouping, high grouping cost, and relatively short cycle life High power limitation of mechanical structure

According to the *Annual Tracking Report of Power Battery Technologies for New Energy Vehicles (2019) (China Industry Technology Innovation Strategic Alliance for Electric Vehicle* 2020*)*, a LFPLFP can reach the energy density of 170 Wh/kg and the volume energy density of 360–390 Wh/l. After grouping, the energy density of the battery system will exceed 140 Wh/kg, and the cycle life will be over 5000 times. A prismatic hard-shell ternary power battery can reach the energy density of 240 Wh/kg (270 Wh/kg for pouch) and the volume energy density of 540–590 Wh/l (590–630 Wh/l for pouch). After grouping, the energy density of the battery system will exceed 170 Wh/kg. In addition, the ternary material system is gradually changing from NCM523 and NCM622 to NCM712 and NCM811 in pursuit of higher energy density. In terms of price, the prices of a ternary power battery system and a LFPLFP power battery system are above RMB 1.05/(Wh) and RMB 0.95/(Wh), respectively.

At present, there are three types of batteries according to packaging: prismatic hard-shell, ALF pouch, and cylindrical. See Table 1.1 for the advantages and disadvantages of the three types (Wang and Xia 2017).

The energy density of prismatic batteries of the NCM523 and NCM622 systems adopted in China is between 200 and 240 Wh/kg. While that of LFP batteries is usually between 140 and 180 Wh/kg, and some products can reach 200 Wh/kg. As for pouch batteries, many enterprises have developed battery cells with energy density of 300 Wh/kg based on NCM811 cathode material and silicon–carbon anode material. The energy density of ternary cells produced by LG Chem of Korea can reach about 250 Wh/kg. Cylindrical power battery products have a tendency to increase in size, from 18650 (i.e., 18 mm in diameter and 65 mm in height) to 21700. At present, the energy density of cylindrical batteries produced in China is generally 230–260 Wh/kg. The energy density of 21700 cylindrical batteries made of Ni–Co–Al (NCA) cathode material and Si–C anode material by Panasonic can reach 270 Wh/kg (China Industry Technology Innovation Strategic Alliance for Electric Vehicle 2020).

Generally speaking, current power battery products mainly adopt LFP and ternary materials. With the increasing demand for driving range, the ternary material system with high nickel content is generally favored by battery enterprises, but their disadvantage in safety have also become a major obstacle in their development process. In terms of overall technology, the *Technology Roadmap for Energy Saving and New Energy Vehicles 2.0*, which was revised and compiled under the leadership of China-SAE, specifies that by 2035, China's power battery technology for new energy vehicles should be in an leading position of the world in general, platforms of multi-material system power battery, module and system products shall be formed to continuously improve the energy density of power batteries, and the energy density of universal, commercial and high-end energy-type power batteries should exceed 300 Wh/kg and 250 Wh/kg respectively to significantly improve the durability and reliability of power batteries (China Society of Automotive Engineers 2020).

1.2 Thermal Management and Thermal Safety of Power Batteries

Performance, life and safety of lithium-ion batteries are closely related to the temperature of the batteries. Too high battery temperature will speed up the side reaction, accelerate the aging (approximately, the increase of every 15 °C in temperature will cause the service life to be reduced by half), and even cause safety accidents. While too low temperature will cause significant decline in power and capacity. If the power is not limited, lithium-ions may be precipitated, causing irreversible attenuation and potential safety hazards. Usually, the suitable operating temperature of lithium-ion batteries is 10–30 °C, while the operating temperature range of vehicles is 30–50 °C. The thermal environment around batteries in vehicles is often uneven, which poses a severe challenge to the thermal management of battery packs. The large-scale and grouped application of power batteries makes the heat dissipation capacity of batteries (packs) much lower than the heat generation capacity. Especially for HEVs and PHEVs characterized by high rate discharge, it is necessary to design complicated heat dissipation systems. When battery cells are used in parallel (the pole pieces inside the cells are also connected in parallel), the uneven temperature will cause thermoelectric coupling, that is, the battery cell (or part) with high temperature will share more current, resulting in uneven state of charge, thus accelerating the deterioration of the battery pack. Therefore, the thermal management technology of vehicle power battery systems is one of the key technologies to ensure their performance, life and safety. Table 1.2 shows an overview of battery pack thermal management systems of several electric vehicles (including HEVs, PHEVs and BEVs) (Zhang et al. 2012).

In addition, lithium-ion batteries also have thermal safety problems, which may lead to a decline in battery performance, affecting the performance and driving range of electric vehicles, and may even lead to safety accidents, resulting in casualties and major economic losses (Wu et al. 2015; Tröltzsch et al. 2006; Shao and Feng 2018).

Table 1.2 Overview of battery pack thermal management systems of several electric vehicles

Country	Model	Type	Battery connection method	Thermal management system and others
US	GM Volt	PHEV	It consists of 288 cells, which are arranged in 7 modules each containing 36 cells and 2 modules each containing 18 cells. The electrical connection is a hybrid structure, which is relatively independent of the mechanical arrangement. The electrical connection can be equivalent to 96 cells connected in series and 3 groups connected in parallel	It adopts liquid cooling method, with the cooling liquid being a mixture of 50% water and 50% glycol. The metal cooling fins are separated between the cells, and the cooling liquid circulates in the cooling fins in a closed way. The thickness of the cooling fins between the cells is only about 1 mm. When the temperature is too low, the heating coil can heat the cooling liquid and raise the temperature of the battery
			If the nominal voltage and total capacity of a battery pack are kept unchanged, it is composed of about 250 battery cells	
	Enwel Th!nk City	BEV	In the case of 28 kWh, the battery consists of 432 cells. Using the parallel-series structure, the electrical connection is equivalent to 108 cells in series and 4 groups in parallel	It adopts forced air cooling method, with a hollow aluminum heat conducting grooves arranged at the heads of every two parallel connected cells, which is connected with the ventilation and diversion groove of the whole battery pack
	Tesla Roadster	BEV	It consists of 6,831 18650 lithium-ion batteries. Where, 69 are connected in parallel to form a brick, then 9 groups are connected in series to form a sheet, and finally 11 sheets are stacked in series	It adopts liquid cooling method, with the cooling liquid being a mixture of 50% water and 50% glycol

(continued)

Table 1.2 (continued)

Country	Model	Type	Battery connection method	Thermal management system and others
Japan	Toyota Prius Prius, PHV	HEV, PHEV	–	It adopts forced air cooling method, with the ventilation fans operating in four modes: off, low speed, medium speed and high speed. The battery temperature control system determines the operation modes of the battery fans
	Nissan Leaf	BEV	It consists of 192 cell connected in a mixed structure, that is, 48 modules connected in series, and each module containing 2 serial and 2 parallel groups	The battery pack is sealed (heating option is available in cold areas)
	Mitsubishi iMiEV, Minicap	BEV	IMiEV is composed of 88 cells connected in series	Adopting forced air cooling method

Table 1.3 shows safety accidents caused by thermal runaway of lithium-ion power batteries according to incomplete statistics in recent years, and Fig. 1.2 shows phones corresponding to some safety accidents.

The causes of accidents listed in Table 1.3 are mostly battery collision, internal short circuit, overheating, overcharge, etc., all of which may lead to thermal runaway.

Table 1.3 Safety accidents caused by thermal runaway of lithium-ion power batteries according to incomplete statistics

No.	Time	Site	Description
1	2013.6	Hong Kong	Electric bus catching fire during rapid charging
2	2015.4	Shenzhen	Electric bus catching fire during charging
3	2016.4	Pudong, Shanghai	Spontaneous combustion of electric passenger vehicle
4	2016.5	Zhuhai	Electric bus catching fire due to short circuit of battery
5	2017.5	Beijing	Spontaneous combustion of electric bus in the parking lot
6	2019.4	Shanghai	Spontaneous combustion in parking lot after rapid charging of electric passenger car
7	2019.8	Hunan	Electric bus catching fire when charging
8	2020.10	Beijing	Spontaneous combustion of electric vehicle

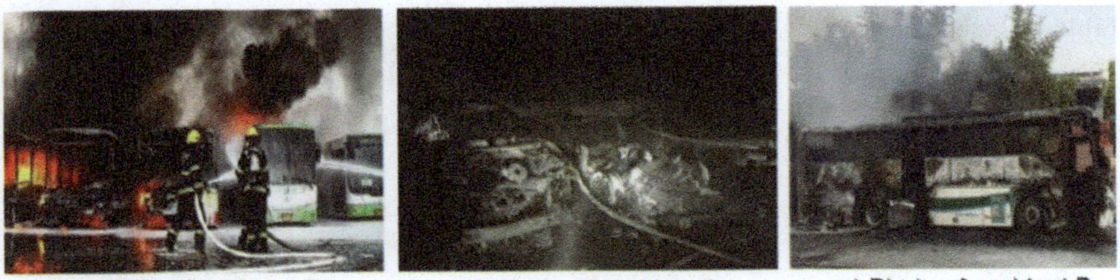

a) Photo of accident 5 b) Photo of accident 6 c) Photo of accident 7

Fig. 1.2 Photos of safety accident scenes of lithium-ion power batteries in recent years

After the thermal runaway of a battery cell occurs, the generated heat will be transferred to the adjacent cells, causing the spread of thermal runaway. Thermal runaway and heat spread lead to fire and even explosion of battery modules, threatening the property and safety of the passengers. *Technology Roadmap for Energy Saving and New Energy Vehicles 2.0* proposes the goal to limit the fire accident rate of new energy vehicles below 0.01 times/10,000 vehicles by 2035 (China Society of Automotive Engineers 2020). In 2020, China issued the *Fuel Cell Electric Vehicles Safety Requirements* and the *Electric Vehicles Traction Battery Safety Requirements*, which were implemented in 2021. The *Electric Vehicles Traction Battery Safety Requirements* is a mandatory national standard.

Therefore, the research on thermal safety of lithium-ion power battery is crucial, and the solutions mainly include: the improvement of cathode and anode, separator and electrolytes (as for battery materials), preventing thermal runaway battery cells from transferring heat to adjacent cells and resulting in thermal runaway of the whole battery pack (as for the prevention and control of thermal propagation), and so on (China Industry Technology Innovation Strategic Alliance for Electric Vehicle 2020).

1.3 Research Methods of Power Battery Thermal Management

In order to apply lithium-ion batteries to power battery systems of electric vehicles, many issues need to be considered, and one important issue is the thermal management of lithium-ion power batteries. Research on thermal management of power batteries for electric vehicles mainly involves the following three aspects: heat dissipation of power battery packs; low temperature heating of power battery packs: temperature field distribution of power battery packs.

As for the first aspect, because a battery pack generates heat during charge and discharge, if the heat is not dissipated in time, it will lead to an increase in the battery pack temperature and accelerate the decline in the life of the battery pack. If the temperature is too high, it may even cause the battery pack to catch fire or even explode. Therefore, the research on heat dissipation of power battery pack is very important. As for the second aspect, with the gradual popularization of electric vehicles, people pay more and more attention to the low-temperature performance of power batteries, and problems such as difficulty in charging battery packs at low temperature, degradation of discharge capacity and shortening of driving range are gradually exposed, so it is also necessary to study the low-temperature heating of power batteries. As for the third aspect, because there are many battery cells in a battery pack, if the temperature field distribution of the battery pack is uneven, the inconsistency of the battery cells will increase, thus affecting the overall performance of the whole battery pack. Therefore, the research on the temperature field distribution of the battery pack is also an important aspect of thermal management research of power battery packs.

1.3.1 Heating Methods of Power Battery Packs

The research progress of battery pack heating is relatively slow, and it is technically more difficult to realize battery pack heating than battery pack cooling in electric vehicles. With the gradual popularization of electric vehicles, heating problems of battery packs cannot be evaded. When a battery is at a low temperature (below 10 °C), its charge and discharge performance will greatly decrease (Zhang et al. 2003), as shown in Fig. 1.3. This is because at low temperature, the polarization of the battery electrodes is serious, the internal resistance of the battery increases significantly, and the active substances in the electrolyte cannot be fully utilized. The following describes the current development of battery pack heating technologies from two aspects: internal heating and external heating.

1. Internal heating

Stuart and Hande (2004) studied the internal heating of applying alternating current to batteries at low temperature. Firstly, the internal heating effect of 60 Hz alternating current was studied. By applying 60 Hz alternating current to lead–acid batteries

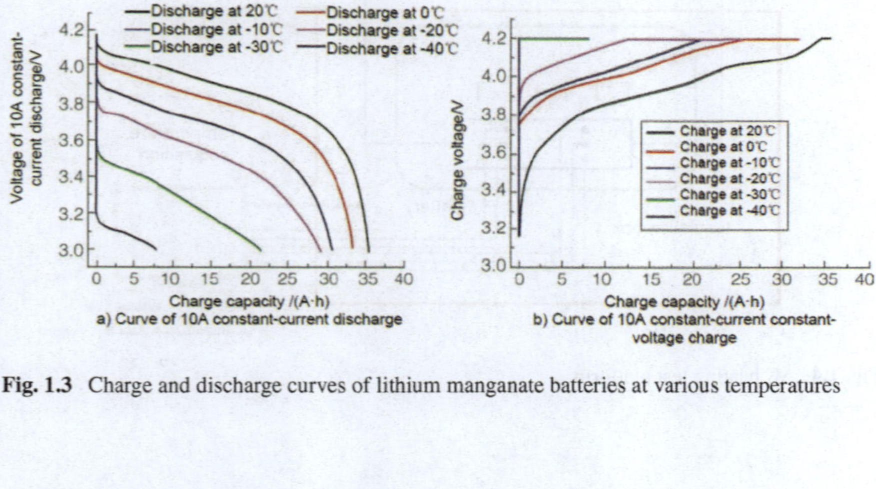

Fig. 1.3 Charge and discharge curves of lithium manganate batteries at various temperatures

with different SOC at low temperature, it was found that the greater the amplitude of alternating current, the more obvious the internal heat generation was. It was proposed that after applying 60 Hz alternating current with the amplitude of 100 A to a lead–acid battery, the battery can rapidly heat up within 5 min and realize normal charge and discharge. The high-frequency alternating current of 10–20 kHz was also applied to 16 Panasonic Ni–MH battery packs. The experimental results showed by applying 10–20 kHz alternating current with the amplitude of 60–80 A to Ni–MH battery packs at 30–20 °C, the battery packs could quickly restore to normal charge and discharge state within a few minutes.

Hande and Stuart (2004) adopted inverter circuits to heat the batteries at high frequency, and the results showed that the temperature of the batteries can rise to room temperature in a few minutes. He also claimed that the charge and discharge performance of the batteries is improved and the internal resistance is reduced after heating, but he has not modeled and analyzed this heating method, nor has he analyzed the influence of this method on the capacity and life of the batteries. Zhao et al. (2011) claimed that a battery generates more heat when discharged than when charged by comparing the heat generation processes during charge and discharge. By using this characteristic, the battery can be heated at low temperature by combining high pulse discharge with low pulse charging.

Zhang et al. (2015) adopted sinusoidal alternating current to heat batteries at low temperature, and claimed that within a certain range, the higher the amplitude and the lower the frequency of sinusoidal alternating current, the faster the temperature rise of the battery. The research adopted 18650 batteries (see Fig. 1.4 for the experiment platform). When the amplitude of sinusoidal alternating current is 7 A (2.25C), the frequency is 1 Hz and the external convection heat transfer coefficient is 15.9 W/(m^2 k), the battery can rise from 20 to 5 °C within 15 min, and the temperature distribution of the battery remains homogeneity.

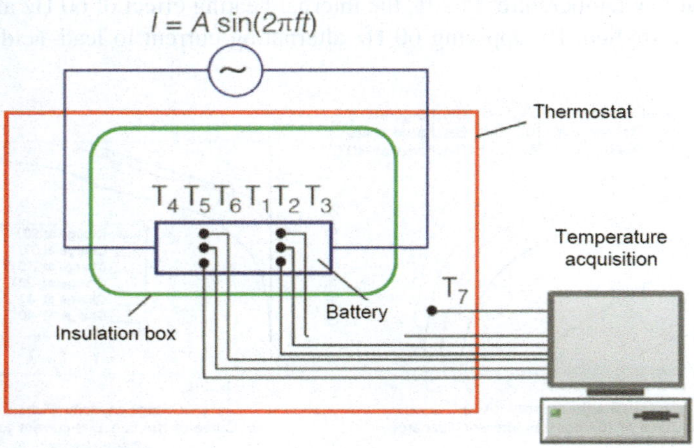

Fig. 1.4 AC heating test platform

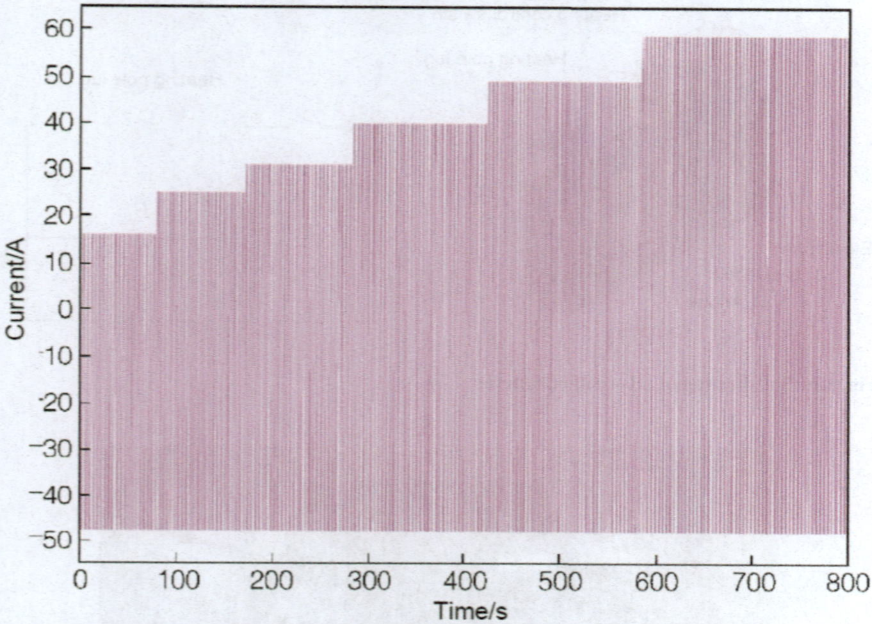

Fig. 1.5 Changes of charge and discharge currents of stepwise charging preheating method

Ruan et al. (2014) also put forward a stepwise segmented charging technique, which, as shown in Fig. 1.5, is to conduct low-depth charge and discharge of a battery pack at low temperature while maintaining the current of low-depth discharge unchanged and increasing the current of low-depth charge continuously. During this process, the battery will generate heat internally and finally reach the required temperature. The experimental results show that the battery temperature can rise from 10 to 0 °C after 15 min. Because of the low depth, the charge and discharge will hardly affect the battery capacity or life.

As a special internal heating method, the all-weather self-heating battery (Fig. 1.6) proposed by the research team of Professor Wang Chaoyang of Pennsylvania State University (Wang et al. 2016) has attracted wide attention in recent years. All-climate batteries are made by embedding nickel foil in traditional lithium-ion batteries as the heating source. When low-temperature preheating is needed, the heating control switch is closed to form a self-heating current loop to generate a large amount of heat inside the battery and realize rapid preheating of the power battery. At the cell level, the results show that the battery cells can be heated from −30 to 0 °C within 30 s, and the energy consumption is less than 5% of itself. At the level of industrial production and application of batteries, Beijing Institute of Technology and RiseSun MGL Co. Ltd. have launched a research project on technical solutions of all-weather battery cells and self-heating, and made great breakthroughs and achievements in the development of all-weather power battery systems (see Fig. 1.7). In the parking

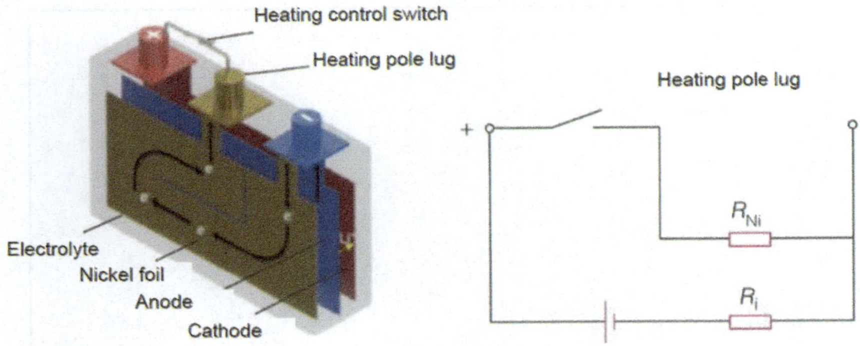

Fig. 1.6 Self-heating of all-weather batteries

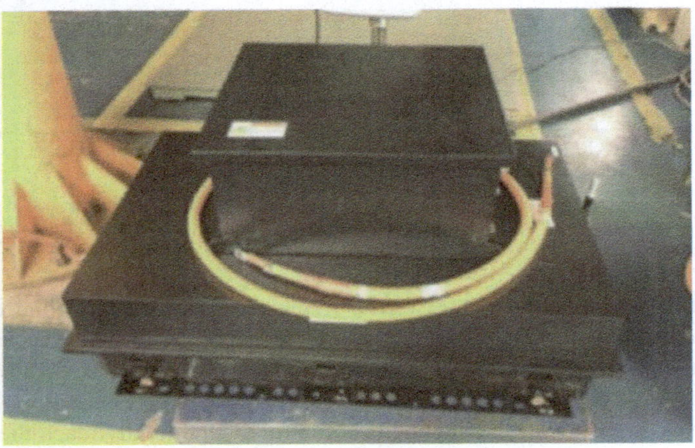

Fig. 1.7 Prototype of all-weather power battery systems

heating test in cold regions, the rapid self-heating start-up in 6 min was realized, and the temperature rise rate exceeded 5 °C/min. The energy consumption of battery heating during low-temperature start-up was not higher than 5%, which showed the great potential of this technology in future industrial application (Wang and Sun 2019).

2. External heating

(1) Liquid or gas heating

Liquid or gas heating means to heat the battery by filling heated liquid or gas into the battery box. The electric vehicle VOLT introduced by General Motors adopts liquid to heat and dissipate heat from the battery packs (see Figs. 1.8 and 1.9) (Matthe and Turner 2011). The use of liquid heating requires higher sealing and insulation of a

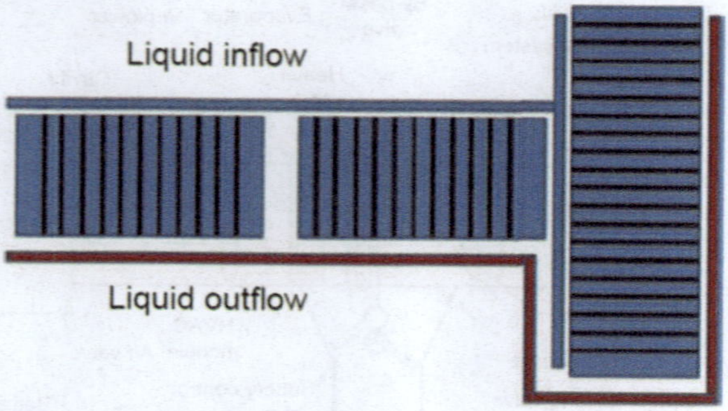

Fig. 1.8 Flow direction of heat transfer liquid of VOLT

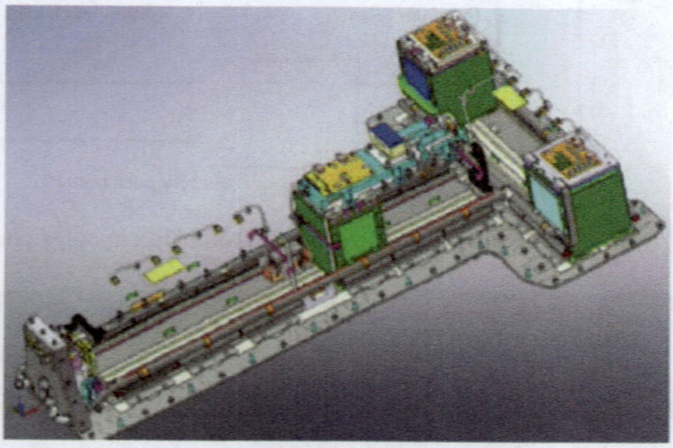

Fig. 1.9 Battery module structure of VOLT

battery box, which will increase the complexity of the whole battery box. There are still many reliability problems to be solved. Although gas heating (see Fig. 1.10) has no special requirements for sealing and insulation, it has the disadvantages of slow heating speed and high heating energy consumption (United Automotive Electronic Systems Co., Ltd. 2009; Matthe and Turner 2011).

(2) Heating plate heating

Heating plate heating refers to adding an electric heating plate at the top or bottom of a battery pack. When heating, the electric heating plate is electrified, part of the heat of the heating plate is directly transferred to the battery by heat conduction, and the other part heats the battery by convection through the surrounding heated air. Ma (2010) studied the bottom heating of power battery packs (see Fig. 1.11 for the heating system). The results show that the heating time was long when the heating plate was used. After heating, the temperature distribution of the battery pack was uneven and high temperature difference occurred.

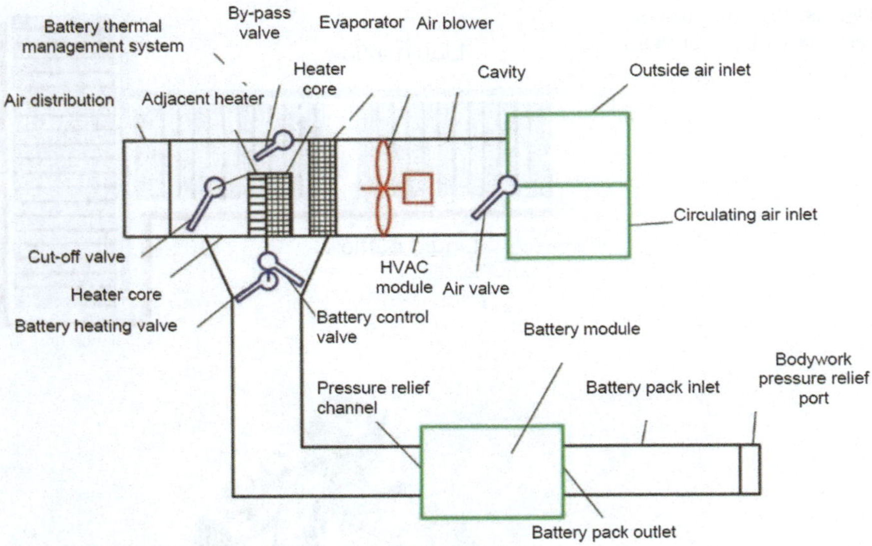

Fig. 1.10 Battery gas heating management system of GM global technology operations

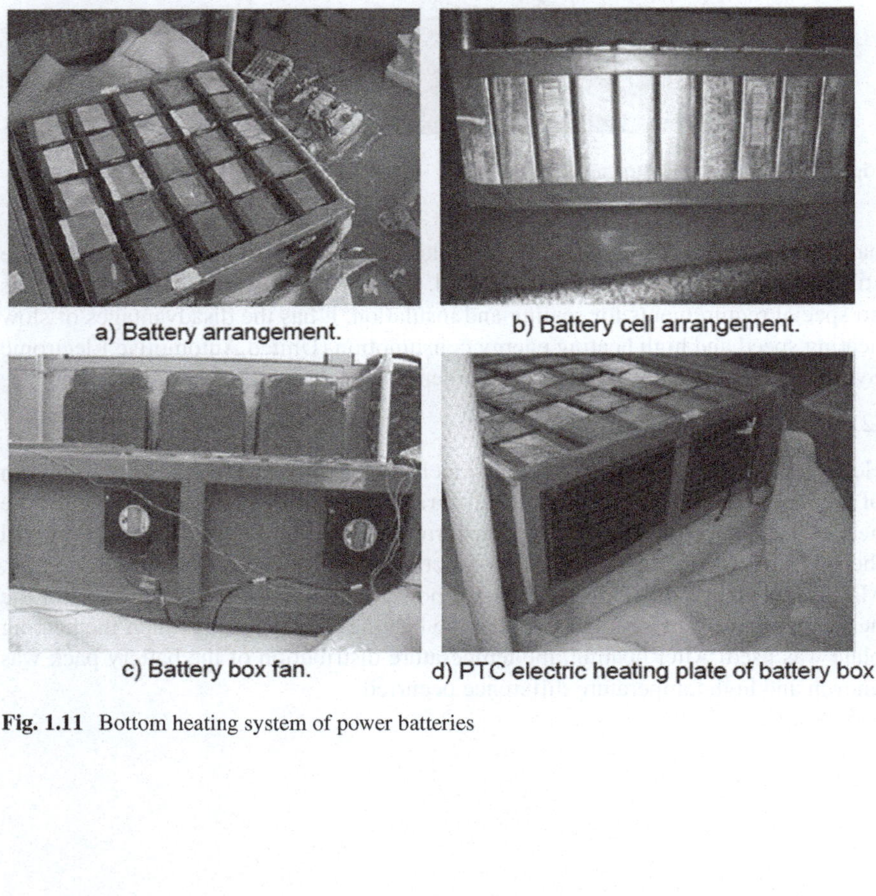

Fig. 1.11 Bottom heating system of power batteries

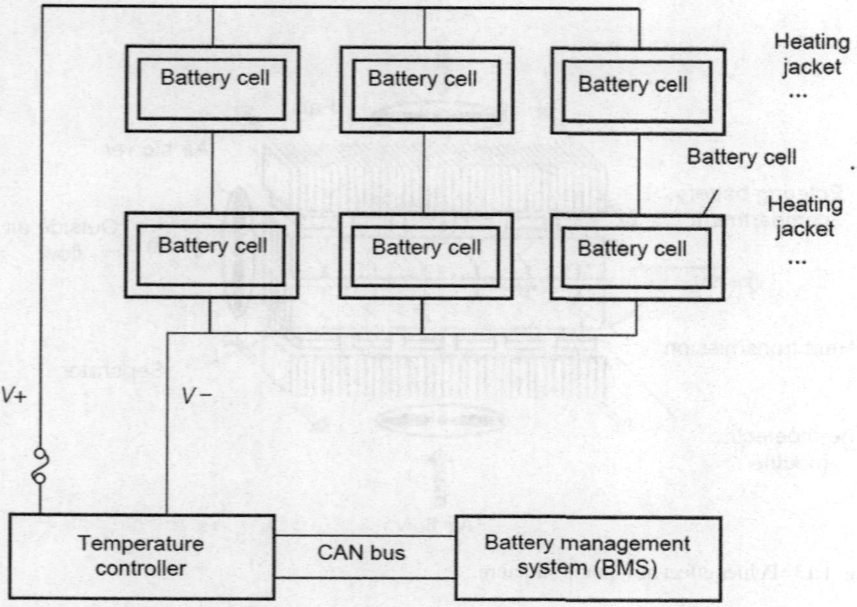

Fig. 1.12 Heating scheme of the heating jacket of a vehicle brand

(3) Heating jacket heating

Heating jacket heating refers to adding a heating jacket made of a resistive material to each battery cell. This heating mode (see Fig. 1.12) can make battery cells in a battery pack heated evenly with less energy loss (Chery Automobile Co. Ltd. 2010). However, in summer, heating jackets will cause such problems as heat dissipation difficulties of batteries.

(4) Peltier effect heating method

Peltier effect means that when the current flows through the interface of two different conductors, it will absorb heat from the outside or release heat to the outside. With the special property of Peltier effect, two functions of heating and cooling can be realized by changing the direction of current, and the intensity of heating and cooling can be accurately controlled by changing the current. It is an active thermal management system for battery packs. Peltier effect has been used in electronic devices to some extent, but there are few researches on applying Peltier effect to power batteries.

Alaoui and Salameh (2001, 2003, 2004) studied the application of Peltier effect in electric vehicles. Alaoui and Salameh (2001) made an electric heating device with Peltier effect, and tested the device. The experimental results showed that the device had the advantages of simple structure, high temperature control precision and low energy consumption. Alaoui and Salameh (2003, 2004) provide the device structure for applying Peltier effect to vehicles (see Fig. 1.13), as well as arrangement schemes

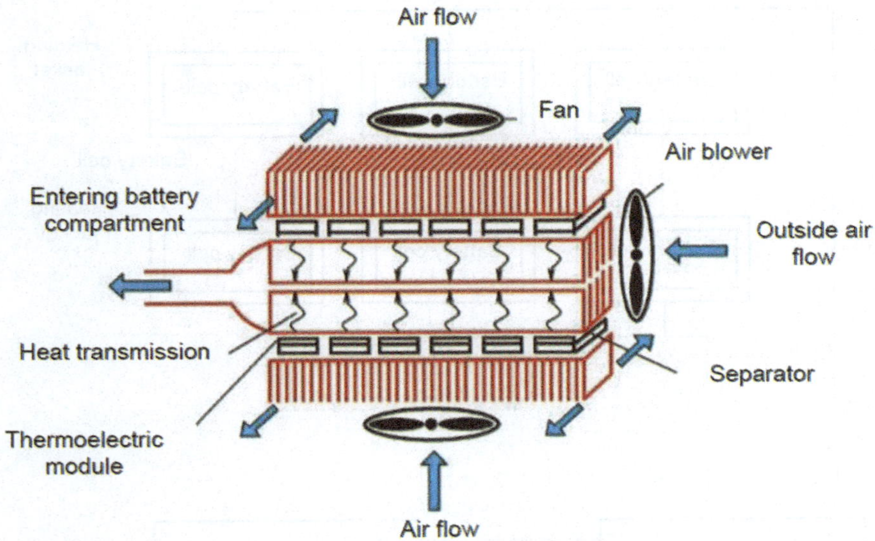

Fig. 1.13 Peltier effect heat pump structure

of Peltier effect on vehicles (see Fig. 1.14). The device can not only heat and cool batteries, but also replace vehicle air conditioners.

Kras et al. (2009), Kras and Aebi (2010) developed an active battery pack thermal management system based on Peltier effect, which was assembled on the electric vehicle SAM EVII. The lithium-ion power battery had a capacity of 7 kWh, and the thermal management system could effectively cool and heat the battery, but the specific structure of the thermal management system was not given.

Ji and Wang (2013) made a comparative study on different heating methods from the following four aspects:

① The energy loss of the battery during heating.
② Heating time.
③ Influence of heating on the battery system and the battery itself.
④ Cost of the heating system.

This paper selects the following four heating methods for comparison:

① DC internal heating of the battery, that is, discharging the battery at constant current or constant voltage, so as to generate heat inside the battery.
② Self-heating of the battery, specifically, the battery supplies power to convection heating device for self-heating.
③ MPH (Mutual Pulse Heating), that is, the whole battery pack being divided into two parts, which are connected by DC/DC devices in the middle, and one part is discharged to charge the other part, and then the other part is discharged to charge this part, so that the temperature of the whole battery pack gradually rises during the alternate charge and discharge of the two parts.

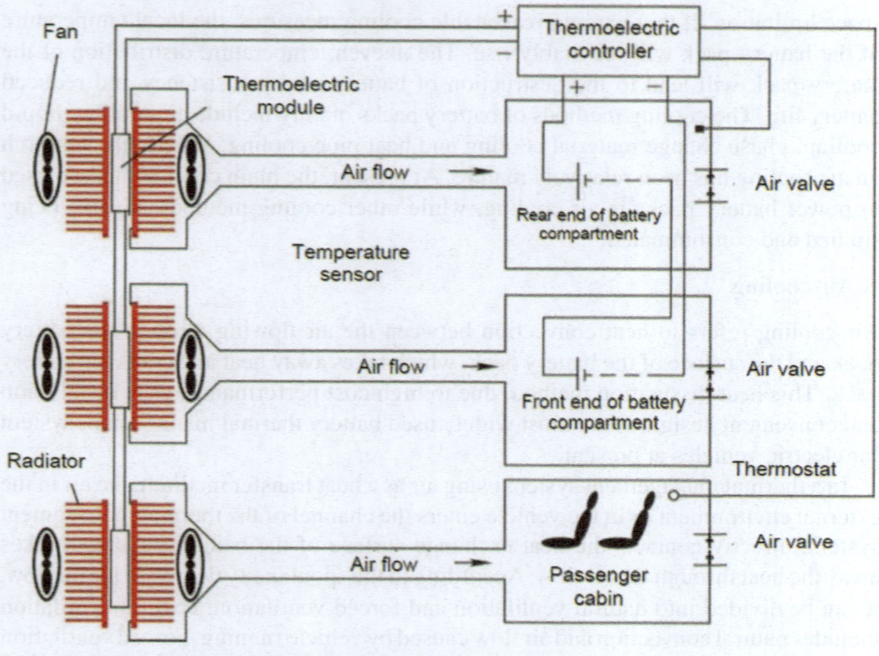

Fig. 1.14 Peltier effect thermal management system

④ AC electric heating, that is, applying AC with certain frequency and amplitude to the cathode and anode of the battery, so that the battery can be heated by itself.

After research, this paper concluded that: Method ②, battery self-heating, requires the shortest heating time; Method ③, MPH, needs the least energy and heats evenly. The AC used in Method ④ AC heating can be converted from commercial power without consuming the internal energy of the battery, and the heating uniformity of this method is good; Method ①, DC internal heating of the battery, requires no design for additional heating devices, but it has low heating efficiency, long heating time and larger capacity loss of batteries.

1.3.2 Heat Dissipation Methods of Power Battery Packs

The heat dissipation problem of electric vehicle battery packs has attracted researchers' attention for a long time. As early as 1979, Chen and Gibbard (1979) put forward the thermal management problems of lead–acid power battery packs. When a battery pack is charged and discharged at a high rate or working in a high temperature environment, the layout of the power battery pack is compact due to the space limitation. If there are no reasonable cooling measures, the local temperature of the battery pack will inevitably rise. The uneven temperature distribution of the battery pack will lead to the destruction of battery pack consistency and reduced battery life. The cooling methods of battery packs mainly include air cooling, liquid cooling, phase change material cooling and heat pipe cooling. Where, the research on air cooling has been relatively mature. At present, the main cooling method used in power battery packs is air cooling, while other cooling methods are still being studied and consummated.

1. Air-cooling

Air cooling refers to heat convection between the air flowing through the battery pack and the surface of the battery pack, which takes away heat and cools the battery pack. This heat dissipation method, due to high cost performance, easy installation and convenient design, is the most widely used battery thermal management system for electric vehicles at present.

In a thermal management system using air as a heat transfer medium, the air in the external environment or in the vehicle enters the channel of the thermal management system, directly contacts the heat exchange surface of the battery pack, and takes away the heat through the air flow. According to the spontaneity degree of the air flow, it can be divided into natural ventilation and forced ventilation. Natural ventilation includes natural convection and air flow caused by vehicle running. Forced ventilation is mainly driven by fans, and the instantaneous power of the fans is determined by the control circuit of the thermal management system.

A schematic diagram of external circulation and internal circulation of air cooling, and cooling modes and processes of active cooling and passive cooling are shown in Fig. 1.15. In passive air cooling, air is introduced from the outside, which dissipates heat to the battery and then is directly discharged by a fan. Active cooling is the internal air circulation, which is discharged by the fan and then returned to the cooling/heating device of the vehicle for new circulating heat dissipation.

2. Liquid cooling

Thermal management systems using liquid as heat transfer medium are mainly divided into contact type and non-contact type. The contact type adopts highly insulating liquids such as silicon-based oil and mineral oil, which can directly soak the battery pack in heat transfer liquid; Conductive liquids such as water, glycol or coolant are used in non-contact mode, and the battery pack cannot be in direct contact with the heat transfer liquid. At this time, distributed airtight pipes should be arranged inside the battery pack, and the heat transfer liquid flows through the pipes and takes away the heat. The material and tightness of the pipes ensure the electrical insulation between the conductive liquid and the battery. The liquid flow in contact or non-contact liquid cooling system is mainly driven by oil pumps/water pumps, etc.

Compared with gas, liquid has higher heat capacity and thermal conductivity coefficient, so at the same volume and flow rate, the cooling effect of liquid is obviously better than that of air. However, although the effect of liquid cooling is

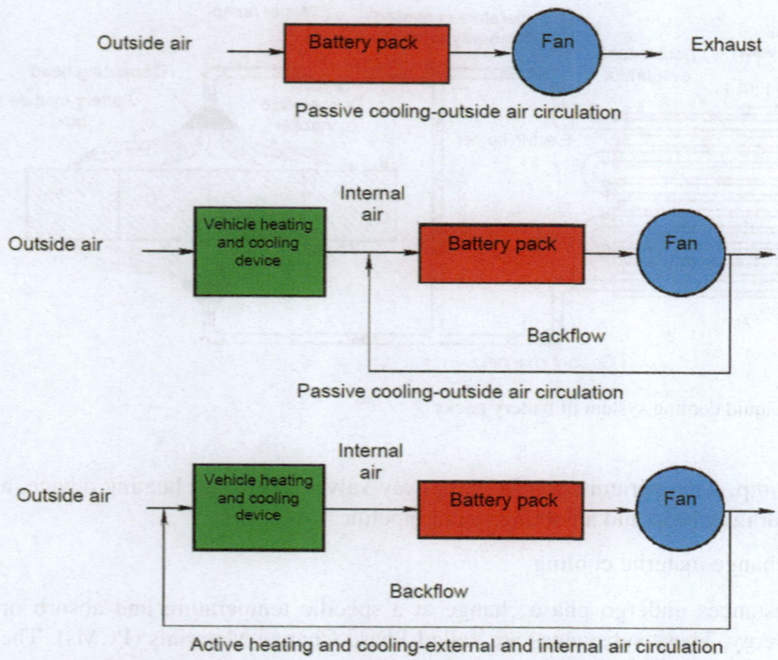

Fig. 1.15 Air cooling method

better than that of air cooling, problems such as sealing, insulation, reliability, energy density reduction of the battery pack and cost must be considered when adopting liquid cooling. The insulating oil of heat transfer medium in a contact liquid cooling system has high viscosity, which requires high oil pump power to maintain the required flow rate. In a non-contact liquid cooling system, distributed closed flow channels need to be designed inside the battery pack, which increases the overall mass of the battery pack and reduces the heat transfer efficiency between the battery surface and the heat transfer medium.

Pesaran et al. (1999) and Pesaran (2001) discussed these problems in detail. Pesaran et al. (1999) claims that if the cooling liquid is in direct contact with the battery, the cooling liquid must be insulated, such as mineral oil. Because of the high viscosity, the flow rate of oil is relatively low, which reduces the cooling effect and causes high energy consumption of the pump. If water is used for cooling, because water is conductive, it can only be cooled in a non-contact way, which makes the structure of the cooling system complicated, and the heat generated by the battery can only be transmitted to the cooling water through the water jacket, which reduces the cooling effect. Wu and Zhang (2008) puts forward a liquid cooling system for power battery packs, which can get better control effect through simulation (see Fig. 1.16). The system comprises a battery module, a battery module box, a sleeve evaporator,

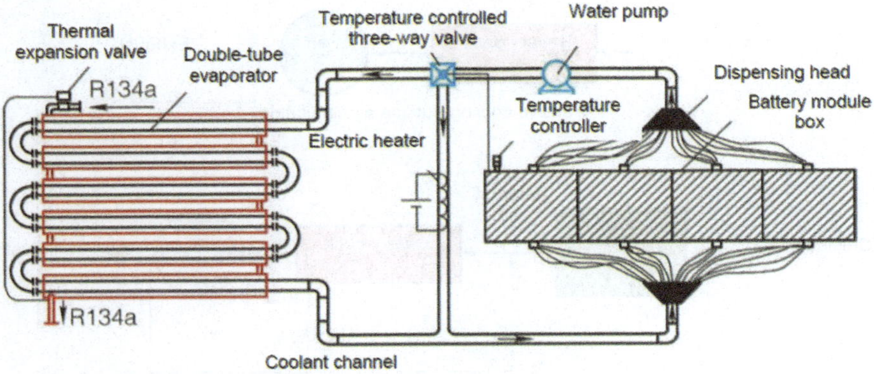

Fig. 1.16 Liquid cooling system of battery packs

a water pump, a temperature control three-way valve, an electric heating device, a liquid separating head and a cooling liquid pipeline.

3. Phase change material cooling

Some substances undergo phase change at a specific temperature and absorb or release energy. These substances are called Phase Change Materials (PCMs). The phase change temperature can be adjusted near the upper limit of the suitable working range of the battery by adjusting the types and composition ratios of the phase change materials and additives. When this kind of phase-change material is used to wrap a battery pack, when the battery temperature rises to the phase-change temperature, the phase change material will absorb a large amount of latent heat, so that the battery temperature can be maintained within the suitable working range of the battery, and the battery pack can be effectively prevented from overheating.

A thermal management system using PCMs as the heat transfer medium has the advantages of simple overall structure, high system reliability and high safety, and has been widely used in the cooling system of electronic devices. In 1994, Rafalovich et al. (1994) used phase change materials to cool lead–acid batteries. It was proven through numerical simulation and experiments that PCMs can make lead–acid batteries work normally in a wide temperature range.

Al-Hallaj and Selman (2000, 2002), Khateeb et al. (2004, 2005), Kizilela et al. (2008) have made a series of studies on PCMs as cooling materials for lithium-ion power batteries. Al-Hallaj and Selman (2000, 2002) demonstrated through simulation that it is completely feasible to use PCMs as cooling materials for passive thermal management systems of lithium-ion power batteries. Khateeb et al. (2004, 2005) took electric scooters as the research objects, using 18650 lithium-ion batteries instead of the lead–acid batteries of the original scooters, and provided the calculation method to determine the number of PCMs needed for each battery cell. Meanwhile, through comparative experiments, it was found that, due to the low thermal conductivity coefficient of PCMs, if PCMs were used for cooling alone, most of the heat generated

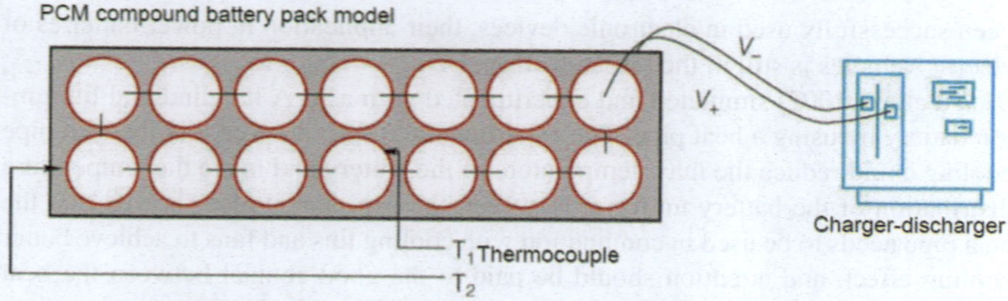

Fig. 1.17 Cooling test equipment for phase change materials

during battery discharge could not be dissipated into the air, which would lead to high temperature difference between battery cells at different positions in the battery pack. Moreover, when the battery pack was continuously charged and discharged, it was prone to temperature accumulation. By adding aluminum foam into the PCMs, the thermal conductivity coefficient of the PCMs can be significantly improved, and the temperature distribution of the battery pack can be uniform. Kizilela et al. (2008) compared the cooling effect of forced cooling with that of using PCMs, and the related test equipment is shown in Fig. 1.17. In order to improve the thermal conductivity coefficient of the PCM, graphite was added into the phase change material. The simulation results showed that the cooling effect of PCMs was obviously better than that of forced cooling. Moreover, under the condition of 45 °C ambient temperature and heavy current discharge, the cooling system using PCMs could control the temperature of the battery pack within a safe range, and the temperature distribution of the battery pack was uniform.

Duan and Naterer (2010) adopted electric heating tubes to simulate battery heat generation, and studied the whole phase change process of phase change materials and the temperature changes at different positions in phase change materials. The experimental results showed that the phase change material could control the temperature of the electric heating tube within a set range and had a good cooling effect.

When used as passive cooling systems for power batteries, PCMs have their unique advantages: no need for cooling fans, exhaust fans, condensers, design of cooling routes, etc., and no requirements for energy consumption of some active cooling systems. Despite the above advantages, PCMs also have disadvantages that cannot be ignored: if thermal management system adopts a PCM as the cooling material, the sealing problem must be considered, and the volume of the battery box will also increase, and its energy density will decrease, which is a big disadvantage for electric vehicles.

4. Heat pipe cooling

Heat pipe cooling was put forward by an American named R. S. Goller in 1942. In 1967, heat pipes were first used in aerospace and achieved success. Later, many electronic devices began to adopt heat pipes for cooling. Although heat pipes have been successfully used in electronic devices, their application in power batteries of electric vehicles is still in the research stage.

Wu et al. (2002) simulated and experimented with a 12 A h cylindrical lithium-ion battery by using a heat pipe. The experimental results showed that the heat pipe cooling could reduce the max. temperature of the battery and make the temperature distribution of the battery uniform. However, the experiment also showed that the heat pipe needs to be used in conjunction with cooling fins and fans to achieve better cooling effect, and attention should be paid to the good contact between the heat pipe and the battery. Zhang et al. (2009) adopted three methods, natural convection, forced convection and heat pipe, to cool SC Ni–MH batteries, and the heat pipe cooling system is shown in Fig. 1.18. By comparing the results of the three cooling methods, the conclusion was drawn that the cooling effect of heat pipe is better, with more uniform temperature distribution of the battery pack.

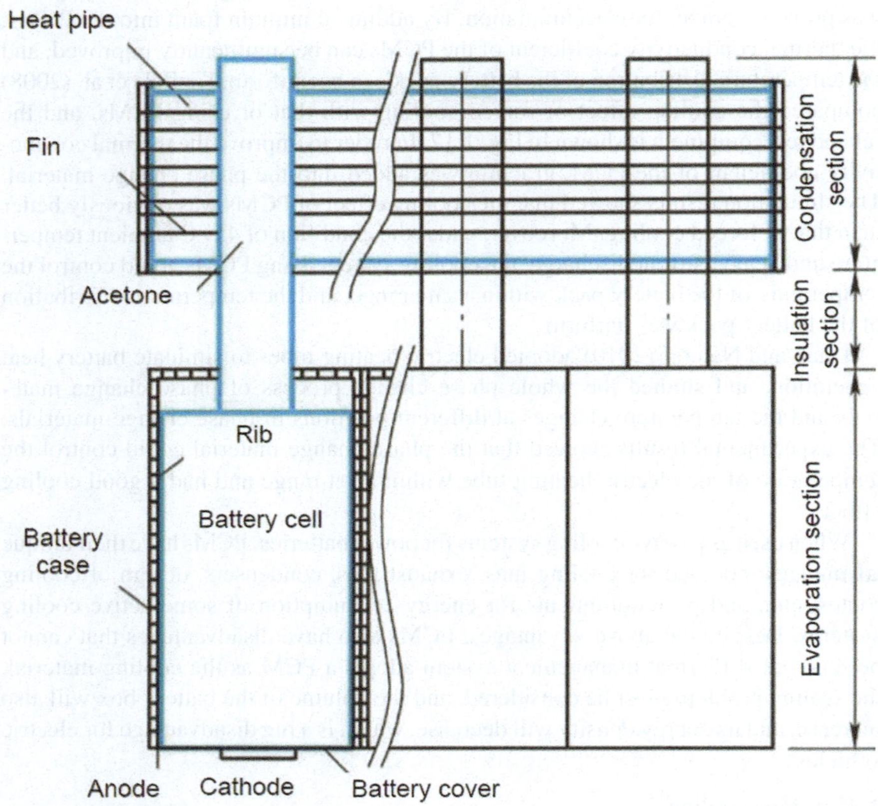

Fig. 1.18 Structural diagram of gravity heat pipe battery cooling system

1.4 Current Research on Thermal Characteristics Modeling of Batteries

1.4.1 Research on Heat Generation Models of Power Batteries

In the low rate charge and discharge environments, the heat generated by a battery mainly includes electrochemical reaction heat, ohmic heat and polarization heat. However, in high rate charge and discharge, if the heat dissipation is poor, it will cause heat accumulation and increase the battery temperature, which may be accompanied by side reaction heat besides the above. In addition, in cases of abuse such as high temperature, overcharge, overdischarge, short circuit, extrusion of foreign objects, nail penetration, etc., there will be more severe side reaction heat generation. To solve the heat generation problems of power batteries, researchers have carried out a lot of research from the perspectives of experiment, mechanism and modeling.

Newman et al. (1985) proposed a battery heat generation rate model based on the theory of energy conservation. The model can well simulate the complicated heat generation and temperature distribution in batteries in normal charge and discharge. Bernardi's equation to calculate heat generation rate is as follows:

$$Q = I_L(E_0 - U_L) - I_L T \frac{\mathrm{d}E_0}{\mathrm{d}T} \tag{1.1}$$

Where, Q is the total heat production of the battery; I_L is the current; E_0 is the open circuit voltage; U_L is the working voltage; T is the temperature in kelvin. The first term on the right of the equal sign is irreversible heat, including ohmic heat and polarization heat, and the second term is reversible heat of electrochemical reaction.

Electrochemical reaction heat means that when lithium-ions are intercalated and deintercalated into and out of cathode and anode materials, the battery reaches another equilibrium state from one equilibrium state, which shows that the battery overcomes the reaction energy barrier and absorbs heat or releases heat. According to the thermodynamics reversibility of electrochemistry, it can be seen that the reaction heat is reversible heat.

Ohmic heat means that the battery will have ohmic resistance at the interface between current collector and SEI film, and irreversible Joule heat will be generated when a current passes through it. The magnitude of ohmic heat is related to the thickness of SEI film and ohmic resistance of each interface.

Polarization heat means that when the charge and discharge current is high and exceeds the maximum current that the electrolyte system can provide, the limit diffusion current, there will be concentration overpotential, that is, polarization. The heat generated due to the existence of overpotential is polarized heat.

The above heat generation is the normal heat generation of batteries in the normal charge and discharge cycles. Because there is no decomposition and side reaction of the battery material, there is no severe heat generation. However, in the working condition of high current and poor heat dissipation, a large amount of heat will accumulate inside the battery. When the temperature reaches the initial temperature of side reaction, the battery material will decompose rapidly to generate heat, which leads to thermal runaway.

Side reaction refers to the violent chemical reaction in which the material structure and concentration inside the battery change when the temperature or voltage reaches a certain level. Generally, side reactions will produce a large amount of heat and gas, which will consume cathode and anode materials and electrolyte, and make the battery structurally damaged and out of control. The side reactions mainly include the decomposition of SEI film, Lithium deposition on the anode and reacts with electrolyte, decomposition of electrolyte and decomposition of cathode material, etc. Nobusato (Noboru 2001) added the influence of side reaction heat generation when studying the thermal characteristics of batteries, and improved the heat generation model of batteries:

$$Q_{\mathrm{t}} = Q_{\mathrm{r}} + Q_{\mathrm{p}} + Q_{\mathrm{s}} + Q_{\mathrm{ohm}} = I_L T\left(\frac{\mathrm{d}E_0}{\mathrm{d}T}\right) + Q_{\mathrm{p}} + Q_{\mathrm{ohm}} + Q_{\mathrm{s}} \tag{1.2}$$

where, Q_{t} is the total heat production of the battery; Q_{r} is the reversible heat; Q_{p} is the polarization heat; Q_{ohm} is the ohmic heat; Q_{s} is the heat of side reaction.

In addition, there are many methods to model the thermal characteristics of batteries, such as one-dimensional, two-dimensional or three-dimensional thermal models of batteries according to different dimensions (Hallaj and Selman 2002; Khateeb et al. 2005; Kim and Pesaran 2006), and battery thermal models can be divided into electrochemical–thermal coupling models and electro-thermal coupling models according to different modeling principles (Hallaj et al. 1999; Funahashi et al. 2002). In practical research, there are various practical methods that can be used to model the thermal characteristics of batteries.

Wang (2013) established a prediction model for the internal temperature of battery cells according to the principles of electrochemical model and heat transfer. Joule heat, electrode reaction, heat conduction and convection heat transfer were mainly considered in the model. The accuracy of the above model is verified by the constant rate discharge experiment of batteries.

Kim et al. (2007) established a heat abuse model based on a three-dimensional thermal model for lithium-ion batteries, studied all the reactions that may cause heat generation in the batteries, coupled the heat generation of these reactions into the above model, and then analyzed the internal temperature field of the batteries through experiments. The results showed that the above model has high accuracy.

Zhu et al. (2013) established a battery thermal model based on the Porous Electrode and Concentrated Solution Theory, which can accurately predict the heat generation rate, heat dissipation rate and the temperature rise of the cell in the battery pack. As for battery heat generation rate, the author studied the reaction heat, Joule heat and polarization heat of batteries, and studied the influence of current and SOC on the heat generation rate of the batteries during charge and discharge. Then, the author studied the heat dissipation model of the battery, and studied the heat dissipation under natural convection and forced convection. Finally, the author concluded that: ① the model was in good agreement with the experimental results; ② battery SOC had great influence on reversible heat. When SOC was the same, reversible heat and irreversible heat generated during constant current charge and discharge were the same; ③ When charging and discharging at a constant current, reversible heat takes effect, and when charge and discharge, reversible heat will cancel each other out, so reversible heat will have no effect.

Smyshlyaev et al. (2011) adopted the form of two-dimensional partial differential equation when establishing the thermal model of battery. The model was simplified on the basis of the CFD/FEM model, and had good compatibility with CFD model. In addition, it can estimate and track the parameters and states of the model. Simulation results show that this model can shorten the calculation time.

Gambhire et al. (2015) proposed a Reduced Order Model, which can be used to study the electrochemical and thermal characteristics of batteries. The model adopted lumped heat balance equation. In order to consummate the model, the author also adopted the distributed heat balance equation, which included many parts of a battery cell that can generate heat. The improved model can not only be used in electric vehicles in real time, but also be beneficial to the design and development of lithium-ion batteries.

Liu and Li (2013) put forward an integrated method based on experimental data to predict the temperature distribution of lithium-ion batteries online, which is highly practical.

Rad et al. (2013) developed a battery heat generation model, and considered that there are two factors for battery heat generation:

① The influence of overpotential based on temperature and current, which includes joule heat;
② The influence of entropy based on battery SOC.

In particular, the paper emphasized the influence of entropy that could not be ignored. The accuracy of the model could be improved by integrating the influence of entropy based on SOC into the model. This paper also studied the convection heat dissipation of batteries, and claimed that only when the convection heat transfer coefficient reached 80 W/(m^2 K) could the heat dissipation requirements be met.

Damay and Forgez (2013) analyzed the geometric and physical characteristics of prismatic lithium-ion power battery, obtained related thermophysical parameters of the battery through experiments, and then used the parameters in the simulation of the battery thermal model. The correctness of the model was proved by comparing the simulation results with the experimental results.

On the whole, the modeling of battery thermal characteristics involves many disciplines such as electrochemistry, materials science, heat transfer and thermodynamics, and it also needs the help of mathematical modeling, engineering numerical calculation and CFD thermodynamic and fluid simulation.

1.4.2 Research on the Modeling of Thermal Runaway of Power Batteries

In order to clearly understand the relationship between temperature and heat generation rate in each stage of side reaction when thermal runaway occurs in lithium-ion batteries, it is necessary to establish a battery thermal runaway model to simulate the thermal characteristics of the battery under abuse, so as to provide design basis and simulated analysis means for battery safety early warning and protection design. So far, scholars at home and abroad have done a lot of research on thermal runaway of batteries.

Hatchard et al. (2001) established a thermal runaway model of concentrated mass of lithium-ion batteries when studying the thermal runaway of lithium-ion batteries at high temperature. Three Arrhenius side reaction equations were added to the model to describe the reaction rate and heat generation rate of SEI film decomposition, anode material side reaction and cathode material side reaction. Then, the high temperature hot box experiment of lithium-ion battery in the range of 145–175 °C was carried out. The accuracy of the model was verified by comparing the experimental data with the simulation data. In addition, by changing experimental conditions and model simulation conditions, it was claimed that the model can be applied to batteries of other materials and sizes.

Kim et al. (2007) established a three-dimensional heat abuse model for lithium-ion batteries, which was used to simulate the thermal runaway phenomenon of batteries at high temperature. Some simulation results are shown in Fig. 1.19. It is worth noting that the three-dimensional model includes side reaction models. The side reaction theory refers to the research of T. D. Hatchard, and is also expressed by Arrhenius side reaction equation. The three-dimensional model can describe the temperature distribution of battery under high temperature abuse experiment, and reflect whether the battery has thermal runaway by temperature. After studying the results of batteries with different sizes under the same abuse conditions, it is found that batteries with large sizes are more prone to thermal runaway.

Feng Xuning et al. of Tsinghua University (Feng 2016; Feng et al. 2015) studied the thermal runaway propagation of a 25 A h ternary lithium-ion battery pack, and established a three-dimensional model of thermal runaway propagation of the battery pack. Six batteries connected in series were utilized in the experiment, and nail penetration experiment was conducted to induce the thermal runaway of the battery pack for verifying the accuracy of the model. The built three-dimensional model can simulate the temperature distribution during thermal runaway expansion, as shown in Fig. 1.20. The errors between the simulated battery temperature curve and the measured data were within the acceptable range, and the battery simulation with different parameters also attain the requirements. Through experiments and model analysis, it was found that reducing the internal energy of the battery, increasing the convective heat transfer coefficient to 70 W/(m^2 K) or adding heat-resistant and heat-resistant materials between the batteries could effectively reduce the harm of heat runaway propagation.

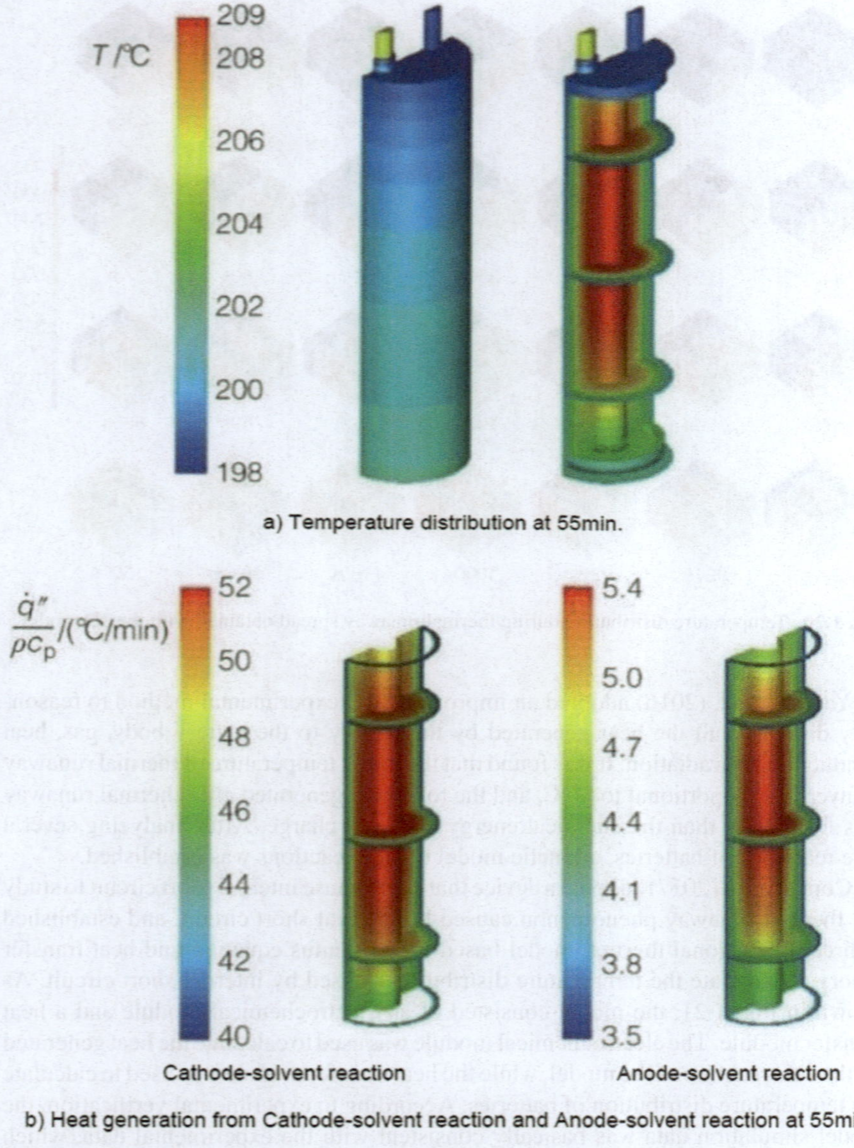

Fig. 1.19 Three-dimensional simulation of 18650 batteries in oven test at 155 °C

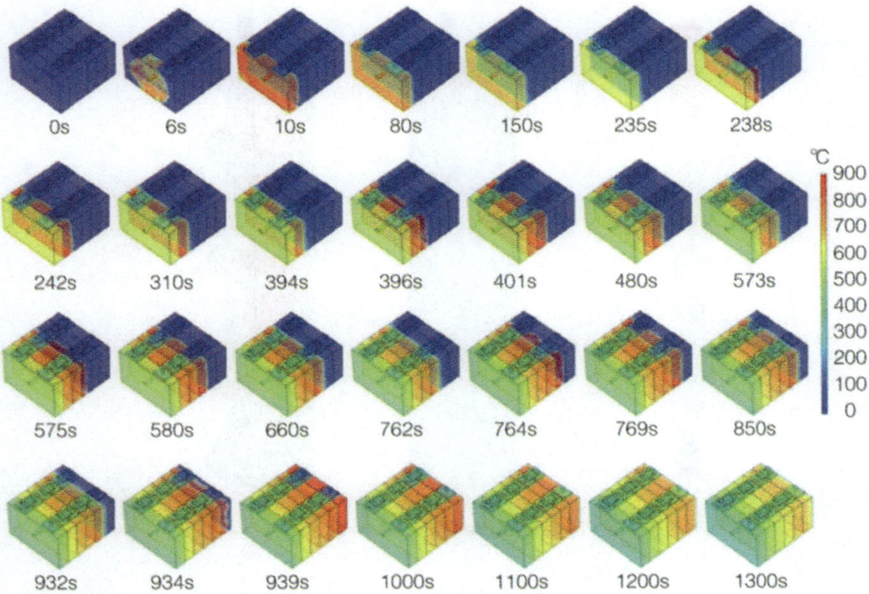

Fig. 1.20 Temperature distribution during thermal runaway spread obtained with the 3D model

Yayathi et al. (2016) adopted an improved ARC experimental method to reasonably distribute all the heat generated by the battery to the battery body, gas, heat conduction and radiation. It was found that the initial temperature of thermal runaway is inversely proportional to SOC, and the total heat generated after thermal runaway was far greater than the chemical energy after full charge. After analyzing several side reactions of batteries, a kinetic model of side reactions was established.

Coman et al. (2017) adopted a device that could cause internal short circuit to study the thermal runaway phenomenon caused by internal short circuit, and established a three-dimensional thermal model based on Arrhenius equation and heat transfer theory to simulate the temperature distribution caused by internal short circuit. As shown in Fig. 1.21, the model consisted of an electrochemical module and a heat transfer module. The electrochemical module was used to calculate the heat generated by the side reaction of the model, while the heat transfer module was used to calculate the temperature distribution of batteries. According to experimental verification, the model simulation data was basically consistent with the experimental data, which could accurately reflect the temperature distribution of the batteries.

At present, the research on thermal runaway model of lithium-ion battery mainly focuses on the change of heat generation rate of battery side reactions with the increase of temperature under the abuse of high temperature. For other abuse cases, such as overcharging, overdischarging, short circuit, nail penetration and extrusion, there is a lack of reasonable modeling analysis. Most of the theoretical basis for the study of side reactions is Arrhenius equation, that is, the relationship between

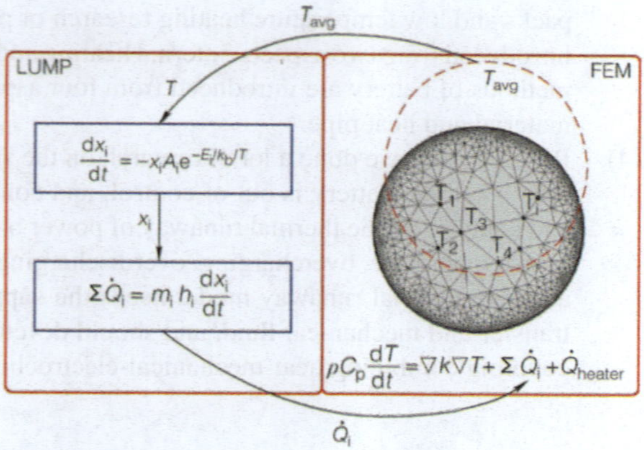

Fig. 1.21 Coupling modeling of thermal runaway based on lumped electrochemical model and three-dimensional finite element heat transfer model

chemical reaction rate and temperature. Three-dimensional thermal diffusion equation is the theoretical basis of battery temperature distribution. There is a lack of modeling and analysis for the variation of voltage and current of batteries under abuse, and the reactions inside the battery electrochemistry are not really involved, and the coupled simulation of electrochemistry and heat is not achieved. At present, the thermal runaway model of lithium-ion battery needs further improvement, and the parameters of battery such as temperature, voltage, current and impedance should be reflected in the model. To build an accurate model, the support of many disciplines such as electrochemistry, heat transfer and mechanical fluid are needed, and the future development direction should be the multi-physical coupled modeling with the combination of heat-mechanical-electrochemistry.

Summary

In this chapter, the thermal management status of power batteries is elaborated, including the research status of new energy vehicles, power batteries, thermal safety of power batteries, and thermal management and modeling of batteries, as follows:

(1) With the double guarantee of policy support and technological progress, new energy vehicles have developed rapidly in recent years, but at the same time, some safety problems can not be ignored. On this basis, researchers have done a lot of research on thermal management and thermal safety management of power batteries.

(2) Researchers have done a lot of research from the perspectives of experiment, mechanism and modeling, and the battery heat generation rate model proposed by D. Bernardi has been widely used. The modeling of battery thermal characteristics involves many disciplines, such as electrochemistry, materials science, heat transfer and thermodynamics. At the same time, mathematical modeling, engineering numerical calculation and CFD thermodynamic and fluid simulation are needed.

(3) Thermal management technology of electric vehicle power battery mainly involves the following two aspects: heat dissipation research of power battery

packs and low temperature heating research of power battery pack. Heating is introduced from two aspects: internal heating and external heating, and cooling methods of battery are introduced from four aspects: air, liquid, phase change material and heat pipe.

(4) Researchers have done a lot of research on the side reactions of various materials when the battery is out of control, and conducted experimental research and analysis on the thermal runaway of power battery from the perspectives of high temperature, overcharging, overdischarging and internal short circuit. An accurate thermal runaway model needs the support of electrochemistry, heat transfer and mechanical fluid, and should develop towards the multi-physical coupling combining heat-mechanical-electrochemistry.

References

Alaoui C, Salameh ZM (2001) Solid state heater cooler: design and evaluation. In: 2001 large engineering systems conference on power engineering. Conference proceedings, Halifax, NS, Canada, pp 139–145

Alaoui C, Salameh ZM (2003) Modeling and simulation of a thermal management system for electric vehicles. In: The 29th annual conference of the IEEE industrial electronics society, Roanoke, VA, United States, pp 887–890

Alaoui C, Salameh ZM (2004) A novel thermal management for electric and hybrid vehicles. IEEE Trans Veh Technol 54(2):468–476

Al-Hallaj S, Selman JR (2000) A novel thermal management system for electric vehicle batteries using phase-change material. J Electrochem Soc 147(9):3231–3236

Al-Hallaj S, Selman JR (2002) Thermal modeling of secondary lithium batteries for electric vehicle/hybrid electric vehicle applications. J Power Sources 110:341–348

An P, Qi L (2006) Application and development of lithium-ion secondary batteries. Acta Scientiarum Naturalium Universitatis Pekinensis 42(1)

Brown S, Pyke D, Steenhof P (2010) Electric vehicles: the role and importance of standards in an emerging market. Energ Policy 38(7):3797–3806

Chen XG (2020) How to "climb a hill over a hurdle." Energy Conserv Environ Prot 5:18–19

Chen CC, Gibbard HF (1979) Thermal management of battery systems for electric vehicles and utility load leveling. In: Proceedings of the 14th intersociety energy conversion engineering conference. American Chem. Soc., IEEE, American Nuclear Soc, Boston, MA, USA, pp 725–729

Chery Automobile Co. Ltd. (2010) A lithium-ion power battery heating device. China: CN101710631A

China Industry Technology Innovation Strategic Alliance for Electric Vehicle (2020) Annual tracking report on power battery technology for new energy vehicles (2019). China Industry Technology Innovation Strategic Alliance for Electric Vehicle, Beijing

China Society of Automotive Engineers (2020) Technical roadmap for energy saving and new energy vehicles 2.0. China Society of Automotive Engineers, Beijing

Coman PT, Darcy EC, Veje CT et al (2017) Modelling lithium-ion battery thermal runaway triggered by an internal short circuit device using an efficiency factor and Arrhenius equationtions. J Electrochem Soc 164(4):A587–A593

Dai YN, Yang B, Yao YC, et al (2005) Development of lithium-ion batteries. Batter Bimon 35(3)

Damay N, Forgez C (2013) Thermal modeling and experimental validation of a large prismatic lithium-ion battery. Ind Electron Soc 4694–4699

Duan X, Naterer GF (2010) Heat transfer in phase change materials for thermal management of electric vehicle battery modules. Int J Heat Mass Transf 53:5176–5182

Feng X, He X, Ouyang M et al (2015) Thermal runaway propagation model for designing a safer battery pack with 25Ah $LiNi_xCo_yMn_zO_2$ large format lithium-ion battery. Appl Energy 154:74–91

Feng XN (2016) Mechanism, modeling and prevention of thermal runaway in automotive lithium-ion power batteries, Tsinghua University

Funahashi A, Kida Y, Yanagida K, et al (2002) Thermal simulation of large-scale lithium secondary batteries using a graphite-coke hybrid carbon negative electrode and $LiNi_{0.7}Co_{0.3}O_2$ positive electrode. J Power Sources 104:248–252

Gambhire P, Ganesan N, Basu S, Hariharan KS, Kolake SM, Song T, Oh D, Yeo T, Doo S (2015) A reduced order electrochemical thermal model for lithium-ion batteries. J Power Sources 290:87–101

GM Global Technology Operations. Vehicle heating, ventilation, air conditioning and battery thermal management. 101386285A

Hallaj SA, Maleki H, Hong JS et al (1999) Thermal modeling and design considerations of lithium-ion batteries. J Power Sources 83:1–8

Hallaj SA, Selman JR (2002) Thermal modeling of secondary lithium batteries for electric vehicle/hybrid electric vehicle applications. J Power Sources 110:341–348

Hande A, Stuart T (2004) Effects of high frequency AC currents on cold temperature battery performance. In: Proceedings of the 2nd IEEE India International Congress on Power Electronics (IICPE 2004), Mumbai, India

Hatchard TD, Macneil DD, Basu A et al (2001) Thermal model of cylindrical and prismatic lithium-ion batteries. J Electrochem Soc 148(7):A755–A761

Intelligence Research Group (2018) Report on special investigation and investment prospect evaluation of China's lithium-ion battery market from 2018 to 2024. Intelligence Research Group, Beijing

Ji Y, Wang CY (2013) Heating strategies for lithium-ion batteries operated from subzero temperatures. Electrochim Acta 107:664–674

Khateeb SA, Amiruddin S, Farid M et al (2005) Thermal management of lithium-ion battery with phase change material for electric scooters: experimental validation. J Power Sources 142:345–353

Khateeb SA, Farid MM, Selman JR, Al-Hallaj S, et al (2004) Design and simulation of a lithium-ion battery with a phase change material thermal management system for an electric scooter. J Power Sources 128:292–307

Khateeb SA, Amiruddina S, et al (2005) Thermal management of lithium-ion battery with phase change material for electric scooters: experimental validation. J Power Sources 142:345–353

Kim GH, Pesaran A, Spotnitz R (2007) A three-dimensional thermal abuse model for lithium-ion batteries. J Power Sources 170:476–489

Kim GH, Pesaran A (2006) Battery thermal management system design modeling. In: The 22nd international battery, hybrid and fuel cell electric vehicle conference and exhibition. Japan Automobile Research Institute, Yokohama, Japan

Kizilela R, Lateefa A, Sabbaha R, et al (2008) Passive control of temperature excursion and uniformity in high-energy lithium-ion battery packs at high current and ambient temperature. J Power Sources 183:370–375

Kras B, Aebi A (2010) Improvement of low temperature performance of SAM EV-II lithium-ion battery pack by applying active thermal management based on Peltier elements. In: Proceedings of the 25th world battery, hybrid and fuel cell electric vehicle symposium, Shenzhen, China, pp 1–5

Kras B, Ciosek M, Makomaski K (2009) Thermal management of lithium polymer based battery pack for urban BEV. In: Proceedings of the 24th world battery, hybrid and fuel cell electric vehicle symposium, Stavanger, Norway, pp 1–4

Liu Z, Li H (2013) Integrated modeling for intelligent battery thermal management. International conference, pp 2522–2527

Ma X (2010) Research on temperature characteristics and heating management system of lithium-ion batteries for electric vehicles. Beijing Institute of Technology, Beijing, pp 47–52

Matthe R, Turner L (2011) VOLTEC battery system for electric vehicle with extended range. SAE 4(1):1944–1962

Newman, Bernardi, Pawlikowski (1985) A general energy-balance for battery systems. J Electrochem Soc 132(1):5–12

Noboru S (2001) Thermal behavior analysis of lithium-ion batteries for electric and hybrid vehicles. J Power Sources 99(1–2):70–77

Pesaran AA (2001) Battery thermal management in EVs and HEVs: issues and solutions [EB/OL]. http://www.nrel.gov/vehiclesandfuels/energystorage. Accessed 06 Feb 2001

Pesaran AA, Burch S, Keyser M (1999) An approach for designing thermal management systems for electric and hybrid vehicle battery packs. In: Proceeding of the 4th vehicle thermal management systems conference and exhibition, London, UK, pp 1–16

Qin W, Chen X (2015) Top ten strategic industries expect breakthroughs—detailed explanation of technology roadmap in key areas of the new Made in China 2025. Equip Manuf

Rad MS, Danilov DL, Baghalha M, Kazemeini M, Notten PHL (2013) Adaptive thermal modeling of lithium-ion batteries. Electrochim Acta 102:183–195

Rafalovich A, Longardner W, Keller G, et al (1994) Thermal management of electric vehicle's batteries using phase change materials. In: Winter annual meeting of the American Society of Mechanical Engineers, Chicago, IL, USA, pp 1–4

Ruan H, Jiang J, Sun B, Wu N, Shi W, Zhang Y (2014) Stepwise segmented charging technique for lithium-ion battery to induce thermal management by low-temperature internal heating. In: 2014 IEEE transportation electrification conference and expo, ITEC Asia-Pacific 2014-coference proceedings

Rui C, Tang CM, Wang XW, Zhang MR (2013) Influence of graphite on the safety of lithium-ion batteries for power tools. Electr Tool 05:16–19

Shao XT, Feng H (2018) Research progress of life cycle safety evolution of lithium-ion batteries. Chem Eng Des Commun 44(12):193

Smyshlyaev A, Krstic M, Chaturvedi N, et al (2011) PDE model for thermal dynamics of a large lithium-ion battery pack. In: American control conference, pp 959–964

Stuart TA, Hande A (2004) HEV battery heating using AC currents. J Power Sources 129:368–378

Su BH (2019) Research on the development trend of new energy vehicles. Auto Time 16:88–89

Tröltzsch U, Kanoun O, Tränkler H-R (2006) Characterizing aging effects of lithium-ion batteries by impedance spectroscopy. Electrochim Acta 51(8):1664–1672

United Automotive Electronic Systems Co., Ltd. (2009) Automotive battery thermal management systems and working methods. China 101577354A

Wang T (2013) Development of a one-dimensional thermal-electrochemical model of lithium-ion battery. Ind Electron Soc 6709–6714

Wang WW, Sun FC (2019) Key technologies and prospects of new energy vehicles in the whole climate. Strat Study of CAE 21(03):47–55

Wang F, Xia J (2017) Design and manufacturing technologies of EV power battery systems. Science Press, Beijing

Wang CY, Zhang GS, Ge SH et al (2016) Lithium-ion battery structure that self-heats at low temperatures. Nature 529:515–518

Wu ZJ, Zhang GQ (2008) Liquid cooling systems of Ni-MH batteries for hybrid electric vehicles. J Guangdong Univ Technol 25(4):28–31

Wu MS, Liu KH, Wang YY, et al (2002) Heat dissipation design for lithium-ion batteries. J Power Sources 109:160–166

Wu K, Zhang Y, Zeng YQ, Yang J (2011) Study on the safety performance of lithium-ion batteries. Prog Chem Z1:401–409

Wu C, Zhu C, Ge Y et al (2015) A review on fault mechanism and diagnosis approach for lithium-ion batteries. J Nanomater 2015:1–9
Yayathi S, Walker W, Doughty D et al (2016) Energy distributions exhibited during thermal runaway of commercial lithium-ion batteries used for human spaceflight applications. J Power Sources 329:197–206
Yu XS (2020) The international competitiveness of China's lithium-ion battery industry has obviously improved. China Information World. 2020-07-20(014)
Zhang JB, Lu LG, Li Z (2012) Key technologies and fundamental academic issues for traction battery systems. J Automot Saf Energy 3(02):87–104
Zhang GQ, Wu ZJ, Rao ZH et al (2009) Experimental study on cooling effect of power battery heat pipes. Chem Ind Eng Prog 28(7):1165–1168
Zhang J, Ge H, Li Z, Ding Z (2015) Internal heating of lithium-ion batteries using alternating current based on the heat generation model in frequency domain. J Power Sources 273:1030–1037
Zhang SS, Xu K, Jow TR (2003) The low temperature performance of lithium-ion batteries. J Power Sources 115:137–140
Zhao J, Yang WZ, Zhao JM (2000) Application and development of lithium-ion batteries. Batter Ind 01:31–36
Zhao XW, Zhang GY, Yang L, Qiang JX, Chen ZQ (2011) A new charging mode of lithium-ion batteries with $LiFePO_4$/C composites under low temperature. J Therm Anal Calorim 104(2):561–567
Zhu C, Li X, Song L, Xiang L (2013) Development of a theoretically based thermal model for lithium-ion battery pack. J Power Sources 223:155–164XXX

Chapter 2
Analysis on Charge and Discharge Temperature Characteristics of Lithium-ion Batteries

The influences of temperature on the characteristics of lithium-ion batteries are mainly reflected in battery capacity, internal resistance, charge and discharge power and so on. High temperature and low temperature have different influences on the battery characteristics, low temperature mainly causes the battery performance to deteriorate or even fail to be used normally, while high temperature mainly considers the runaway behavior of the battery when heated. Therefore, this chapter takes prismatic ALF batteries as the research object, aiming at the low temperature condition, through the charge and discharge experiments of batteries in different temperature environments, studies the influence of temperature on the charge and discharge capacities, voltage working platforms, ohmic resistance, AC impedance and charge and discharge powers of batteries. At the same time, under natural heat dissipation, the thermal characteristics of the battery during charge and discharge are studied and analyzed.

2.1 Structure and Working Principle of Lithium-ion Batteries

2.1.1 Structure of Lithium-ion Batteries

A lithium-ion battery refers to a secondary battery system in which two different compounds capable of reversibly intercalating and deintercalating lithium-ions are used as the cathode and anode of the battery respectively (Zheng 2007). A lithium-ion battery is mainly composed of cathode, anode, electrolyte and separator. Figure 2.1 shows the structural schematic diagram of a cylindrical and a prismatic lithium-ion battery. Lithium-ion battery cathode adopts the lithium-ion intercalation compounds, such as lithium cobalt oxide ($LiCoO2$), lithium nickel oxide ($LiNiO2$), lithium

J. Li, *Modeling and Simulation of Lithium-ion Power Battery Thermal Management*,
Key Technologies on New Energy Vehicles,
https://doi.org/10.1007/978-981-19-0844-6_2

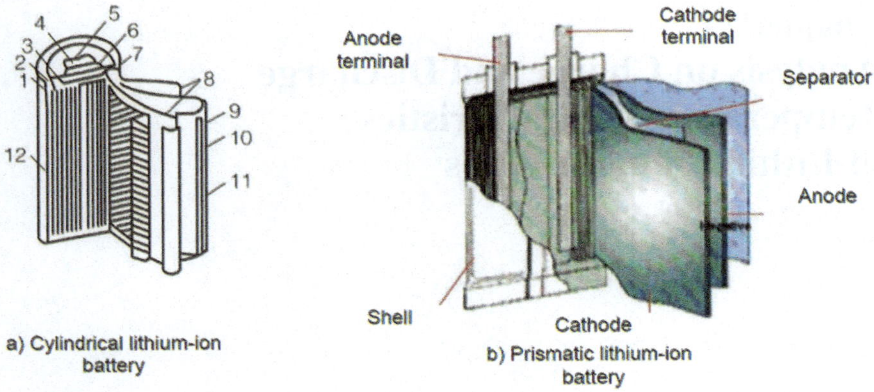

Fig. 2.1 Lithium-ion battery structure. 1-insulator, 2-gasket, 3-PTC element, 4-cathode terminal, 5-vent hole, 6-explosion-proof valve, 7-cathode, 8-separator, 9-anode, 10-anode lead, 11-cathode, 12-shell

manganese oxide (LiMn2O4), lithium iron oxide (LiFePO4) and vanadium oxide. The anode is divided into carbon-based materials and non-carbon-based materials, and the carbon-based materials mainly include graphitized carbon materials and amorphous carbon materials. Non-carbon-based materials mainly include titanium oxides, nitrides, silicon-based materials, tin-based oxides and nano-oxides (Wu et al. 2006). Traditional battery electrolytes are aqueous solutions, such as lead-acid batteries. Because the working voltage of lithium-ion battery is high (3 –4.2 V), non-aqueous electrolyte is adopted, which mainly includes liquid, solid and molten salt (Huang et al. 2008).

Microscopically, a lithium-ion battery is formed by winding or stacking several cells in sandwich structure, as shown in Fig. 2.2. Each cell consists of five parts: positive and negative current collectors, positive and negative active materials and a separator. And the electrolyte and solvent are distributed around the sandwich structure.

Positive and negative current collectors: Most of the anodes are made of aluminum foil, and the cathodes are made of copper foil. Their function is to collect the current from the cathode and anode, so that the active materials can be uniformly distributed and support the overall structure of the battery.

Positive and negative active materials: Because of their porous properties, they are also named porous electrodes. Carbon is dominant in cathode materials. Lithium salts such as lithium iron phosphate, lithium manganate and NMC are the main components of the cathode. Where, the solid phase transfer coefficient and lithium concentration are two indexes that affect the charge–discharge rate and capacity, and the energy density is its volume fraction.

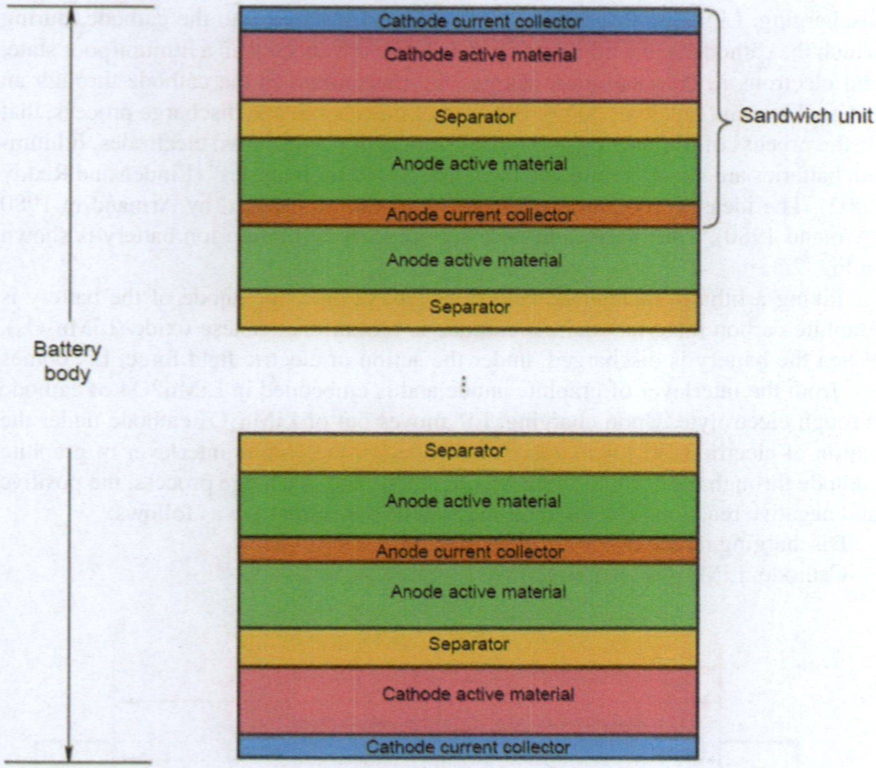

Fig. 2.2 Microstructure of lithium-ion battery

Separator: its function is to separate the cathode and the anode to prevent internal short circuit. Polyolefin materials such as polyethylene and polypropylene with pores are its main components, and only lithium-ions can pass through the separator.

Electrolyte: it includes electrolyte lithium salt, organic solvent and additives. the lithium salt is generally $LiPF_6$ with high conductivity, good thermal stability and no toxicity, while the organic solvents are mostly EC, PC and EMC, etc.

2.1.2 Working Principle of Lithium-ion Battery

In the charging process of a lithium-ion battery, Li^+ is removed from the positive compound and adsorbed by the carbon substance in the anode, and the anode is in a low-potential lithium-rich state, while the cathode is in a high-potential lithium-poor state. Electrons are transferred to the anode as compensation charges through an external circuit, so that the anode charges are kept in balance. Similarly, during discharging, Li^+ is removed from the anode and inserted into the cathode, during which the cathode is in a lithium-rich state while the anode is in a lithium-poor state, and electrons as compensation charges are transmitted to the cathode through an external circuit (Liu et al. 2009). Because of this charge and discharge process, that is, the process of lithium-ion going back and forth between two electrodes, lithium-ion batteries are vividly compared to as "rocking chair batteries" (Linden and Reddy 2002). The idea of "rocking chair battery" was first proposed by Armand in 1980 (Armand 1980). The charge and discharge process of lithium-ion battery is shown in Fig. 2.3.

Taking a lithium manganate battery as an example, the anode of the battery is graphite carbon material, and the cathode is lithium manganese oxide ($LiMn_2O_4$). When the battery is discharged, under the action of electric field force, Li^+ comes out from the interlayer of graphite anode and is embedded in LiMn2O4 of cathode through electrolyte. Upon charging, Li^+ moves out of $LiMn_2O_4$ cathode under the action of electric field force, and is embedded in the carbon interlayer of graphite cathode through electrolyte. In the whole charge and discharge process, the positive and negative reactions and the total reaction of the battery are as follows:

Discharging reaction:

Cathode: $LiMn_2O_4 \rightarrow Li_{1-x}Mn_2O_4 + xLi^+ + xe^-$.

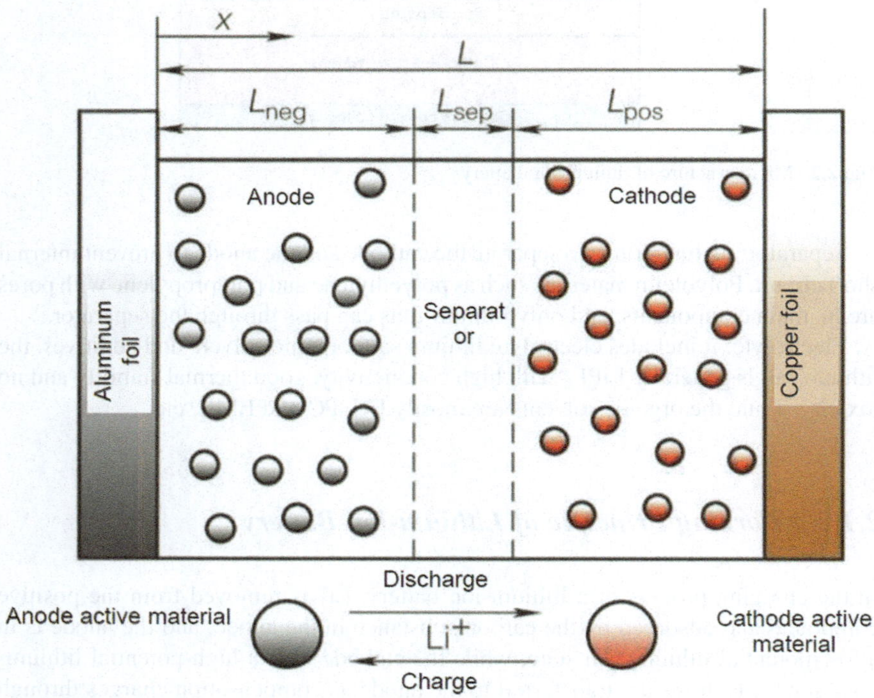

Fig. 2.3 Charge and discharge process of lithium-ion battery

Anode: $C + xLi^{+} + xe^{-} \rightarrow Li_xC$.
Total battery reaction: $LiMn_2O_4 + C \rightarrow Li_{1-x}Mn_2O_4 + Li_xC$.
Discharge reaction:
Cathode: $Li_{1-x}Mn_2O_4 + xLi^{+} + xe^{-} \rightarrow LiMn_2O_4$.
Anode: $Li_xC \rightarrow C + xLi^{+} + xe^{-}$.
Total battery reaction: $Li_{1-x}Mn_2O_4 + Li_xC \rightarrow LiMn_2O_4 + C$.

2.2 Influence of Temperature on Charge and Discharge Performance of Lithium-ion Batteries

Temperature is an important factor affecting the performance of lithium-ion batteries, so it is a key element in the research of battery thermal characteristics and thermal management to clarify the influence of temperature on battery charge and discharge performance. This section will take a lithium-ion power battery as an example, starting from the battery temperature characteristic experiment, and analyze the concrete influence of temperature on the battery charge and discharge voltage, capacity and internal resistance.

2.2.1 Experimental Platform for Battery Charge and Discharge Temperature Characteristics

The structural block diagram of a battery performance test experimental platform is shown in Fig. 2.4. The whole platform is composed of battery charge and discharge devices, thermostat, temperature measurement module, data acquisition system and electrochemical workstation.

The battery charge and discharge devices are Digatron EVT500-500 developed for lithium-ion battery pack test and Qingtian HT-V5C200D200-4 developed for battery cell test. Digatron EVT500- 500 can reach the maximum charge and discharge current of 500 A, and the maximum voltage of 500 V. Its main working modes include constant current mode, constant voltage mode, constant power mode and constant internal resistance mode, and dynamic working conditions such as DST and FUDS can be called. EVT500-500 is connected with the upper control computer through the CAN interface, and the upper control computer sets the working mode of the EVT500-500 through the test system "BTS-600" and records the current and voltage values of the battery pack in real time. During the test, the upper control computer can adjust the parameters such as charge and discharge currents, charge and discharge time, cut-off voltage and cycle times as required. After the test, the test system can make the experimental data into graphics, texts or tables as required. Kinte HT-V5C200D200-4 can reach the max. voltage of 5 V and the max. charging/discharging current of 200A. This device is only used for testing battery cells with high testing

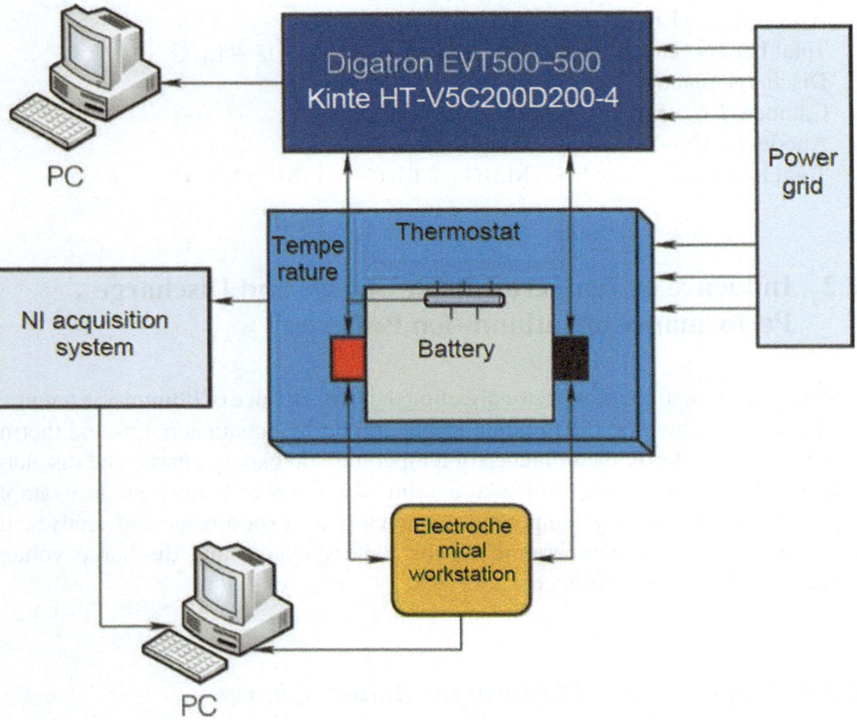

Fig. 2.4 Block diagram of experimental platform for battery performance test

accuracy, and its main functions are similar to those of EVT500-500. See Table 2.1 for main parameters of Digatron EVT500-500 and Kinte HT-V5C200D200-4.

The function of the thermostat is to provide the ambient temperature required for testing. In the testing process, the tested battery is placed in a temperature box with a set temperature, and the battery reaches the set temperature by standing for a certain time, thus simulating the real state of the battery in different temperature environments. Then, the battery in the thermostat is tested for charge and discharge, and the charge and discharge performance of the battery at different temperatures

Table 2.1 Parameters of EVT500-500 and HT-V5C200D200-4 test systems

Parameters	EVT500-500	HT-V5C200D200-4
Max. charge and discharge current	500 A	200 A
Current measurement error	±0.5%	±0.05%
Max. voltage	500 V	5 V
Voltage measurement error	±0.5%	±0.05%

is obtained. With the dimensions of 600 mm × 600 mm × 730 mm, the thermostat can provide a temperature range from 40 to 80 °C, meeting the requirements of the experiment.

The temperature measurement module is used to measure the temperature change of the battery surface and lugs during charge and discharge. The temperature measurement module has 16 channels and adopts PT100 temperature sensor. The voltage signal output by the temperature sensor is transmitted to the upper computer through the data acquisition system, and the software installed in the upper computer converts the voltage signal into a specific temperature value and saves and displays the data in real time.

The electrochemical workstation is used to measure the AC impedance spectrum of the battery and the impedance value at fixed frequency.

2.2.2 *Charge and Discharge Characteristics of Lithium-ion Batteries at Room Temperature*

The lithium manganate battery is taken as the research object, and its appearance is shown in Fig. 2.5. This battery is a pouch battery, and its shell is made of ALF. See Table 2.2 for its basic parameters.

Fig. 2.5 Cell appearance

Table 2.2 Basic parameters of lithium manganate battery

Parameters of lithium manganate battery	Specific numerical value
Rated capacity	35A·h
Rated voltage	3.7 V
Dimensions	300 mm × 168 mm × 15 mm
Mass	1.02 kg

1. Static charge and discharge characteristics

In order to understand the influence of charging current on charge capacity and voltage, and the influence of discharging current on discharge capacity and voltage, static charge and discharge experiments should be carried out on lithium manganate batteries. Static charging methods mainly include constant current-constant voltage charging, constant voltage charging and constant power charging. A constant current-constant voltage charging method is adopted here, that is, firstly, the battery is charged to the upper limit cut-off voltage with a certain constant current *I*, then the battery is trickle charged with the cut-off voltage as a constant voltage, and the charging is stopped when the charging current drops to *I*/10. Static discharge methods mainly include constant current discharging, constant resistance discharging and constant power discharging. A constant current discharge method is adopted here, that is, the battery is discharged with a certain constant current *I*, and the discharging is stopped when the terminal voltage of the battery drops to the lower limit cut-off voltage.

Static discharge experiment is carried out at normal temperature. Before the constant current discharging experiment, the battery is charged at constant current and constant voltage at a rate of 1/3C at first, and then stand for 2 h after being fully charged. After standing, the battery is discharged at constant current of 10 A, 35 A, 70 A and 140 A respectively, with a cut-off voltage of 3 V. The relationships between cell voltage and capacity at different discharge rates are shown in Fig. 2.6.

It can be seen from Fig. 2.6 that with the increase of discharge rate, the terminal voltage of the battery drops rapidly and the discharge capacity decreases. Comparing the constant-current discharge of 140A (4C rate) with the constant-current discharge of 10A, the terminal voltage of the battery decreased by 8.23% on average with the

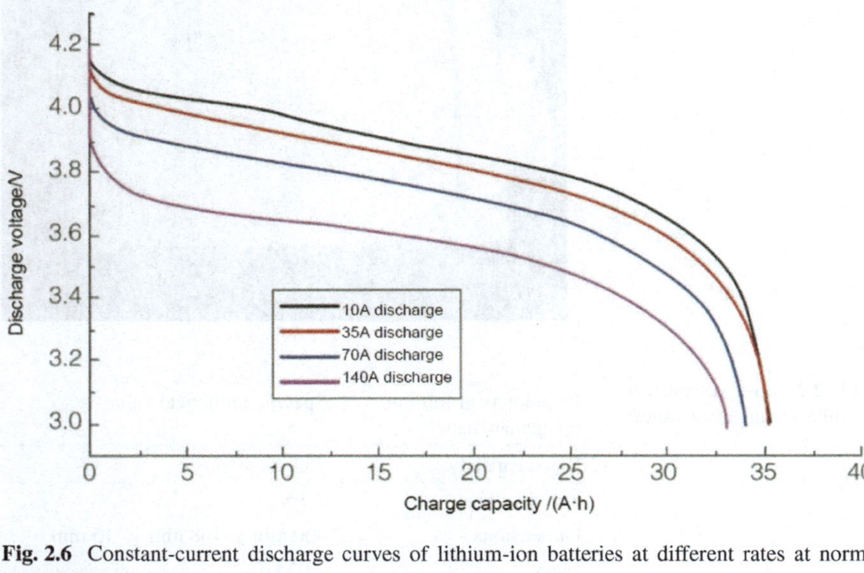

Fig. 2.6 Constant-current discharge curves of lithium-ion batteries at different rates at normal temperature

max. decrease of 14.11%, and the discharge capacity decreased by 6.49%. Compared with the voltage drop degree, the discharge capacity drops slightly. If constant current discharge is carried out at 70A, the capacity will only decrease by 3.71%. It can be found through discharge experiments at different rates that when the capacity of the battery is released to about 20%, the terminal voltage of the battery drops rapidly to the cut-off voltage, whether it is discharged at a high rate or a low rate, which indicates that the battery polarization is serious at the end of discharge. Moreover, related research shows that the discharge depth of the battery has great influence on the cycle life. Therefore, in practical use, deep discharge of batteries should be avoided as far as possible (Doerffel and Sharkh 2006). The purpose of discharging the battery at high current is to meet the demand of high power in vehicle operation. However, if a battery is discharged at high rate for a long time, the discharge energy of the battery will decrease. When a battery is discharged at the constant current of 10A, the discharge energy of the battery is 135.46 W·h, while when the battery is discharged at the constant current of 140A, the discharge energy of the battery is 117.48 W·h, decreasing by 13.27%. Therefore, batteries should avoid long-term high current discharge.

According to Peukert's theory, at the same temperature, when a battery is discharged at a constant current, the current I, the discharge time t and the discharge capacity C of the battery satisfy the following relationship:

$$I^n t = C \tag{2.1}$$

where, n is the time constant of the battery.

If C_s is defined as the discharge capacity at standard current I_s and C_q is defined as the discharge capacity at another current I_q, then according to Peukert theory, we have,

$$I_q^n t_q = I_s^n t_s = C \tag{2.2}$$

Equation (2.2) can be rewritten as:

$$I_q t_q I_q^{n-1} = I_s t_s I_s^{n-1} \tag{2.3}$$

Finally:

$$C_q = C_s \left(\frac{I_s}{I_q} \right)^{n-1} \tag{2.4}$$

From the above deduction, it can be seen that the closer the time constant n of the battery is to 1, the less the discharge capacity of the battery is affected by the discharge current, and the better the stability of the discharge capacity of the battery. For this battery, if the standard discharge current is 10 A, the standard discharge capacity is 35.33 A h. According to Eq. (2.4), when a battery is discharged at 35 A, 70 A and 140 A, the time constants of the battery are 1.003, 1.019 and 1.026, respectively, which are very close to 1. This shows theoretically that this battery has good stability and high discharge efficiency during high current discharge, which is of great help to reduce the usage of the battery and save the space inside the vehicle.

A static charging experiment was carried out at normal temperature. Before the constant current-constant voltage charging experiment, the battery was discharged at a constant current with rate of 1/3C, and then shelve for 2 h. After shelving, the battery was charged at constant current with rates of 10 A, 35 A, 70 A and 140 A respectively. When the battery voltage rises to 4.2 V, the battery was charged at a constant voltage of 4.2 V, and when the current dropped to I/10 (I represents a different charging current), the charging was stopped. The relationship between battery charging voltage and charge capacity is shown in Fig. 2.7.

It can be clearly seen from Fig. 2.7 that with the increase of the charging rate, the terminal voltage of the battery rises rapidly, the constant current charge time shortens, and the charge capacity decreases, but the constant voltage charge capacity increases. the specific parameters are shown in Table 2.3. It can be seen from the table that using high-rate charging can shorten the charge time, and the charge capacity is only reduced by 3.5% compared with 35A·h, which indicates that this lithium manganate battery has good high-rate charging characteristics.

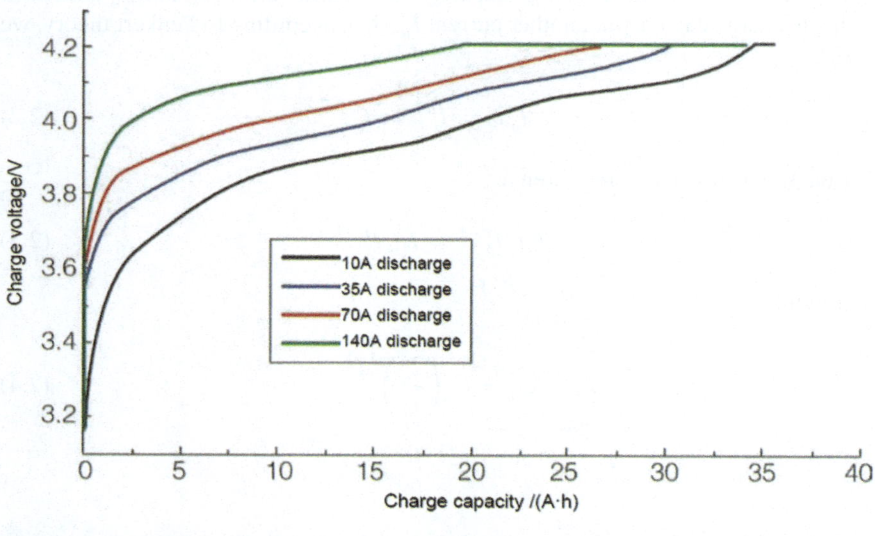

Fig. 2.7 Constant current-constant voltage discharge curves of lithium-ion batteries at different rates at normal temperature

Table 2.3 Comparison of charge capacity and charge time at different charging currents

Charging current /A	10	35	70	140
Constant current charge capacity /(A·h)	34.49	30.44	26.99	19.68
Total charge capacity /(A·h)	35.51	34.23	33.95	33.78
Ratio of total charge capacity to 35 Ah (%)	101	97.8	97	96.5
Remaining time/min	225	66	53	25

2. Dynamic charge and discharge characteristics

Dynamic charge and discharge capability of lithium-ion batteries is an important index to characterize battery performance. According to the high-power charge and discharge requirements of batteries under specific conditions, a compound pulse condition shown in Fig. 2.8 was established, with the maximum discharging current of 280A (8C) and the max. charging current of 175A(5C). Figure 2.9 shows the pulse

Fig. 2.8 Customized compound pulse condition

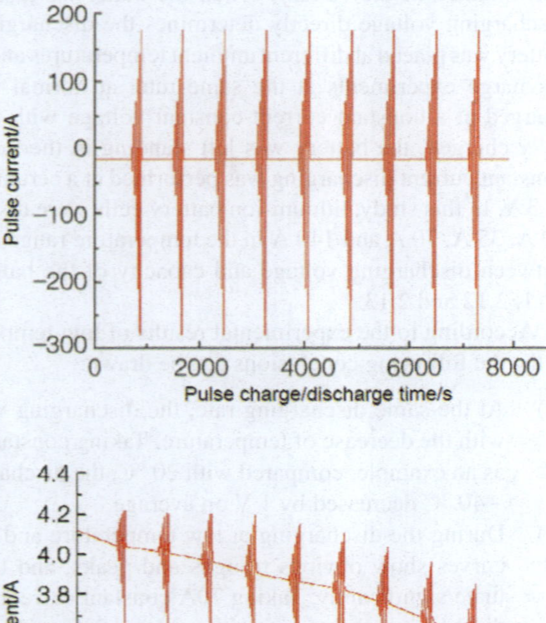

Fig. 2.9 Pulse charge and discharge curves of batteries

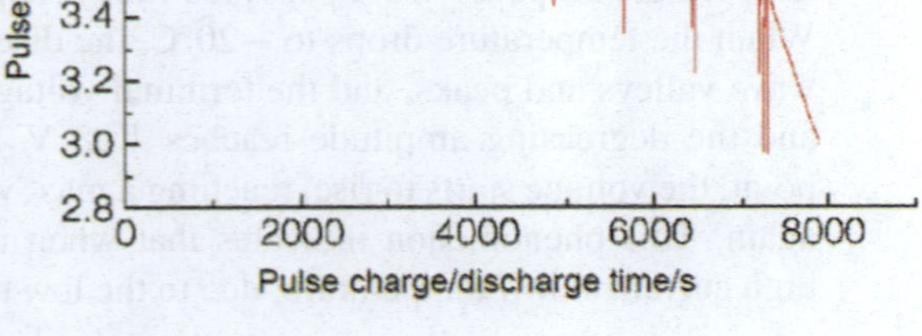

charge and discharge curves of batteries It can be seen from the pulse experiment that when the battery capacity is between 80 and 100%, the max. charging current of the battery is less than 140A. With the decrease of the battery capacity, the charge capacity of the battery is improved. When the battery capacity is greater than or equal to 20%, the battery can be discharged at a high current of 280 A.

2.2.3 *Influence of Temperature on Battery Discharging Voltage*

At present, lithium-ion batteries can normally work in the range of 20–50 °C, but in practical use, most lithium-ion batteries can only ensure the working performance above 0 °C. This section will study and analyze the charge and discharge performance of lithium-ion batteries at low temperature.

The discharging voltage of a battery is an important index to characterize the performance of the battery. When the battery is discharged at the same rate, the discharging voltage directly determines the discharging power of the battery. The battery was placed at different ambient temperatures and subjected to constant current discharge experiments at the same rate: at normal temperature, the battery was charged at a constant current-constant voltage with rate of 1/3C, and after being fully charged, the battery was left standing in thermostat for 5 h; After standing, constant current discharging was performed at a certain rate, with the cut-off voltage of 3 V. In this study, lithium-ion battery cells were discharged at constant current at 10 A, 35 A, 70 A and 140 A in the temperature range of 40 –20°C. The relationship between discharging voltage and capacity of the batteries is shown in Figs. 2.10, 2.11, 2.12 and 2.13.

According to the experimental results of low-temperature discharging of battery cells, the following conclusions can be drawn:

(1) At the same discharging rate, the discharging voltage of a battery decreases with the decrease of temperature. Taking constant current discharging at 10 A as an example, compared with 20 °C, the discharging voltage of the battery at −40 °C decreased by 1 V on average.

(2) During the discharging at low temperature and high current, the discharging curves show obvious troughs and peaks, and the discharging voltages fluctuate significantly. Taking 70A constant current discharge as an example, the discharging curve is normal at 20 and 0°C, without valleys or peaks. When the temperature drops to − 10°C, obvious valleys appear in the discharging curves. When the temperature drops to −20°C, the discharging curves show obvious wave valleys and peaks, and the terminal voltage drops from 4.15 to 3.07 V, and the degreasing amplitude reaches 1.08 V. After dropping to the lowest point, the voltage starts to rise, reaching a max. value of 3.35 V, and then drops again. This phenomenon indicates that when the battery is discharged at a high current at low temperature, due to the low temperature in the initial stage

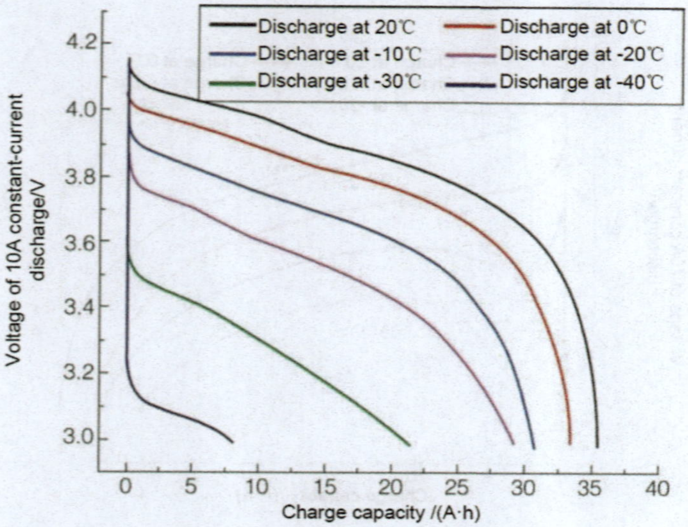

Fig. 2.10 Curves of 10A constant current discharge at different temperatures

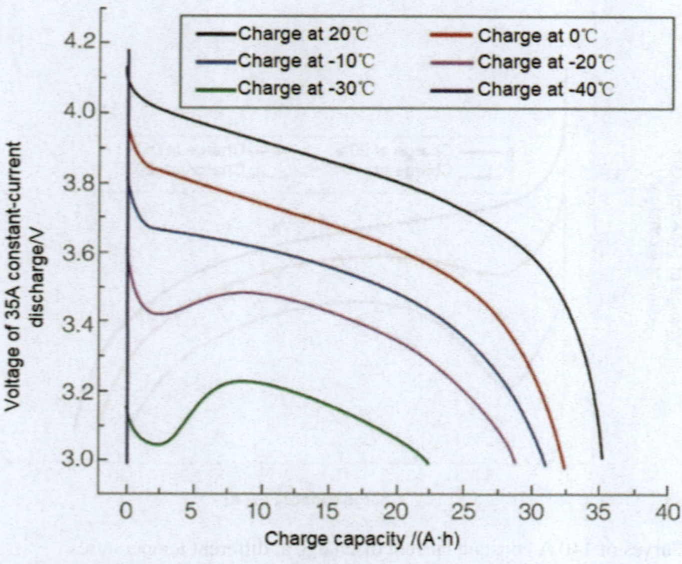

Fig. 2.11 Curves of 35 A constant current discharge at different temperatures

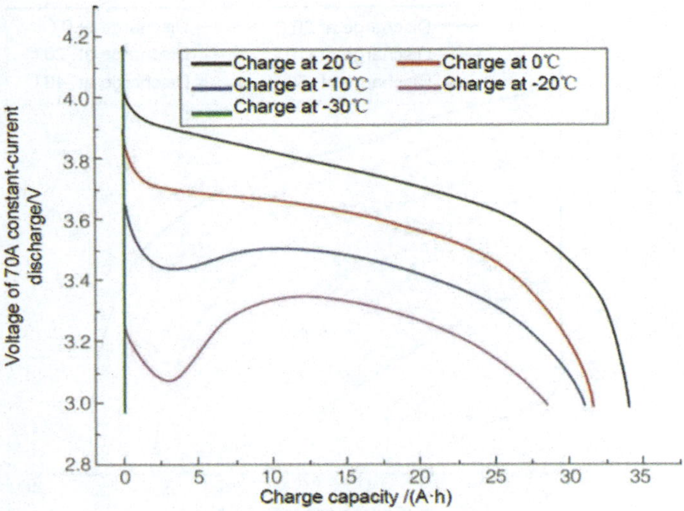

Fig. 2.12 Curves of 70 A constant current discharge at different temperatures

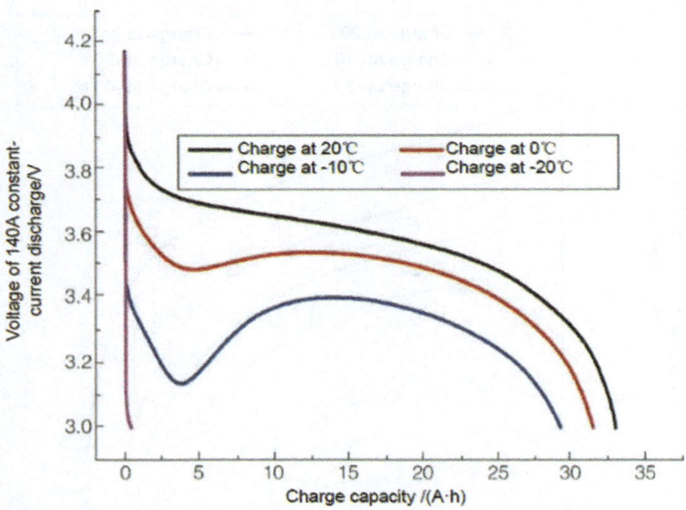

Fig. 2.13 Curves of 140 A constant current discharge at different temperatures

of discharge, the active substances of the battery cannot be fully utilized, the electrode polarization is serious, the internal resistance of the battery is high, and the discharging voltage of the battery drops rapidly in the initial stage of discharging. During the discharging, the current flows through the battery, and Joule heat generated by the internal resistance of the battery makes the temperature of the battery rise rapidly, and the active substances of the battery are activated, so the discharge voltage of the battery starts to rise. With the decrease of battery capacity, the discharging voltage of the battery starts to drop again.

2.2.4 *Influence of Temperature on Battery Discharge Capacity*

At different temperatures, the discharge capacity of the battery will change. In order to study the influence of temperature on a prismatic ALF battery, a battery was discharged at constant current at different discharge rates within the temperature range of 40–20°C. The change of discharge capacity is shown in Table 2.4.

It can be seen from Table 2.4 that under the same discharge rate, with the decrease of ambient temperature, the discharge capacity of the battery decreases rapidly. Discharging at 10A, for example: when the temperature is 20°C, the discharge capacity is 35.33 Ah. When the temperature drops to −30°C, the discharge capacity drops to 21.12 Ah, down by 40.22%. When the temperature drops to −40°C, the discharge capacity is only 7.81 Ah, decreased by 77.89%.

2.2.5 *Influence of Temperature on Battery Charge Capacity*

By studying the discharging characteristics of batteries at different temperatures, it can be seen that with the decrease of temperature, the discharge performance of batteries decreases greatly. This section will study the Influence of low temperature on battery charge performance. The battery was placed at different ambient temperatures and charged at constant current-constant voltage at the same rate: at normal temperature, the battery was discharged at a constant current rate of 1/3C, with a

Table 2.4 Discharge capacity of a battery at different temperatures and different discharge rates (unit: Ah)

Ambient temperature (°C)	Discharge rate			
	10A constant current discharge	35A constant current discharge	70A constant current discharge	140A constant current discharge
20	35.33	35.19	34.01	33.02
0	33.32	32.28	32.47	31.51
−10	30.63	30.96	31.01	29.31
−20	29.07	29.41	28.44	0.38
−30	21.12	22.21	0.06	0
−40	7.81	0.02	0	0

cut-off voltage of 3 V, and after the discharging, the battery was left standing in a temperature-setting thermostat for 5 h; After standing, constant current-constant voltage charging was performed. The relationship between the voltage and charge capacity of batteries charged at constant current and constant voltage of 10 A, 35 A and 70 A is shown in Figs. 2.14, 2.15 and 2.16.

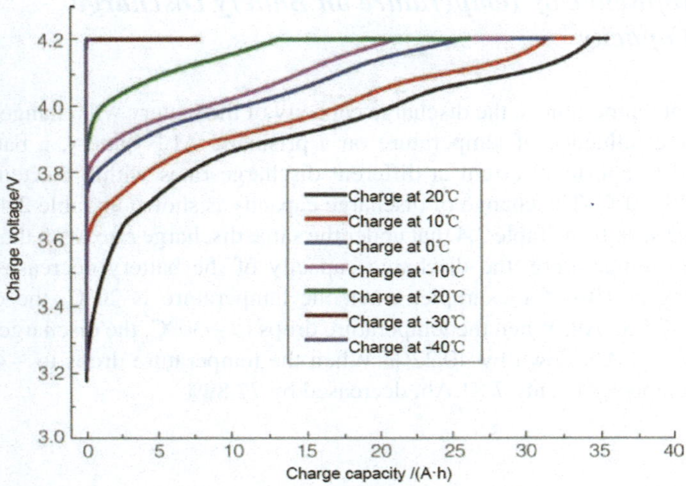

Fig. 2.14 Charging curves of batteries at 10 A constant current—constant voltage at different temperatures

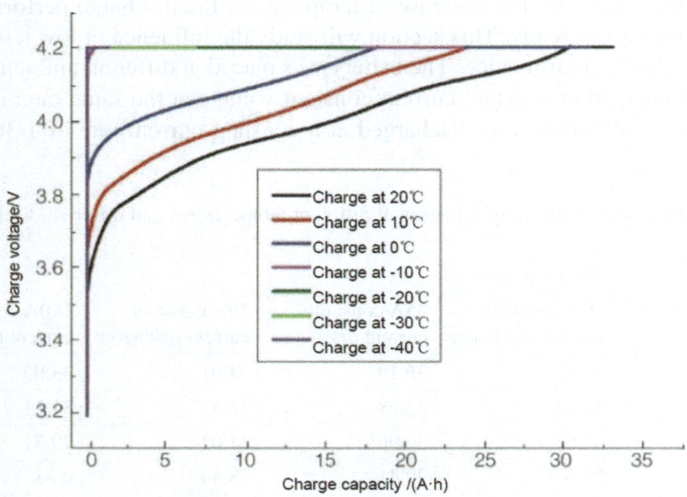

Fig. 2.15 Charging curves of batteries at 35 A constant current—constant voltage at different temperatures

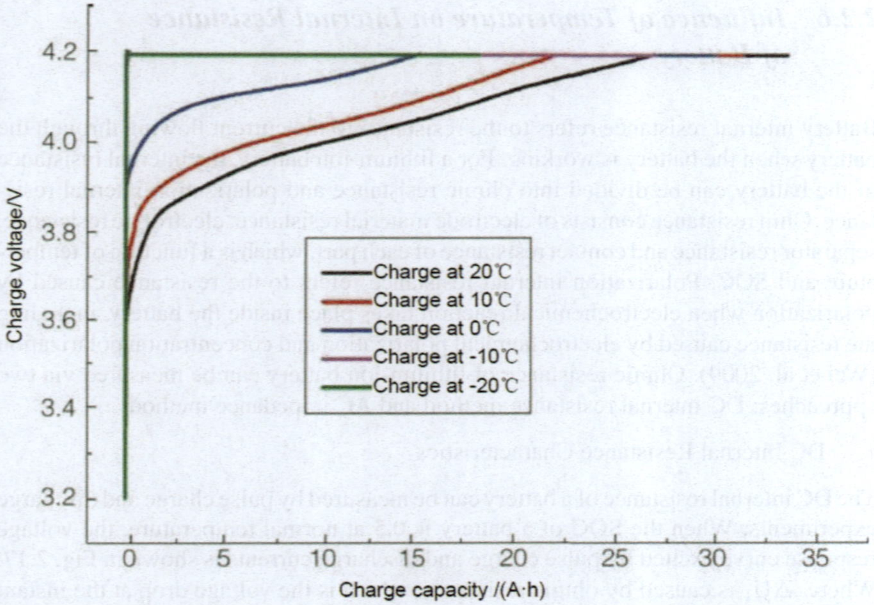

Fig. 2.16 Charging curves of batteries at 70 A constant current—constant voltage at different temperatures

It can be seen from Figs. 2.14, 2.15 and 2.16 that the charge performance of the battery decreases significantly at low temperature. Battery charging at low temperature has the following two characteristics:

(1) When the charging current is the same, the charging voltage increases with the decrease of temperature. Especially when charging with high current, there is no constant current charging process at all below 0 °C. At the moment when the charging current is loaded, the terminal voltage of the battery quickly rises to the cut-off voltage of 4.2 V, and directly enters the constant voltage charging stage.

(2) With the decrease of temperature, constant-current charge time and charge capacity decrease rapidly, while constant-voltage charge time and charge capacity increase, while total charge capacity decreases. When charging a battery at the same current, the time taken to charge the same capacity increases.

2.2.6 *Influence of Temperature on Internal Resistance of Battery*

Battery internal resistance refers to the resistance to the current flowing through the battery when the battery is working. For a lithium-ion battery, the internal resistance of the battery can be divided into ohmic resistance and polarization internal resistance. Ohm resistance consists of electrode material resistance, electrolyte resistance, separator resistance and contact resistance of each part, which is a function of temperature and SOC. Polarization internal resistance refers to the resistance caused by polarization when electrochemical reaction takes place inside the battery, including the resistance caused by electrochemical polarization and concentration polarization (Wei et al. 2009). Ohmic resistance of lithium-ion battery can be measured via two approaches: DC internal resistance method and AC impedance method.

1. DC Internal Resistance Characteristics

The DC internal resistance of a battery can be measured by pulse charge and discharge experiments. When the SOC of a battery is 0.5 at normal temperature, the voltage response curve excited by pulse charge and discharge currents is shown in Fig. 2.17. Where, ΔU_1 is caused by ohmic resistance, which is the voltage drop at the instant when the battery starts discharging, so the ohmic resistance of discharge can be calculated with Eq. (2.5). ΔU_3 is caused by ohmic resistance, which is the boost value at the instant when the battery starts charging. In the same way, the ohmic resistance of charging can be calculated. ΔU_2 and $\Delta U4$ are the voltage changes caused by the polarization internal resistance of the battery.

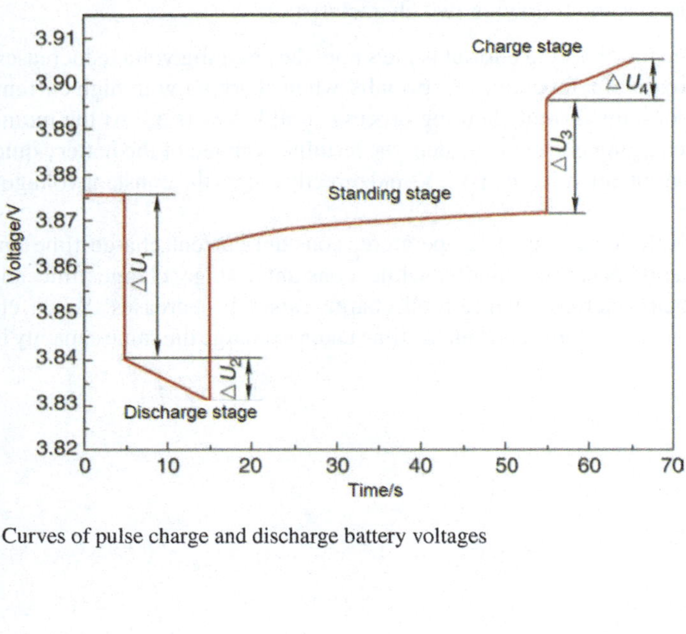

Fig. 2.17 Curves of pulse charge and discharge battery voltages

$$R = \frac{\Delta U_1}{|I|} \tag{2.5}$$

The charging-discharging ohmic resistance curves of a battery at temperature of 20–20 °C and SOC of 0.1 –1.0 are shown in Figs. 2.18 and 2.19, from which the relationship between DC internal resistance and temperature and SOC can be concluded:

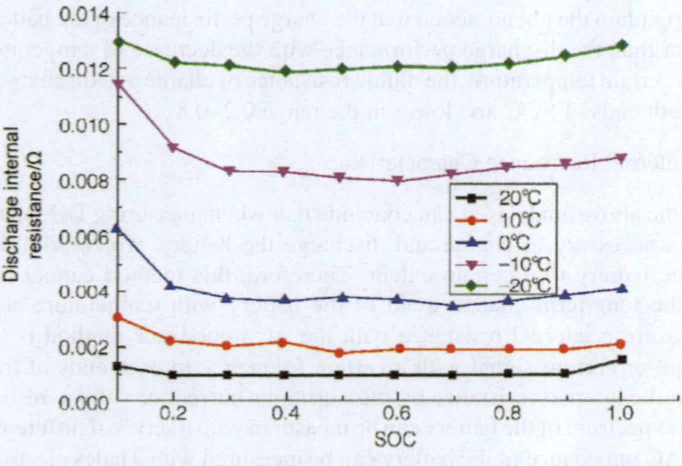

Fig. 2.18 Relationship between ohmic resistance and SOC when discharging at different temperatures

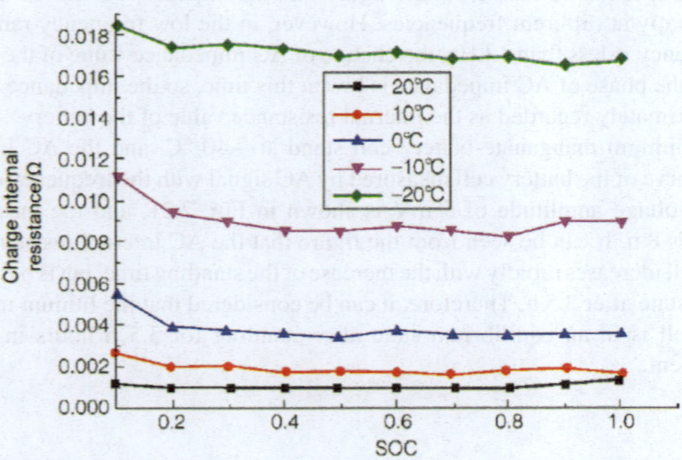

Fig. 2.19 Relationship between ohmic resistance and SOC when charging at different temperatures

(1) With the decrease of the temperature, the internal resistance of charge and discharge increases rapidly, and increases greatly below 0 °C. At 20 °C. The ohmic resistance of charging is obviously higher than the internal resistance of discharging. The average discharging ohmic resistance at −20 °C is 140.05% and the average charging ohmic resistance at −10 °C is 190.53%. This can explain the reasons why the constant current discharging voltage of the battery drops rapidly with the decrease of temperature and the constant current charging voltage rises rapidly with the decrease of temperature. It can also explain the phenomenon that the charge performance of the battery decays faster than the discharge performance with the decrease of temperature.
(2) At a certain temperature, the ohmic resistance of charge and discharge is higher at both ends of SOC and lower in the range 0.2–0.8.

2. AC Internal Resistance Characteristics

Based on the above analysis, it can conclude that when measuring DC internal resistance, it is necessary to charge and discharge the battery, which will change the state of the battery to a certain extent. Therefore, this method cannot be used to measure the long-term change trend of the battery with temperature at a certain SOC. Measuring internal resistance with the AC impedance method is to apply a low voltage or current signal with a certain frequency to both ends of the battery, and obtain the internal resistance by measuring its current or voltage response. The impedance spectrum of the battery can be measured with a series of different frequencies. The AC impedance of the battery can be measured with Thales electrochemical workstation. The AC impedance spectrum of a battery measured at room temperature of 20 °C with 1–100 kHz bandwidth and 5 mV disturbance voltage is shown in Fig. 2.20. It can be seen from the figure that the impedance value of the battery varies greatly at different frequencies. However, in the low frequency range where the frequency is less than 1 kHz, the change of AC impedance value of the battery is low, and the phase of AC impedance is low at this time, so the impedance value can be approximately regarded as the internal resistance value of the battery.

Let a lithium manganate battery cell stand at −40 °C, and the AC impedance change curve of the battery cell measured by AC signal with the frequency of 260 Hz and the voltage amplitude of 5 mV is shown in Fig. 2.21, and the measurement duration is 8 h. It can be seen from the figure that the AC internal resistance of the battery cell increases rapidly with the increase of the standing time, but is basically in a constant state after 3.5 h. Therefore, it can be considered that the lithium manganate battery cell is in an equilibrium state after standing for 3.5 h hours in a certain environment.

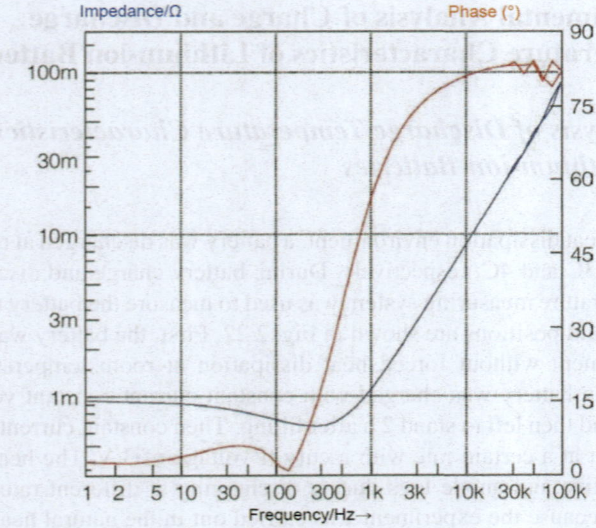

Fig. 2.20 AC impedance spectrum of a battery at normal temperature

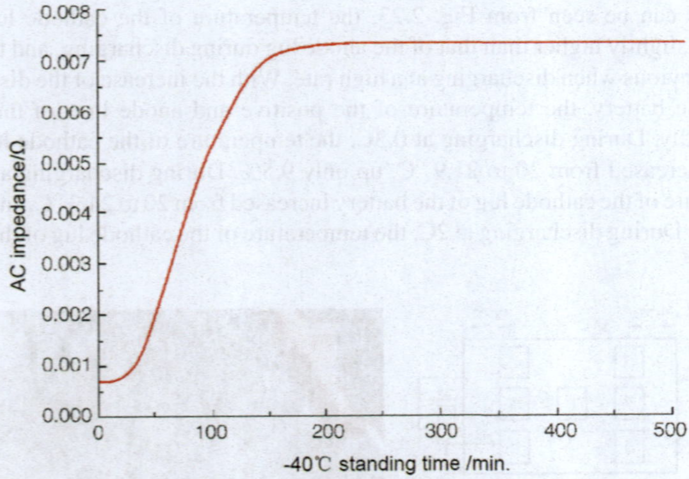

Fig. 2.21 AC impedance curve of a battery during standing at −40 °C

2.3 Experimental Analysis of Charge and Discharge Temperature Characteristics of Lithium-ion Batteries

2.3.1 Analysis of Discharge Temperature Characteristics of Lithium-ion Batteries

In the natural heat dissipation environment, a battery was discharged at rates of 0.3C, 0.5C, 1C, 2C, 3C and 4C, respectively. During battery charge and discharge, a 16-channel temperature measuring system was used to measure the battery temperature.

The labels and positions are shown in Fig. 2.22. First, the battery was suspended in an environment without forced heat dissipation at room temperature. Before discharging, the battery was charged with constant current-constant voltage at the rate of 1/3C, and then left to stand 2 h after filling; Then constant current discharging was carried out at a certain rate with a cut-off voltage of 3 V. The heat generation curves of positive and anode lugs during discharging at different rates are shown in Fig. 2.23. Because the experiment was carried out in the natural heat dissipation environment, the room temperature was slightly different in different time periods. In order to facilitate the comparative study, the initial temperature of the battery was unified at 20 °C in the drawing process.

As we can be seen from Fig. 2.23, the temperature of the cathode lug of the battery is slightly higher than that of the anode lug during discharging, and this trend is more obvious when discharging at a high rate. With the increase of the discharging rate of the battery, the temperature of the positive and anode lugs of the battery rises rapidly. During discharging at 0.3C, the temperature of the cathode lug of the battery increased from 20 to 21.9 °C, up only 9.5%. During discharging at 1C, the temperature of the cathode lug of the battery increased from 20 to 24.3 °C, an increase of 21.5%. During discharging at 2C, the temperature of the cathode lug of the battery

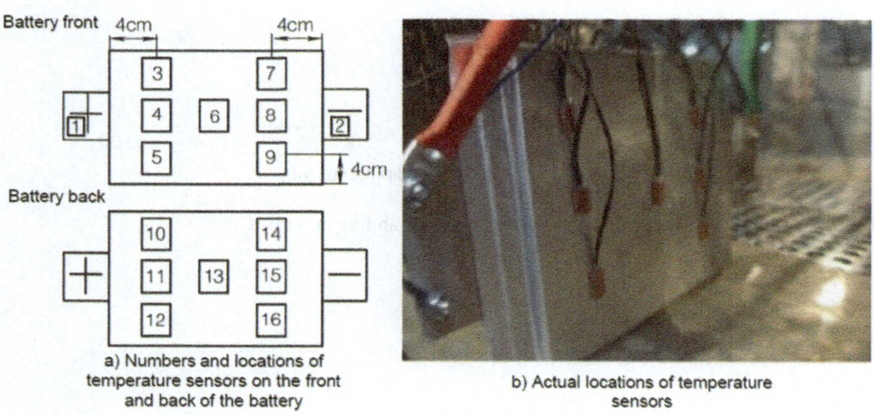

Fig. 2.22 Labels and positions of battery cell temperature sensors

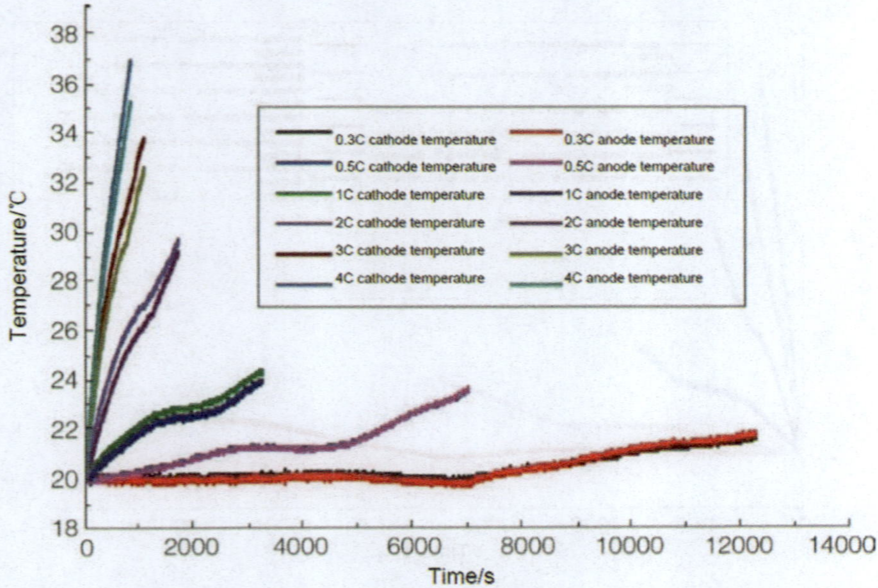

Fig. 2.23 Temperature change curves of positive and anode lugs of a battery with different discharge rates

increased from 20 to 29.6 °C, an increase of 48%. When the battery was discharged at 4C, the temperature of the cathode lug of the battery increased from 20 to 36.96 °C, an increase of 84.8%. Therefore, when the battery is discharged at a high rate in a high temperature environment, corresponding heat dissipation measures must be taken, otherwise, the battery will be prone to performance decline, shortened service life and even a dangerous state of thermal runaway due to overheating.

When discharging a battery at different rates, the average temperature rise curves of the front and back sides of the battery cell are shown in Fig. 2.24. It can be seen from the figure that at different discharge rates, the temperature rise of the battery body has the same trend as that of the positive and anode lugs: the temperature rises rapidly in the initial stage of discharge, rises slowly in the middle stage, and rises rapidly again in the later stage of discharging.

2.3.2 Analysis of Charge Temperature Characteristics of Lithium-ion Batteries

As with the discharge temperature rise experiment, in the charge temperature rise experiment, the battery was suspended in an environment without forced heat dissipation. First, the battery was discharged at a constant current rate of 1/3C, with a cut-off voltage of 3 V. After the discharge, it was left to stand for 2 h, and then

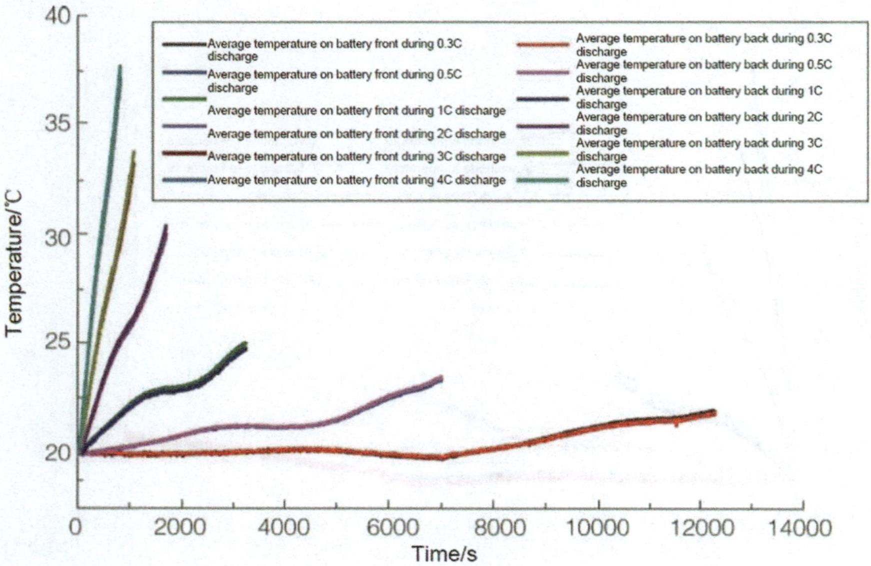

Fig. 2.24 Average temperature change curves of front and back sides of a battery at different discharge rates

it is charged at constant current-constant voltage rates of 0.3C, 0.5C, 1C, 2C, 3C and 4C respectively. When charging the battery at different rates, the temperature curves of positive and anode lugs are shown in Fig. 2.25. It can be seen from the figure that the temperature difference between the positive and anode lugs of the battery during charging is smaller than that during discharging at the same rate. During constant current charging, the temperature of the positive and anode lugs of the battery rises rapidly. In the constant voltage charging stage, the temperature of the battery lugs begins to decrease, which is mainly due to the continuous decrease of the charging current and the decrease of the heat generation rate of the battery. Therefore, in the constant current-constant voltage charging process, the constant current charging process is an important stage of heat accumulation in the battery. The average temperature curves of the front and back sides of the battery cell during charging at different rates are shown in Fig. 2.26. It can be seen from the figure that the temperature of the front and back of the battery cell is almost equal, and the temperature rise of the battery cell itself has the same trend as that of the positive and anode lugs.

Summary

This chapter mainly introduces the structure and working principle of lithium-ion batteries, and studies the related characteristics of a prismatic ALF battery, which lays a foundation for the subsequent research of thermal management systems of

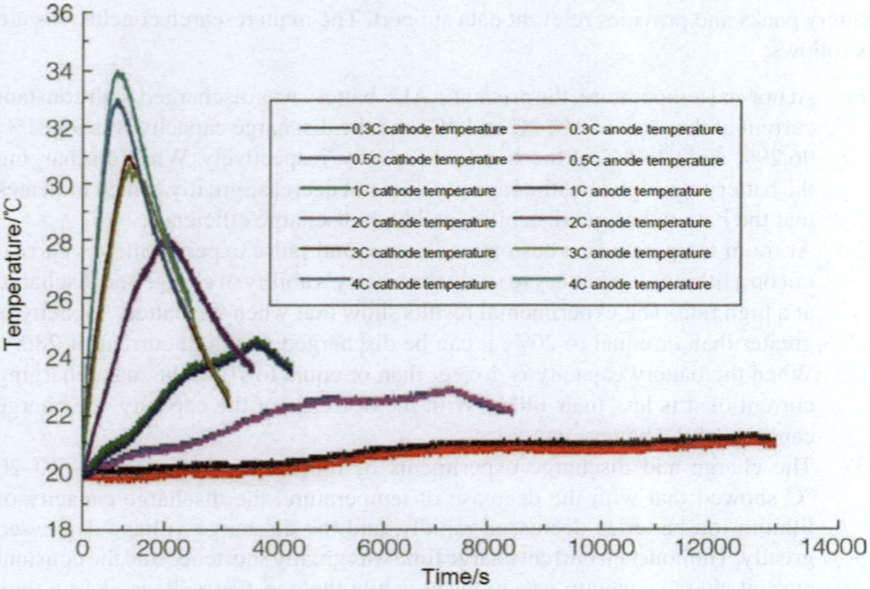

Fig. 2.25 Temperature curves of positive and anode lugs of a battery cell at different charging rates

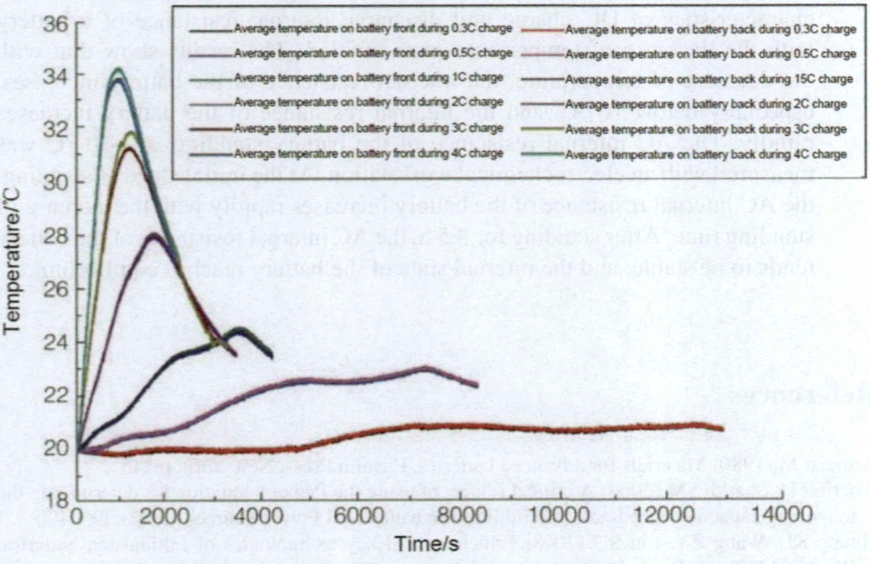

Fig. 2.26 Average temperature curves of front and back sides of a battery cell at different charging rates

battery packs and provides relevant data support. The main research conclusions are as follows:

(1) At normal temperature, the prismatic ALF battery was discharged with constant current at the rates of 1C, 2C and 4C, and the discharge capacity was 99.61%, 96.29% and 93.48% of the standard capacity, respectively. When discharging the battery at a high rate, the capacity does not decrease greatly, which indicates that the battery has good stability and high discharge efficiency.
(2) At room temperature, a customized compound pulse experiment was carried out on a lithium-ion battery to study the battery's ability of charge and discharge at a high rate. The experimental results show that when the battery capacity is greater than or equal to 20%, it can be discharged at a high current of 280A. When the battery capacity is greater than or equal to 70%, the max. charging current of it is less than 140A. With the decrease of the capacity, the charge capacity of the battery increases.
(3) The charge and discharge experiments of lithium-ion batteries at −40–20 °C showed that with the decrease of temperature, the discharge capacity of lithium-ion batteries decreased rapidly, and the discharge voltage decreased greatly. The constant current charge time was greatly shortened and the constant current charge capacity was reduced, while the constant voltage charge time was prolonged and the total charge capacity was reduced.
(4) Through the dynamic measurement of pulse charge and discharge, the change characteristics of DC charge and discharge internal resistance of a battery with the decrease of temperature were studied. The results show that with the decrease of temperature, the internal resistance of the battery increases, especially below 10 °C, and the internal resistance of the battery increases rapidly. The AC internal resistance of the battery standing at −40 °C was measured with an electrochemical workstation. At the initial stage of standing, the AC internal resistance of the battery increases rapidly with the increase of standing time. After standing for 3.5 h, the AC internal resistance of the battery tends to be stable, and the internal state of the battery reaches equilibrium.

References

Armand M (1980) Materials for advanced batteries. Plenum Press, New York, p 145

Doerffel D, Sharkh SM (2006) A critical review of using the Peukert equation for determining the remaining capacity of lead-acid and lithium-ion batteries. J Power Sources 155(2):395–400

Huang KL, Wang ZX, Liu SQ (2008) Principles and key technologies of lithium-ion batteries. Chemical Industry Press, Beijing

Linden D, Reddy TB (2002) Handbook of batteries, 2nd. McGraw Hill, New York

Liu L, Wang HL, Zhi G (2009) Working principles and main materials of lithium-ion batteries. Sci Technol Inf 23:454

Wei XZ, Xu W, Shen D (2009) Identification of internal resistance of lithium-ion batteries and its application in life estimation. Chin J Power Sources 3(3):217–220
Wu YP, Dai XB, Ma JQ et al (2006) Lithium-ion batteries: application and practice. Chemical Industry Press, Beijing
Zheng HH (2007) Lithium-ion battery electrolytes. Chemical Industry Press, Beijing

Chapter 3
Electrothermal Coupling Modeling of Lithium-ion Batteries

When designing a thermal management system of power batteries, it is often necessary to establish a thermal model of power batteries to simulate and analyze the changes of battery temperature. The calculation of heat generation of lithium-ion batteries is related to the accuracy of the battery thermal model, and it is difficult to measure it accurately in real time during the charge and discharge of the batteries at present. In addition, it is difficult to obtain the internal temperature of the battery cells, which undoubtedly increases the design difficulty of battery thermal management system. In this chapter, the methods for obtaining thermophysical parameters of batteries are introduced, and the electrothermal coupling modeling methods based on Bernardi heat generation rate and electrochemical model of prismatic ALF batteries are systematically expounded. In addition, a radial layered modeling method is proposed for cylindrical batteries.

3.1 Principles of Heat Generation and Heat Conduction of Lithium-ion Batteries

3.1.1 Heat Generation of Lithium-ion Batteries

During normal charge and discharge of lithium-ion batteries, the heat generation mainly includes polarization heat generation, internal resistance Joule heat generation and chemical reaction heat generation.

(1) Internal resistance joule heat

It is mainly generated by internal resistance of batteries. The internal resistance of a battery mainly includes electronic internal resistance (including contact resistance among conductive lugs, current collectors and active materials) and ionic internal

J. Li, *Modeling and Simulation of Lithium-ion Power Battery Thermal Management*, Key Technologies on New Energy Vehicles,
https://doi.org/10.1007/978-981-19-0844-6_3

resistance of the electrolyte (including electrodes and separator). The internal resistance Joule heat is always positive during the charge and discharge of the battery, that is, whether charging or discharging, the internal resistance Joule heat generation will only generate heat but not absorb it, and this part of heat is the main part of the heat generated during the charge and discharge process of the battery.

(2) Chemical reaction heat generation

Because the battery will undergo chemical reactions during charge and discharge, and heat will be generated during the chemical reactions, the reaction heat can be calculated with Eq. (3.1):

$$Q_r = -T\frac{\partial \Delta G}{\partial T} = -T\frac{\partial \Delta H}{\partial T} - T\left[-\frac{\partial (T\Delta S)}{\partial T}\right] = -\sum_i \frac{\Delta H_i}{n_i F} I_i + \sum_j \frac{\Delta S_i}{n_j F} I_j \tag{3.1}$$

where, H is the enthalpy (J); S is the entropy (J/K); G is the Gibbs free energy (J), $\Delta G = \Delta H - T\Delta S$; T is the thermodynamic temperature (K); n is the number of electrons; F is the Faraday constant.

This part of heat is positive in the battery discharging stage and negative in the battery charging stage. When the battery is charged and discharged at constant current at the same rate at normal temperature, the average surface temperature of the battery during discharge is higher than that during charge.

(3) Battery polarization heat generation

The battery will be polarized due to the passing of load current, and heat will be generated in the polarization process, and this part of heat will take a positive value in the charge and discharge process. Cell polarization mainly includes activation polarization and concentration polarization. Activation polarization can drive the electrochemical reaction between electrode and electrolyte interface, while concentration polarization is caused by the concentration difference between products and reactants between the electrolyte–electrode interface and the electrolyte body.

The above is the heat generation during normal use of batteries. In case of thermal runaway, the battery will generate side reaction heat, which will cause damage to the battery and cause danger.

3.1.2 Heat Conduction of Lithium-ion Batteries

The heat conduction model of lithium-ion power batteries is established in rectangular coordinate system, and a micro-cell parallelepiped is randomly taken out from the lithium-ion power batteries to analyze the energy balance of micro-cell body, as shown in Fig. 3.1.

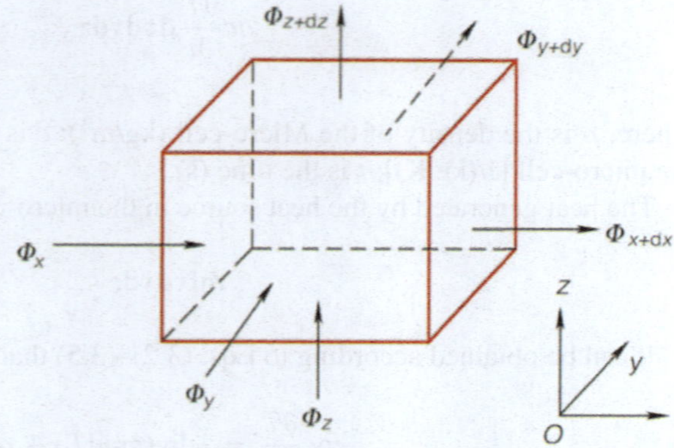

Fig. 3.1 Thermal conduction and heat balance analysis of micro-cells

Any micro-cell in the uniform medium in the rectangular coordinate system will generate heat during the charge and discharge of the battery, and has an internal heat source. Let its value be $\dot{q}$, which indicates the heat energy generated or consumed in unit volume in unit time. According to Fourier's law of heat conduction (Tao 2006; Yang and Tao 2006; Zhao 2008), it can be obtained as follows:

$$\begin{cases} (\Phi_x)_x = -\lambda\left(\frac{\partial T}{\partial x}\right)_x \mathrm{d}y\mathrm{d}z \\ \left(\Phi_y\right)_y = -\lambda\left(\frac{\partial T}{\partial y}\right)_y \mathrm{d}x\mathrm{d}z \\ (\Phi_z)_z = -\lambda\left(\frac{\partial T}{\partial z}\right)_z \mathrm{d}x\mathrm{d}y \end{cases} \tag{3.2}$$

where, $(\Phi_x)_x$ is the value of the component of heat flow in x direction at x point; $\left(\Phi_y\right)_y$ is the value of the component of heat flow in y direction at y point; $(\Phi_z)_z$ is the value of the component of heat flow in z direction at z point.

The heat flux of micro-cell is derived from three surfaces: $x = x + \mathrm{d}x$, $y = y + \mathrm{d}y$ and $z = z + \mathrm{d}z$. In the same way, it is obtained according to Fourier heat conduction law.

$$\begin{cases} (\Phi_x)_{x+\mathrm{d}x} = (\Phi_x)_x + \frac{\partial \Phi_x}{\partial x} = (\Phi_x)_x + \frac{\partial}{\partial x}\left[-\lambda\left(\frac{\partial T}{\partial x}\right)_x \mathrm{d}y\mathrm{d}z\right]\mathrm{d}x \\ \left(\Phi_y\right)_{y+\mathrm{d}y} = \left(\Phi_y\right)_y + \frac{\partial \Phi_y}{\partial y} = (\Phi_y)_y + \frac{\partial}{\partial y}\left[-\lambda\left(\frac{\partial T}{\partial y}\right)_y \mathrm{d}x\mathrm{d}z\right]\mathrm{d}y \\ (\Phi_z)_{z+\mathrm{d}z} = (\Phi_z)_z + \frac{\partial \Phi_z}{\partial z} = (\Phi_z)_z + \frac{\partial}{\partial z}\left[-\lambda\left(\frac{\partial T}{\partial z}\right)_z \mathrm{d}x\mathrm{d}y\right]\mathrm{d}z \end{cases} \tag{3.3}$$

According to the law of conservation of energy, in any time interval, the total heat flux introduced into the micro-cell and the heat generated by the heat source in the micro-cell are equal to the sum of the total heat flux introduced into the micro-cell and the increment of thermodynamic energy (i.e. internal energy) of the micro-cell, where the increment of thermodynamic energy of the micro-cell is:

$$\rho c \frac{\partial T}{\partial t} \mathrm{d}x\mathrm{d}y\mathrm{d}z \tag{3.4}$$

where, ρ is the density of the Micro-cell (kg/m^3); c is the specific heat capacity of the micro-cell [J/(kg K)]; t is the time (s).

The heat generated by the heat source in the micro-cell is:

$$\dot{q}\mathrm{d}x\mathrm{d}y\mathrm{d}z \tag{3.5}$$

It can be obtained according to Eqs. (3.2)–(3.5) that

$$\rho c \frac{\partial T}{\partial t} = \lambda \mathrm{div}(\mathrm{grad}T) + \dot{q} \tag{3.6}$$

When the thermal conductivity coefficient is anisotropic, the differential equation of thermal conductivity is:

$$\rho c \frac{\partial T}{\partial t} = \lambda_x \frac{\partial^2 T}{\partial x^2} + \lambda_y \frac{\partial^2 T}{\partial y^2} + \lambda_z \frac{\partial^2 T}{\partial z^2} + \dot{q} \tag{3.7}$$

When that battery pack is charged/discharged or heated by itself as a heating source, the battery is a typical unsteady heat conductor with an internal heat source. If the battery is heated by an external power source and the internal heat source is zero, the battery is an unsteady heat conductor without internal heat source.

Heat generated by the battery will be exchanged with the outside, mainly including the following types.

(1) Heat conduction

Heat conduction refers to the process of transferring heat within or between objects with temperature gradient by means of molecular thermal motion, and the equation is as follows:

$$q = -\lambda_{\mathrm{n}} \frac{\partial T}{\partial n} \tag{3.8}$$

where, q is the heat flux density (W/m^2); λ_n is the conductivity coefficient [W/(m K)]; $\partial T/\partial n$ is the temperature gradient (K/m) in the n direction; – means that the direction of temperature rise is opposite to the direction of heat transfer.

(2) Heat convection

Heat convection refers to the process of heat transfer between parts with temperature differences in liquid or gas by circulating flow. The equation is as follows:

$$q = h(T_1 - T_2) \tag{3.9}$$

where, h is the convective heat transfer coefficient W/m^2 K; T_1 is the solid surface temperature (K); and T_2 is the fluid temperature (K).

(3) Heat radiation

Heat radiation refers to the process in which an object with temperature transfers heat outward in the form of electromagnetic waves. The equation is as follows:

$$q = F\varepsilon\sigma A_1\left(T_1^4 - T_2^4\right) \tag{3.10}$$

where, F is the Correction factor; ε is the Stefan–Boltzmann constant; σ is the Stefan–Boltzmann constant, $\sigma = 5.67 \times 10^{-8}$ W/(m^2 K^4); A_1 is the area of radiation surface of one battery (m^2); T_1 is the temperature (K) of the radiation surface of the battery; T_2 is the temperature (K) of the radiation surface of another surrounding battery.

Because of its compact internal material arrangement and good thermal conductivity, the heat transferred by thermal radiation accounts for a negligible proportion.

3.2 Thermophysical Parameters of Lithium-ion Batteries

The materials of cathode, anode, separator and electrolyte in a battery are all different, and there are even solid, liquid and gaseous substances at the same time. It is extremely difficult to measure the thermophysical parameters of each structure in the battery without damaging the battery structure. The individual thermophysical parameters of each material constituting the battery can be obtained by consulting relevant data and experiments (see Table 3.1), and the overall thermophysical parameters of the battery can be calculated.

3.2.1 Thermal Conductivity Coefficient

The layered structure of a battery cell makes the distribution of thermophysical parameters in the battery show a certain rule. Materials are uniformly distributed in the width x-axis and length y-axis, and stacked in layers in the thickness z-axis. Therefore, in a three-dimensional thermal model, the battery can be treated as an anisotropic material with the same thermal conductivity coefficient in x and y directions and different thermal conductivity coefficient in z direction. Here, the thermal resistance method can be used to calculate the thermal conductivity coefficient of the battery. The thermal resistance method is divided into series thermal resistance method and parallel thermal resistance method, which are respectively expressed by Eqs. (3.11) and (3.12):

Table 3.1 Thermophysical parameters of materials used in a prismatic ALF Battery

Components	Physical property parameters				
	Material	Thickness/μm	Density /(kg/m^3)	Specific heat capacity/[J (/kg K)]	Coefficient of thermal conductivity coefficient/[W (/m K)]
Copper foil	Cu	10	8933	385	398
Cathode material	Lithium manganate compound	55	2840	839	3.9
Diaphragm	PVDF	30	659	1978	0.33
Anode material	Graphite	55	1671	1064	3.3
Aluminum foil	Al	10	2710	903	208
Shell	Aluminum laminated film	318	1636	1377	0.427

$$k = \frac{L_1 + L_2}{L_1/k_1 + L_2/k_2} \tag{3.11}$$

$$k = \frac{A_1}{A_1 + A_2} k_1 + \frac{A_2}{A_1 + A_2} k_2 \tag{3.12}$$

The thermal conductivity coefficient of stacked lithium-ion batteries should be calculated by series thermal resistance method in thickness direction and parallel thermal resistance method in length and width direction. It can be obtained by calculation that

The thermal conductivity coefficient in the x-axis and y-axis directions is

$$k_{\mathrm{T},x} = k_{\mathrm{T},y} = \frac{\sum L_i k_{\mathrm{T},i}}{\sum L_i} \tag{3.13}$$

The thermal conductivity coefficient in z-axis direction is

$$k_{\mathrm{T},z} = \frac{\sum L_i}{\sum L_i / k_{\mathrm{T},i}} \tag{3.14}$$

where, $k_{\mathrm{T},x}$, $k_{\mathrm{T},y}$ and $k_{\mathrm{T},z}$ are the average thermal conductivity coefficient in x-axis, y-axis and z-axis, respectively; L_i is the thickness of each of the five layers in the electrochemical model; $k_{\mathrm{T},i}$ is the thermal conductivity coefficient of materials contained in each of the five layers.

3.2.2 Battery Density

The material inside the battery cell is a mixture of copper, aluminum, positive and negative materials, separator and electrolyte in proportion.

Since each layer of the battery is very thin, it can be considered that the inside of the battery is a uniform substance, so the average density of each part of the material is used as the density of the battery:

$$\rho_{\text{batt}} = \frac{\sum L_i \rho_i}{\sum L_i} \tag{3.15}$$

where, ρ_{batt} is the average density of the battery; ρ_i is the density of materials of each component of the battery; L_i is the thickness of each layer in the battery "sandwich" structure unit.

3.2.3 Specific Heat Capacity of Batteries

The specific heat capacity of a battery is defined as the heat capacity of a unit mass of matter, that is, the heat absorbed or released when a unit mass of matter changes its unit temperature. As the battery is composed of various substances, the specific heat capacity of the battery is calculated with Eq. (3.16), just like the density of the battery.

$$c_{\text{batt}} = \frac{\sum (\rho_i L_i) c_i}{\sum (\rho_i L_i)} \tag{3.16}$$

where, c_{batt} is the average specific heat capacity of the battery; ρ_i is the density of materials of each component of the battery; c_i is the specific heat capacity of each component material of the battery; L_i is the thickness of each layer in the battery "sandwich" structure unit.

3.3 Battery Electrothermal Coupling Model Based on Bernardi Heat Generation Rate

Taking a prismatic ALF battery as an example, Bernardi heat generation rate and Bernardi heat generation rate with current density are compared and calculated, and the heat generation rate model of prismatic ALF battery is determined.

3.3.1 Modeling and Verification of Battery Electrothermal Coupling Model

Because the battery shell is made of ALF, whose thermal radiation to the environment is very low, the thermal model can ignore the thermal radiation between the battery and the surrounding environment. The heat models calculated by the other two heat generation rates are also treated in the same way. Combining Bernardi heat generation rate calculation equation and heat conduction differential equation, a three-dimensional heat generation model of battery cells based on Bernardi heat generation rate can be obtained:

$$\rho c\frac{\partial T}{\partial t} = \lambda_x\frac{\partial^2 T}{\partial x^2} + \lambda_y\frac{\partial^2 T}{\partial y^2} + \lambda_z\frac{\partial^2 T}{\partial z^2} + \beta\frac{I_L}{V_B}\left[(E_0 - U_L) - T\frac{\mathrm{d}E_0}{\mathrm{d}T}\right] \tag{3.17}$$

The initial and boundary conditions are:

$$\begin{cases} T(x, y, z; 0) = T_0 \\ -\lambda\frac{\partial T}{\partial n}|_\Gamma = h(T - T_{\mathrm{amb}})|_\Gamma \end{cases} \tag{3.18}$$

where, T_0 is the initial temperature of the battery cell; T_{amb} is the ambient temperature; β is the correction coefficient of heat generation rate, which is optimized by comparing the simulation results with experimental data. β is 1.13 when discharging and 0.65 when charging. V_B is the cell volume; And I_{L} and U_{L} are the charge and discharge current and voltage of the battery, respectively; E_0 is the open circuit voltage of the battery, and $\mathrm{d}E_0/\mathrm{d}T$ is the temperature influence coefficient of the balanced electromotive force of the battery. Γ is the boundary; $\boldsymbol{n}$ is the normal direction of the boundary Γ.

Because the open-circuit voltage E_0 varies with the capacity and temperature of the battery, this chapter measures the open-circuit voltage of the lithium-ion battery at −40–20 °C and different SOC, and based on relevant experimental data, obtains the function of the open-circuit voltage of the battery with respect to temperature at different SOC by fitting:

$$E_0 = E(\mathrm{SOC}, T) \tag{3.19}$$

When SOC = 1, $-40 \le T \le 20$ °C, the data fitting result is shown in Fig. 3.2, and the fitting function of E_0 about T is shown in Eq. (3.20):

$$E_0 = 1.461 \times 10^{-7}T^3 + 5.298 \times 10^{-6}T^2 + 2.519 \times 10^{-4}T + 4.1534 \tag{3.20}$$

$$\frac{\mathrm{d}E_0}{\mathrm{d}T} = 4.383 \times 10^{-7}T^2 + 1.06 \times 10^{-5}T + 2.519 \times 10^{-4} \tag{3.21}$$

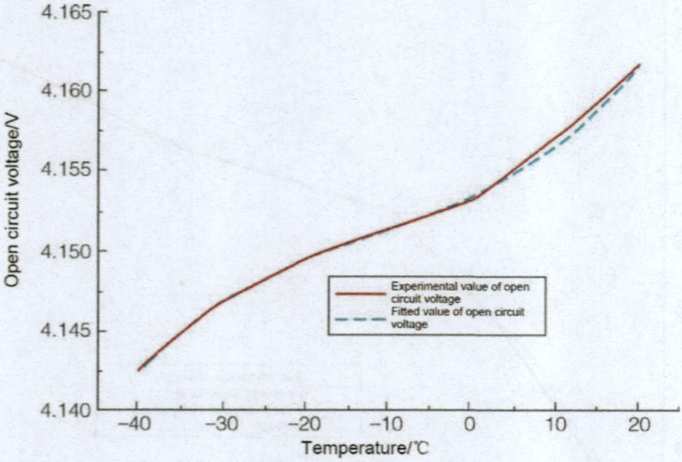

Fig. 3.2 Experimental and fitted values of open circuit voltage when SOC = 1

When SOC = 0.9 and $-40 \leq T \leq 20$ °C, the data fitting result is shown in Fig. 3.3, and the fitting function of E_0 about T is shown in Eq. (3.22):

$$E_0 = 2.547 \times 10^{-7}T^3 + 1.396 \times 10^{-6}T^2 + 2.027 \times 10^{-4}T + 4.068 \quad (3.22)$$

$$\frac{dE_0}{dT} = 7.641 \times 10^{-7}T^2 + 2.792 \times 10^{-6}T + 2.027 \times 10^{-4} \quad (3.23)$$

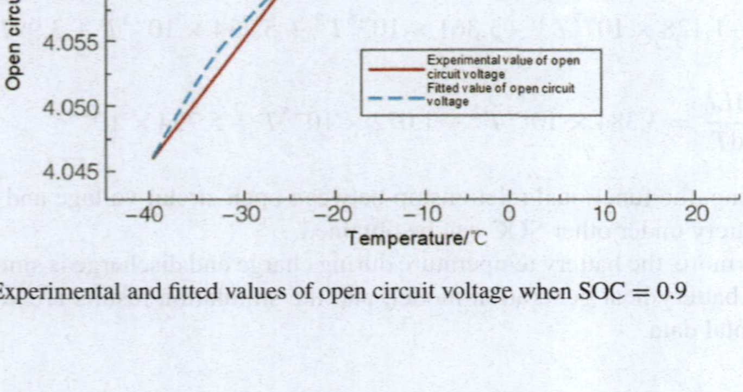

Fig. 3.3 Experimental and fitted values of open circuit voltage when SOC = 0.9

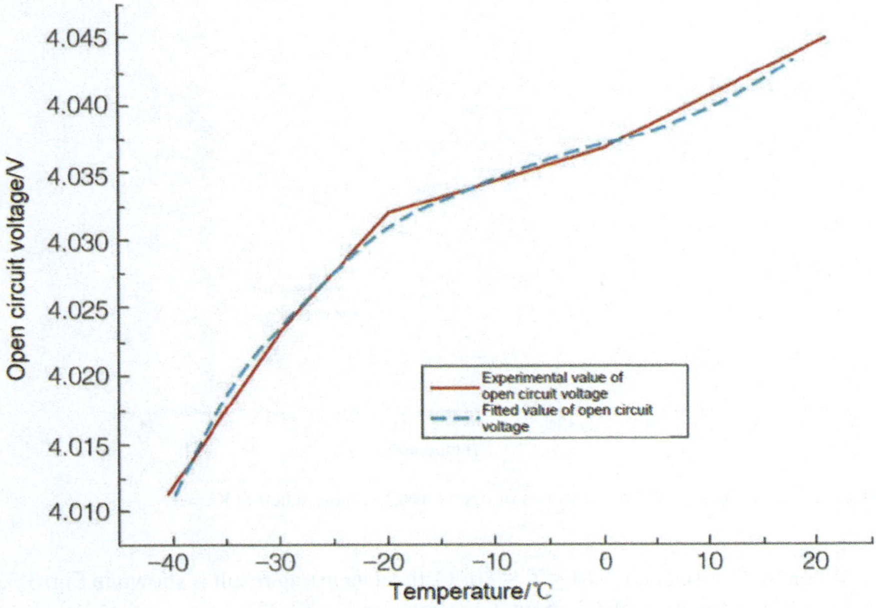

Fig. 3.4 Experimental and fitted values of open circuit voltage when SOC = 0.8

When SOC = 0.8 and −40 ≤ T ≤ 20 °C, the data fitting result is shown in Fig. 3.4, and the fitting function of E_0 about T is shown in Eq. (3.24):

$$E_0 = 2.944 \times 10^{-7}T^3 + 1.289 \times 10^{-6}T^2 + 2.196 \times 10^{-4}T + 4.037 \tag{3.24}$$

$$\frac{\mathrm{d}E_0}{\mathrm{d}T} = 8.832 \times 10^{-7}T^2 + 2.578 \times 10^{-6}T + 2.196 \times 10^{-4} \tag{3.25}$$

When SOC = 0.7 and −40 °C ≤ T ≤ 20 °C, the data fitting is shown in Fig. 3.5, and the fitting function of E_0 about T is shown in Eq. (3.26):

$$E_0 = -1.128 \times 10^{-7}T^3 - 5.361 \times 10^{-6}T^2 + 5.734 \times 10^{-4}T + 3.997 \tag{3.26}$$

$$\frac{\mathrm{d}E_0}{\mathrm{d}T} = 3.384 \times 10^{-7}T^2 + 1.072 \times 10^{-5}T + 5.734 \times 10^{-4} \tag{3.27}$$

By fitting, the functional relationship between open circuit voltage and temperature of battery under other SOC can be obtained.

Furthermore, the battery temperature during charge and discharge is simulated by using the battery heat generation model, and the simulation results are verified by experimental data.

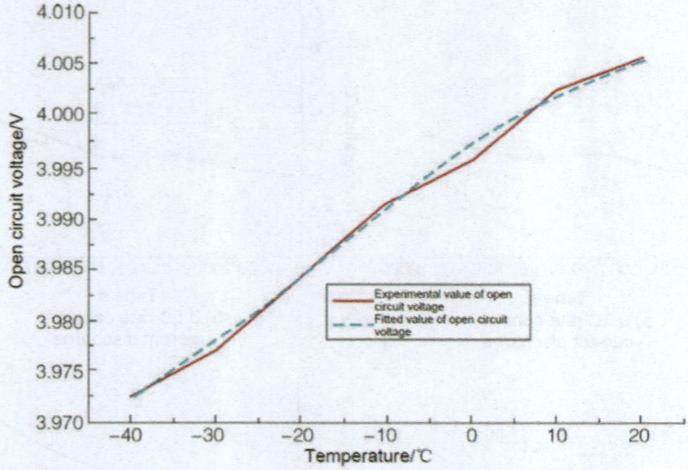

Fig. 3.5 Experimental and fitted values of open circuit voltage when SOC = 0.7

(1) Discharging process

The experimental data of average surface temperature of batteries discharged at rates of 0.3C, 0.5C, 1C, 2C, 3C and 4C are compared with the average surface temperature of batteries simulated by the battery heat generation model, and the latter two models are also compared with the same data. The results are shown in Fig. 3.6.

It can be seen from Fig. 3.6 that the average surface temperature of battery cells calculated by using the three-dimensional heat generation model based on Bernardi heat generation rate is basically consistent with the experimental results, with the average relative error within 1.5% and the max. relative error within 3.5%.

(2) Charging process

The comparison between the experimental and simulated values of the battery surface temperature during charging is shown in Fig. 3.7.

Compared with battery discharge, the error between simulated and experimental values of surface temperature rise during battery charging is increased, the average relative error is kept within 2%, and the max. relative error is increased to within 4.5%.

3.3.2 Modeling and Verification of Electrothermal Coupling Model with Current Density

The electric heating model introduces current density based on Bernardi heat generation rate, and the electric field distribution and thermal field distribution during battery charge and discharge are considered in the model. During charge and discharge,

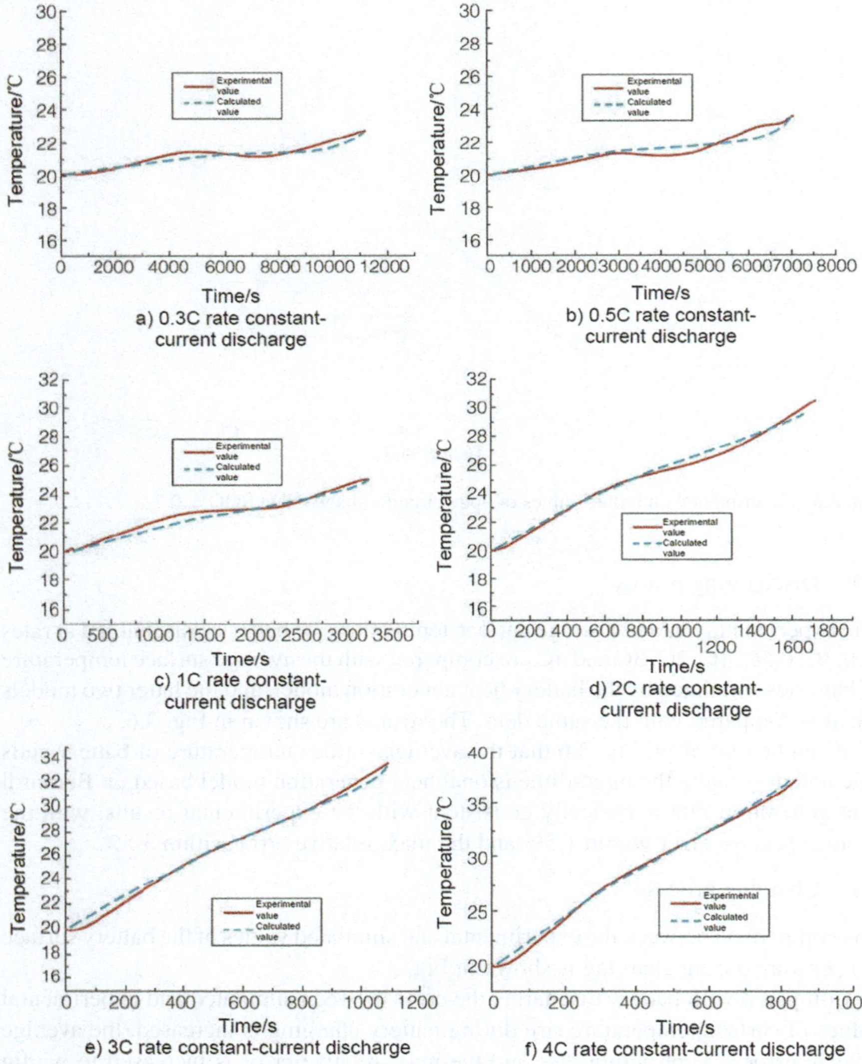

Fig. 3.6 Comparison of surface temperature rise of constant current discharged battery calculated based on Bernardi heat generation rate at normal temperature with experimental values

because the current density distribution of the pole pieces in the battery cell is uneven, the current density distribution of the positive and negative pole pieces can be obtained by establishing the electric field model of the battery pole pieces. Because the pole pieces are very thin, the influence of the thickness direction can be ignored, so a two-dimensional model of the battery pole pieces is established. The two-dimensional geometric model of the positive plate of the battery is shown

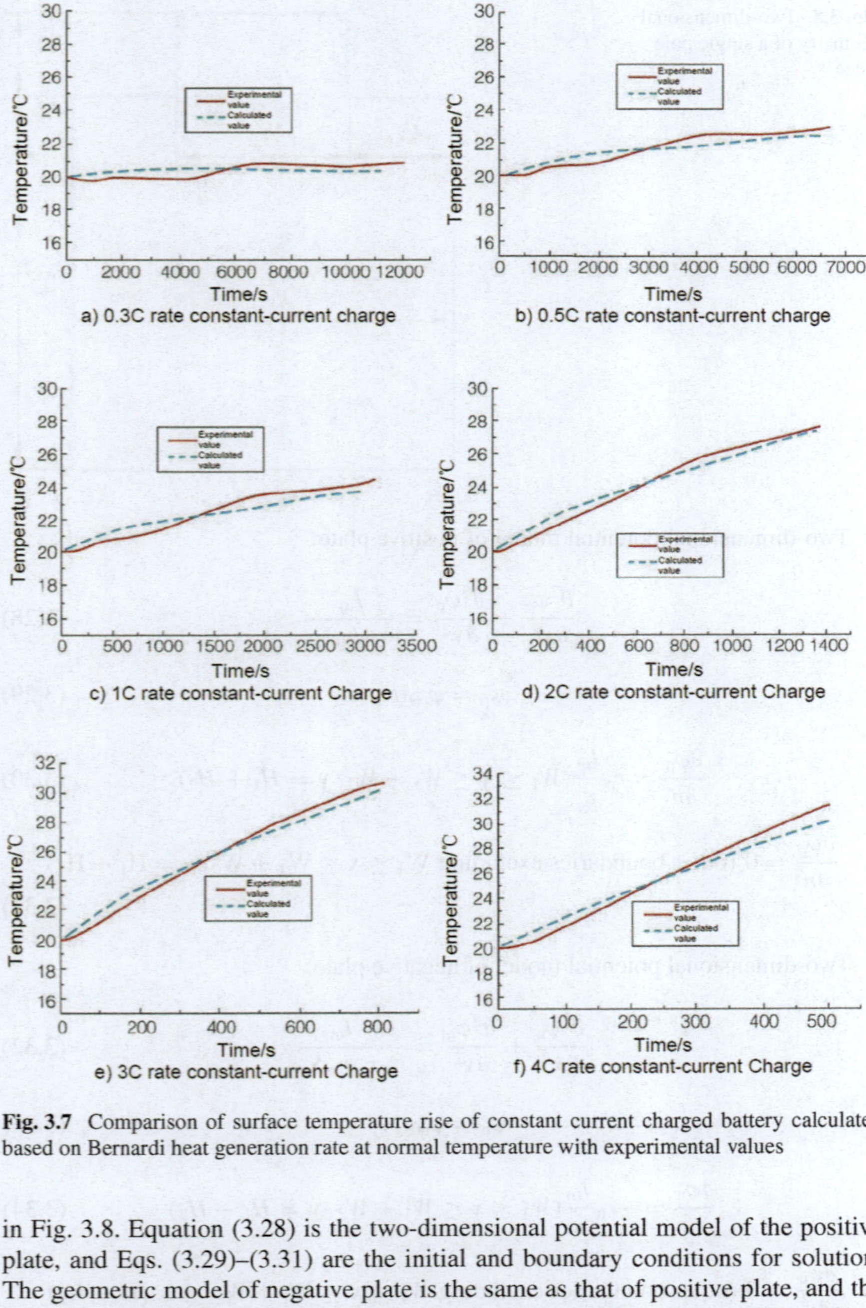

Fig. 3.7 Comparison of surface temperature rise of constant current charged battery calculated based on Bernardi heat generation rate at normal temperature with experimental values

in Fig. 3.8. Equation (3.28) is the two-dimensional potential model of the positive plate, and Eqs. (3.29)–(3.31) are the initial and boundary conditions for solution. The geometric model of negative plate is the same as that of positive plate, and the two-dimensional battery model is the initial condition and boundary condition (Chen 2010) of Eqs. (3.32)–(3.35).

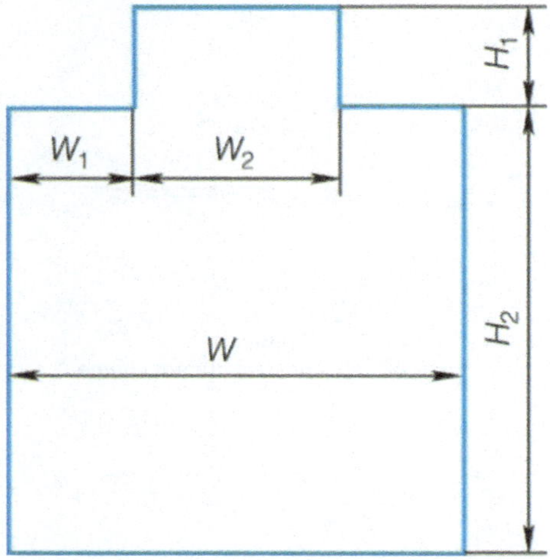

Fig. 3.8 Two-dimensional geometry of a single pole piece

Two-dimensional potential model of positive plate:

$$\frac{\partial^2 \varphi_p}{\partial x^2} + \frac{\partial^2 \varphi_p}{\partial y^2} = \frac{I_{tp}}{\sigma_{cp} h_p S_p} \tag{3.28}$$

$$\varphi_p = \varphi_{p0}(t) \tag{3.29}$$

$$\frac{\partial \varphi_p}{\partial n} = r_p \frac{I_{tp}}{a} (W_1 \le x \le W_1 + W_2, y = H_1 + H_2) \tag{3.30}$$

$$\frac{\partial \varphi_p}{\partial n} = 0 \text{ (other boundaries excluding } W_1 \le x \le W_1 + W_2, y = H_1 + H_2) \tag{3.31}$$

Two-dimensional potential model of negative plate:

$$\frac{\partial^2 \varphi_n}{\partial x^2} + \frac{\partial^2 \varphi_n}{\partial y^2} = -\frac{I_{tn}}{\sigma_{cn} h_n S_n} \tag{3.32}$$

$$\varphi_n = \varphi_{n0}(t) \tag{3.33}$$

$$\frac{\partial \varphi_n}{\partial n} = -r_n \frac{I_{tn}}{a} (W_1 \le x \le W_1 + W_2, y = H_1 + H_2) \tag{3.34}$$

$$\frac{\partial \varphi_n}{\partial n} = 0 \text{ (other boundaries excluding } W_1 \le x \le W_1 + W_2, y = H_1 + H_2) \tag{3.35}$$

where, $\varphi_{\rm p}$ and $\varphi_{\rm n}$ are the potentials of positive plate and negative plate respectively; $\varphi_{\rm p0}$ and $\varphi_{\rm n0}$ are the working voltages of the cathode piece and the anode piece respectively during charge and discharge; $I_{\rm tp}$ and $I_{\rm tn}$ are the currents flowing through the positive plate and negative plate during charge and discharge respectively; $\sigma_{\rm cp}$ and $\sigma_{\rm cn}$ are the conductivity of positive plate and negative plate respectively; $r_{\rm p}$ and $r_{\rm n}$ are the internal resistances of positive plate and negative plate, respectively; $h_{\rm p}$ and $h_{\rm n}$ are the thickness of positive plate and negative plate, respectively. a is the length of the lug; $S_{\rm p}$ and $S_{\rm n}$ are the areas of positive and negative plates, respectively. $\boldsymbol{n}$ is the normal direction outside the boundary.

According to the electric field theory, the current density of positive and negative plates can be solved with Eq. (3.36):

$$\boldsymbol{J} = -\sigma_c \boldsymbol{E} = -\sigma_c \nabla \boldsymbol{\varphi} \tag{3.36}$$

where, $\boldsymbol{J}$ is the current density; $\sigma_{\rm c}$ is the conductivity of positive and negative plates.

The two-dimensional potential model of the positive plate is

$$\begin{aligned} \rho c\frac{\partial T}{\partial t} &= \lambda_x\frac{\partial^2 T}{\partial x^2} + \lambda_y\frac{\partial^2 T}{\partial y^2} + \beta J_p\left[(E_0 - U_L) - T\frac{dE_0}{dT}\right] \\ \frac{\partial^2\varphi_{\rm p}}{\partial x^2} + \frac{\partial^2\varphi_{\rm p}}{\partial y^2} &= \frac{I_{\rm tp}}{\sigma_{\rm cp} h_{\rm p} S_{\rm p}} \\ J_{\rm p} &= -\sigma_c \nabla\varphi_{\rm p} \end{aligned} \tag{3.37}$$

The initial and boundary conditions of the temperature field are

$$\begin{cases} T(x, y, z; 0) = T_0 \\ -\lambda\frac{\partial T}{\partial n}|_\Gamma = h(T - T_{\rm amb})|_\Gamma \end{cases} \tag{3.38}$$

The initial and boundary conditions of the electric field are

$$\varphi_{\rm p} = \varphi_{\rm p0}(t) \tag{3.39}$$

$$\frac{\partial\varphi_{\rm p}}{\partial n} = -r_{\rm p}\frac{I_{\rm tp}}{a}(W_1 \le x \le W_1 + W_2, y = H_1 + H_2) \tag{3.40}$$

$$\frac{\partial\phi_{\rm p}}{\partial n} = 0 \text{ (other boundaries excluding } W_1 < x \le W_1 + W_2, y = H_1 + H_2) \tag{3.41}$$

In the same way, the two-dimensional thermal model of the negative plate can be obtained.

To establish an accurate three-dimensional electric-thermal coupling model of the battery, we must strictly follow the structure of the battery cell. As previously analyzed, the ratio of thickness to length and width of each component of the cell is very low, which makes the calculation amount exceed the acceptable range. Therefore, the battery cell is equivalent to three parts: positive pole piece, negative pole piece and other materials (including packaging, separator and electrolyte, etc.), as shown in Fig. 3.9. The three-dimensional electro-thermal coupling model of the battery is shown in Eq. (3.42), the initial conditions and boundary conditions are shown in Eqs. (3.43)–(3.49), and the parameters in the equations have been explained in the previous contents.

$$\begin{cases} \rho_p c_p \frac{\partial T}{\partial t} = \lambda_{px} \frac{\partial^2 T}{\partial x^2} + \lambda_{py} \frac{\partial^2 T}{\partial y^2} + \lambda_{pz} \frac{\partial^2 T}{\partial z^2} + \beta J_p \left[(E_0 - U_L) - T \frac{dE_0}{dT} \right] \\ \rho_n c_n \frac{\partial T}{\partial t} = \lambda_{nx} \frac{\partial^2 T}{\partial x^2} + \lambda_{ny} \frac{\partial^2 T}{\partial y^2} + \lambda_{nz} \frac{\partial^2 T}{\partial z^2} + \beta J_n \left[(E_0 - U_L) - T \frac{dE_0}{dT} \right] \\ \rho_r c_r \frac{\partial T}{\partial t} = \lambda_{rx} \frac{\partial^2 T}{\partial x^2} + \lambda_{ry} \frac{\partial^2 T}{\partial y^2} + \lambda_{rz} \frac{\partial^2 T}{\partial z^2} \\ \frac{\partial^2 \varphi_p}{\partial x^2} + \frac{\partial^2 \varphi_p}{\partial y^2} + \frac{\partial^2 \varphi_p}{\partial z^2} = \frac{I_{tp}}{\sigma_{cp} h_p S_p} \\ \frac{\partial^2 \varphi_n}{\partial x^2} + \frac{\partial^2 \varphi_n}{\partial y^2} + \frac{\partial^2 \varphi_n}{\partial z^2} = -\frac{I_{tn}}{\sigma_{cn} h_n S_n} \\ J_p = -\sigma_c \nabla \varphi_p \\ J_n = -\sigma_c \nabla \varphi_n \end{cases} \tag{3.42}$$

The initial and boundary conditions of the temperature field are

$$\begin{cases} T(x, y, z; 0) = T_0 \\ -\lambda \frac{\partial T}{\partial n}|_\Gamma = h(T - T_{amb})|_\Gamma \end{cases} \tag{3.43}$$

The initial and boundary conditions of the electric field are

$$\varphi_p = \varphi_{p0}(t) \tag{3.44}$$

Fig. 3.9 Geometric model of electrothermal coupling of battery cells

$$\frac{\partial \varphi_p}{\partial n} = r_p \frac{I_{tp}}{a} (W_1 < x \le W_1 + W_2, y = H_2) \tag{3.45}$$

$$\frac{\partial \varphi_p}{\partial n} = 0 \left(\text{other boundaries excluding } W_1 < x \le W_1 + W_2, y = H_1 + H_2\right) \tag{3.46}$$

$$\varphi_n = \varphi_{n0}(t) \tag{3.47}$$

$$\frac{\partial \varphi_n}{\partial n} = -r_n \frac{I_{tn}}{a} (W_1 < x \le W_1 + W_2, y = H_2) \tag{3.48}$$

$$\frac{\partial \varphi_n}{\partial n} = 0 \text{ (other boundaries excluding } W_1 \le x \le W_1 + W_2, y = H_1 + H_2) \tag{3.49}$$

Furthermore, the model is verified by the battery surface temperature measured by experiments.

(1) Discharging process

Through the model, the temperature field of the battery can be calculated, and the comparison between the experimental value and the simulated value of the battery surface temperature during discharge is shown in Fig. 3.10.

(2) Charging process

The comparison between the experimental value and the simulated value of the battery surface temperature during charging is shown in Fig. 3.11.

The simulation and experimental results of the surface temperature during battery charge and discharge show that the simulation accuracy of the electrothermal coupling model with current density is close to that of Bernardi heat generation rate model, but not higher. Although the electrothermal coupling model with current density takes into account the uneven temperature distribution in the process of battery charge and discharge, and can more accurately reflect the heat generation during battery charge and discharge, it simplifies the battery model when calculating the temperature rise of a battery cell, which reduces the accuracy of the model. In addition, the ratio of the thickness to the length and width of the battery pole piece is very small, so the calculation amount of the monopolar piece model is high, and the calculation amount of the equivalent model of the battery cell will be even higher. Therefore, it is more economical to adopt Bernardi heat generation rate model from the point of model calculation view.

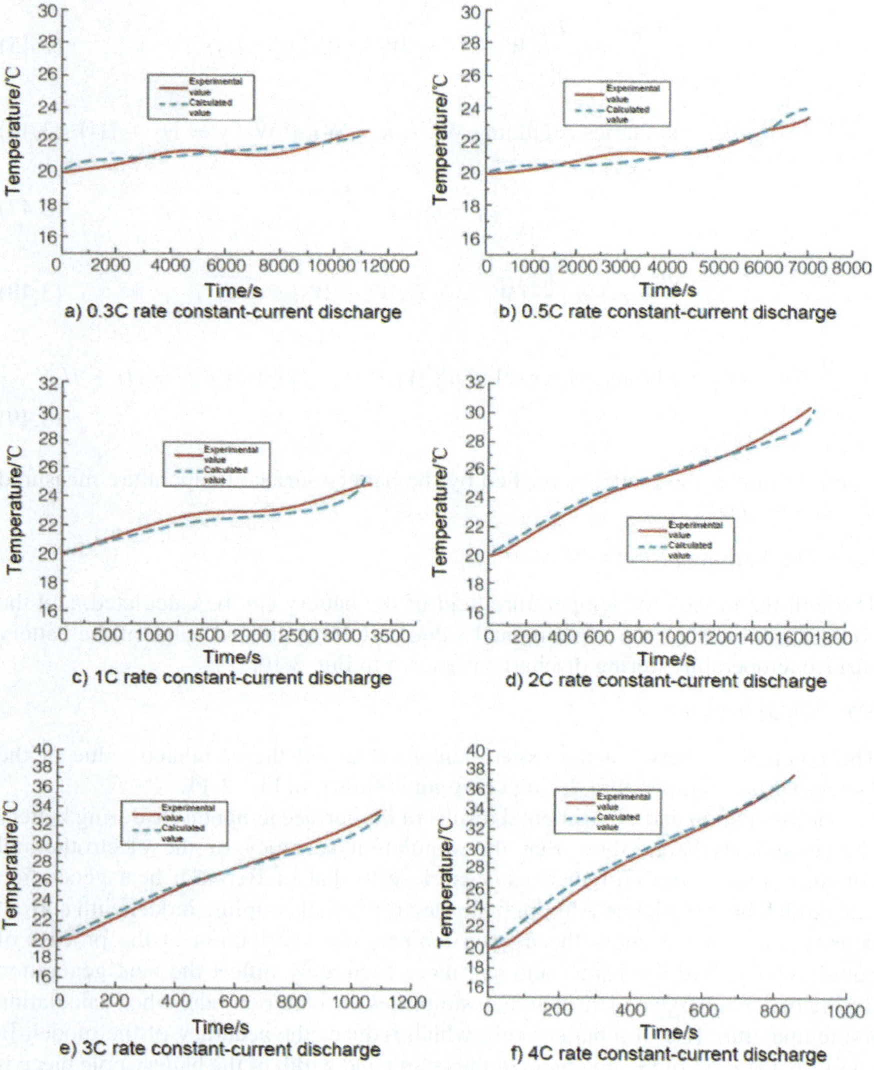

Fig. 3.10 Comparison of surface temperature rise of a constant current discharged battery calculated based on electrothermal coupling model at normal temperature with experimental values

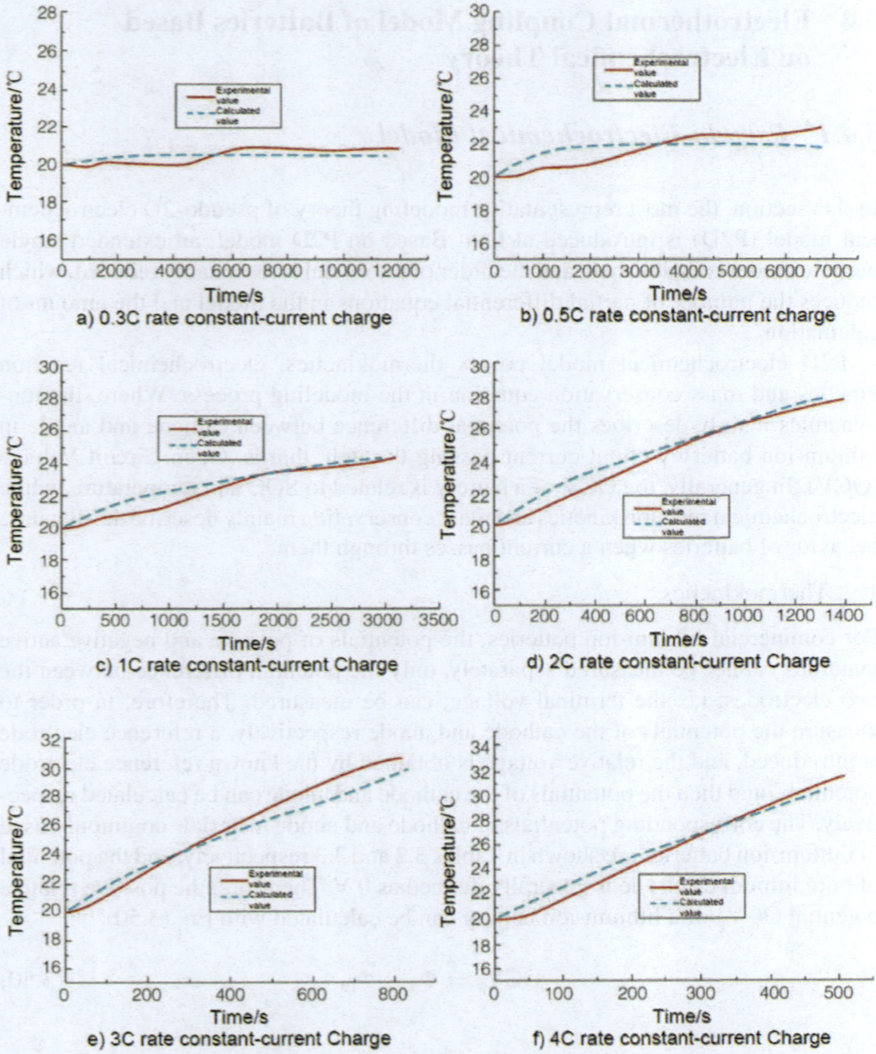

Fig. 3.11 Comparison of surface temperature rise of a constant current charged battery calculated based on electrothermal coupling model at normal temperature with experimental values

3.4 Electrothermal Coupling Model of Batteries Based on Electrochemical Theory

3.4.1 Pseudo-Electrochemical Model

In this section, the most representative modeling theory of pseudo-2D electrochemical model (P2D) is introduced at first. Based on P2D model, an extended single particle model is established, and the order of the model is reasonably reduced, which reduces the number of partial differential equations in the model and the amount of calculation.

P2D electrochemical model covers thermokinetics, electrochemical reaction kinetics and mass conservation equation in the modeling process. Where, thermodynamics mainly describes the potential difference between cathode and anode in lithium-ion battery without current passing through, that is, Open Circuit Voltage (OCV). In generally, the OCV of a battery is related to SOC and temperature, while electrochemical reaction kinetics and mass conservation mainly describe the dynamic behavior of batteries when a current passes through them.

1. Thermokinetics

For commercial lithium-ion batteries, the potentials of positive and negative active materials cannot be measured separately, only the potential difference between the two electrodes, i.e. the terminal voltage, can be measured. Therefore, in order to measure the potentials of the cathode and anode respectively, a reference electrode is introduced, and the relative voltage is obtained by the known reference electrode potentials, and then the potentials of the cathode and anode can be calculated respectively. The corresponding potentials of cathode and anode materials commonly used in lithium-ion batteries are shown in Tables 3.2 and 3.3 respectively, and the potential of pure lithium electrode is generally defined as 0 V. Therefore, the positive relative potential OCV_p of a lithium-ion battery can be calculated with Eq. (3.50):

$$OCV_p = \Phi_p - \Phi_{Li} \tag{3.50}$$

Table 3.2 Potential of common cathode active materials

Cathode material	Potential (vs. Li/Li^+)/V	Specific capacity/(mA h/g)
$LiMn_2O_4$	4.1	100–120
$LiCoO_2$	3.9	140–170
$LiFePO_4$	3.45	170
$Li(NiMnCo)_{1/3}O_2$	3.8	160–170
$LiNi_{0.8}Co_{0.15}Al_{0.05}O_2$	3.8	180–240
Li–sulfur	2.1	1280

Table 3.3 Potential of common anode active materials

Anode material	Potential (vs. Li/Li^+)/V	Specific capacity/(mA h/g)
$Li_4Ti_5O_{12}$	1.6	175
LiC_6	0.1	372
Li–tin	0.6	994
Li metal	0	3862
Li–silicon	0.4	4200

where, OCV_p is the relative potential of cathode of lithium-ion battery; Φ_p is the absolute potential of cathode; Φ_{Li} is the electrode potential of pure lithium.

However, neither the positive absolute potential Φ_p nor the negative absolute potentials Φ_n can be obtained by direct measurement, but the potential difference between Φ_p and Φ_{Li}, that is, OCV_p, can be measured. For a full cell, OCV_{cell} can be calculated with Eq. (3.51):

$$OCV_{cell} = \left(\Phi_p - \Phi_{Li}\right) - (\Phi_n - \Phi_{Li}) = \Phi_p - \Phi_n \tag{3.51}$$

From the point of thermodynamics view, OCV of a battery cell is related to Gibbs free energy [ΔG] of electrode materials. At normal temperature, Gibbs free energy is calculated as follows (Christopher and Wang 2013):

$$\Delta G = \Delta H - T\Delta S \tag{3.52}$$

In the equation, ΔG is the Gibbs free energy, which represents useful work (J) done externally in electrochemical reaction; ΔH is the enthalpy change and represents the total energy generated in electrochemical reaction; $T\Delta S$ is the energy consumed due to heat generation in the reaction process (T is temperature and ΔS is the entropy change).

In general, for a chemical reaction, the starting conditions of the reaction are related to the Gibbs free energy of reactants and products. The smaller the Gibbs free energy is, the more favorable it is for the occurrence of chemical reaction, as shown in Fig. 3.12. For the electrochemical system of lithium-ion battery, the difference of Gibbs free energy of reactants and products corresponding to redox reaction on positive and negative active electrodes is less than zero during the discharge process, so when there is a load outside the battery, the electrochemical reaction will proceed spontaneously. However, in the charging process, the electrochemical reaction can not proceed spontaneously, and extra potentials need to be applied at both ends of the cathode and anode of the battery, so that the charging process can proceed normally.

For the electrochemical system of lithium-ion battery, in order to link Gibbs free energy with potential, the Eq. (3.52) needs to be transformed. Under constant pressure, $H = \overline{U} + p\overline{V}$ therefore, Eq. (3.52) can be rewritten as:

$$\Delta G = \Delta\overline{U} + p\Delta\overline{V} - T\Delta S \tag{3.53}$$

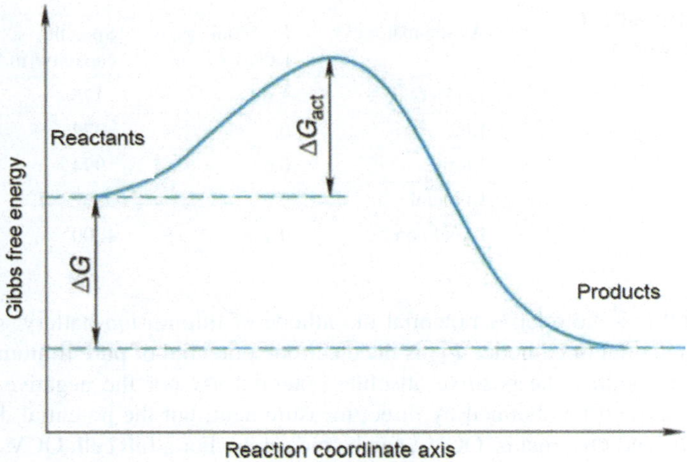

Fig. 3.12 Schematic diagram of Gibbs free energy change from reactant to product

where, $\bar{U}$ (J) is the internal energy of the system (J); p is the pressure (Pa); $\bar{V}$ is the volume (m^3); U and V are used to distinguish from voltage and potential.

In the reaction process of an electrochemical system, the definition of internal energy in the system is as follows:

$$\Delta\overline{U}{=}T\Delta S - p\Delta\overline{V} - n_e FU \tag{3.54}$$

where, n_e is the stoichiometric number of electron transfer during electrochemical reaction; F is Faraday constant (C/mol); U is the open circuit voltage (V).

The relationship between Gibbs free energy and potential can be obtained by substituting Eq. (3.54) into (3.53):

$$\Delta G = -n_e FU \tag{3.55}$$

Under the standard condition, i.e., 1 M electrolyte, $p = 1$ bar, $T = 25$ °C, we have

$$\Delta G^0 = -n_e FU^0 \tag{3.56}$$

But in the non-standard state,

$$\Delta G = \Delta G^0 + \overline{R}T \ln\left(\frac{\Pi \mathrm{act_{prod}}}{\Pi \mathrm{act_{react}}}\right) \tag{3.57}$$

where, $\overline{R}$ is the ideal gas constant [J/(K mol)]; act is related to the concentration of reactants and products, and its calculation equation is:

$$\mathrm{act}_i = \exp\left(\frac{\mu_i - \mu_0}{\overline{R}T}\right) \tag{3.58}$$

where, μ_i is the chemical potential (J/mol) participating in chemical reaction i; μ_0 is the value in the standard state.

Therefore, by substituting Eqs. (3.55) and (3.56) into (3.57), (3.59) can be obtained, that is, Nernst Equation:

$$U = U^0 - \frac{\overline{R}T}{n_e F} \ln\left(\frac{\Pi \mathrm{act}_{\mathrm{prod}}}{\Pi \mathrm{act}_{\mathrm{react}}}\right) \tag{3.59}$$

Therefore, in the electrochemical system of lithium-ion batteries, the open-circuit voltage is related to the concentration of intercalated and deintercalated lithium-ion, and U in Eq. (3.59) is unmeasurable. U_0 is the open-circuit potential in the standard state, which is generally regarded as a constant, but will be affected by temperature. See Tables 3.2 and 3.3 for the standard open-circuit potential of common electrodes. Theoretically, under the condition of known temperature and lithium-ion concentration, the open circuit voltage of a pair of electrodes can be calculated with Eq. (3.59). However, it is very difficult to calculate the activation energy of electrode materials, so empirical values are usually used instead (Verbrugge 1996). There are two ways to obtain the open-circuit voltage of positive and negative materials for a commercial full cell: one is to use the experimental method of a half cell, that is, to separate the positive and negative materials of the full cell, and to form a new "full cell" as cathode and pure lithium electrode respectively, and charge and discharge at a sufficiently small current (generally less than 1/25C). At this time, it is considered that the polarization inside the cell can be ignored, and the terminal voltage of the half cell can be approximated as the open-circuit voltage. For example, Fig. 3.13 shows the OCV curve measured on a half cell composed of graphite and lithium electrode under 1/25C discharge condition. Another method is to use 1/25C rate to

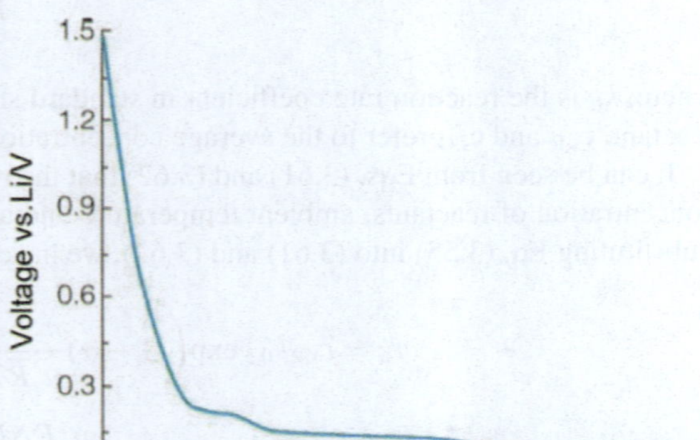

Fig. 3.13 Open-circuit voltage of half cell composed of graphite material and pure lithium material under 1/25C discharging condition

pulse discharge the half cell, and then measure the open circuit voltage when the cell is still in steady state. From 100 to 0% SOC, this method has higher measurement accuracy and more accurate results, but because it takes too long time, most scholars have adopted the first method at present. After determining the open-circuit voltage of the cell in the standing state, the next step is to quantify the influence of ohmic resistance and overpotential in the cell when a current passes through.

2. Reaction kinetics

For a full cell, the theory of reaction kinetics mainly studies the electrochemical reaction between the cathode/anode (respectively) and the electrolyte at the solid–liquid interface. At the solid–liquid interface, there are oxidation and reduction reactions at the same time. In general, the reaction rate is calculated as follows:

$$r = kc \tag{3.60}$$

where, r is the reaction rate [mol/(m^2 s)]; k is the reaction rate coefficient of electrode material (m/s); c is the concentration of lithium-containing active substance (mol/m^3).

When the cell is in a steady state, no current passes through the cell, the oxidation and reduction reactions at the interface are in equilibrium, and the potential difference between the cathode and anode is equal to the open circuit voltage. When a current passes through the battery, the oxidation and reduction reactions at the interface are no longer in equilibrium. The Arrhenius Equation gives the relationship between reaction rate constant k and positive and negative activation energy ΔG_{act} (Forman et al. 2011), so the reaction rate at the interface can be expressed with Eq. (3.61):

$$r_{\mathrm{a}} = c_{\mathrm{s,R}}k_{\mathrm{a}} = c_{\mathrm{s,R}}k_{0,\mathrm{a}}\exp\left(-\frac{\Delta G_{\mathrm{act,a}}}{RT}\right) \tag{3.61}$$

$$r_{\mathrm{c}} = c_{\mathrm{s,O}}k_{\mathrm{c}} = c_{\mathrm{s,O}}k_{0,\mathrm{c}}\exp\left(-\frac{\Delta G_{\mathrm{act,c}}}{RT}\right) \tag{3.62}$$

where, k_0 is the reaction rate coefficient in standard state (if there is more than one reactant, $c_{\mathrm{s,R}}$ and $c_{\mathrm{s,O}}$ refer to the average concentrations of all reactants).

It can be seen from Eqs. (3.61) and (3.62) that the reaction rate is affected by the concentration of reactants, ambient temperature and activation energy of reactants. Substituting Eq. (3.55) into (3.61) and (3.62), we have

$$r_{\mathrm{a}} = c_{\mathrm{s,R}}k_{0,\mathrm{a}}\exp\left((1-\alpha)\frac{n_{\mathrm{e}}F\Delta U}{RT}\right) \tag{3.63}$$

$$r_{\mathrm{c}} = c_{\mathrm{s,O}}k_{0,\mathrm{c}}\exp\left(-\alpha\frac{n_{\mathrm{e}}F\Delta U}{RT}\right) \tag{3.64}$$

where, ΔU is the variation of equilibrium potential of internal electrode potential in standard state, $\Delta U = U - U^0$; α is the proportional coefficient of activation energy to be overcome between cathode and anode, α is usually taken as 0.5.

Therefore, the redox reaction equation for the interface between solid phase and liquid phase during the discharging of the lithium-ion battery is as follows (where, the signs in the redox reaction equation during charging are opposite):

$$r = r_\mathrm{a} - r_\mathrm{c} = c_{\mathrm{s,R}} k_{0,\mathrm{a}} \exp\left[(1-\alpha)\frac{n_\mathrm{e} F \Delta U}{\overline{R} T}\right] - c_{\mathrm{s,O}} k_{0,\mathrm{c}} \exp\left(-\alpha \frac{n_\mathrm{e} F \Delta U}{\overline{R} T}\right) \quad (3.65)$$

According to the relationship $j = n_\mathrm{e} F r$ between the local current density j (A/m^2) at the interface and the reaction rate r, the Eq. (3.65) can be written as:

$$j = nF\left\{c_{\mathrm{s,R}} k_{0,\mathrm{a}} \exp\left[(1-\alpha)\frac{n_\mathrm{e} F \Delta U}{\overline{R} T}\right] - c_{\mathrm{s,O}} k_{0,\mathrm{c}} \exp\left(-\alpha \frac{n_\mathrm{e} F \Delta U}{\overline{R} T}\right)\right\} \quad (3.66)$$

Equation (3.66) is called the Butler–Volmer equation (Xia 2000; Newman and Tiedemann 1974).

For lithium-ion batteries, assuming that the anode is graphite and the cathode is lithium-containing compound of metal N or N oxide, the reactant concentration on the cathode and the anode can be defined as:

$$\text{Anode}: \mathrm{Li}_x\mathrm{C}_6 \rightarrow \mathrm{Li}^+ + \mathrm{C}_6 + \mathrm{e}^- \quad (3.67)$$

$$\text{Cathode}: \mathrm{Li}^+ + \mathrm{N} + \mathrm{e}^- \rightarrow \mathrm{Li}_y\mathrm{N} \quad (3.68)$$

Assuming that $\alpha = 0.5$ and $n_\mathrm{e} = 1$, the reaction rate coefficients of cathode and anode are equal in the internal electrochemical reaction of lithium-ions. Therefore, the redox reaction at the solid–liquid interface of the cathode and anode of the lithium-ion battery can be expressed with Eq. (3.66) respectively.

$$j_\mathrm{p} = F k_{0,\mathrm{p}}\left[c_{\mathrm{s,R,p}} \exp\left(\frac{F \Delta U_\mathrm{p}}{2\overline{R} T}\right) - c_{\mathrm{s,O,p}} \exp\left(-\frac{F \Delta U_\mathrm{p}}{2\overline{R} T}\right)\right] \quad (3.69)$$

$$j_\mathrm{n} = F k_{0,n}\left[c_{\mathrm{s,O,n}} \exp\left(-\frac{F \Delta U_\mathrm{n}}{2\overline{R} T}\right) - c_{\mathrm{s,R,n}} \exp\left(\frac{F \Delta U_\mathrm{n}}{2\overline{R} T}\right)\right] \quad (3.70)$$

In order to better express the reactant concentration on the cathode and the anode, $c_{\mathrm{max},i}$, which represents the max concentration of active material reactant, is introduced, and it is assumed that the max concentration of anode active material $\mathrm{LiC_6}$ or cathode active material LiN is equal to the sum of cathode and anode active materials for lithium intercalation and lithium removal. Therefore, the concentrations of active materials on the cathode and the anode can be defined as:

$$\text{Anode} : c_{\text{max,LiC}_6} = c_{\text{C}_6} + c_{\text{Li}_x\text{C}_6} \tag{3.71}$$

$$\text{Cathode} : c_{\text{max,LiN}} = c_{\text{N}} + c_{\text{Li}_y\text{N}} \tag{3.72}$$

On the basis of the above assumptions, the concentrations of active materials are converted into the concentrations of reactants participating in the redox reaction on each electrode, the following is further defined:

$$c_{\text{s,O,p}} = c_{\text{Li}_y\text{N}} \tag{3.73}$$

$$c_{\text{s,R,p}} = c_{\text{N}} c_{\text{Li}} = \left(c_{\text{max,LiN}} - c_{\text{Li}_y\text{N}}\right) c_{\text{Li}^+} \tag{3.74}$$

$$c_{\text{s,O,n}} = c_{\text{Li}_x\text{C}_6} \tag{3.75}$$

$$c_{\text{s,R,n}} = c_{\text{C}_6} c_{\text{Li}} = \left(c_{\text{max,LiC}_6} - c_{\text{Li}_x\text{C}_6}\right) c_{\text{Li}^+} \tag{3.76}$$

Note that the above reactant concentrations refer to the concentrations of surface active substances participating in electrochemical reactions at the solid–liquid interface. To facilitate further calculation, the following definitions are introduced:

$$c_{\text{Li}_y\text{N}} = c_{\text{s,p}} \tag{3.77}$$

$$c_{\text{max,LiN}} = c_{\text{max,p}} \tag{3.78}$$

$$c_{\text{Li}_x\text{C}_6} = c_{\text{s,n}} \tag{3.79}$$

$$c_{\text{max,LiC}_6} = c_{\text{max,n}} \tag{3.80}$$

where, the lower corner mark s is the surface concentration of the active electrode; The lower corner marks p and n are cathode and anode respectively.

The lithium-ion concentration in the electrolyte is defined as: $c_{\text{Li}^+} = c_{\text{e,p}}$ when it is close to the cathode; $c_{\text{Li}^+} = c_{\text{e,n}}$ when it is close to the anode. Therefore, Eqs. (3.69) and (3.70) can be rewritten as follows:

$$j_{\text{p}} = Fk_{0,\text{p}} \left[\left(c_{\text{max,p}} - c_{\text{s,p}}\right) c_{\text{e,p}} \exp\left(\frac{F\Delta U_{\text{p}}}{2\overline{R}T}\right) - c_{\text{s,p}} \exp\left(-\frac{F\Delta U_{\text{p}}}{2\overline{R}T}\right) \right] \tag{3.81}$$

$$j_{\text{n}} = Fk_{0,\text{n}} \left(\left(c_{\text{max,n}} - c_{\text{s,n}}\right) c_{\text{e,n}} \exp\left(-\frac{F\Delta U_{\text{n}}}{2\overline{R}T}\right) - c_{\text{s,n}} \exp\left(\frac{F\Delta U_{\text{n}}}{2\overline{R}T}\right) \right) \tag{3.82}$$

where, ΔU is calculated from the standard equilibrium potential.

Therefore, when no current passes through the battery, that is, the local current density $j = 0$, the redox reaction rate at the interface between the cathode and anode is the same. It can be obtained according to Eqs. (3.81) and (3.82) that

$$\left(c_{\max,p} - c_{s,p}\right)c_{e,p}\exp\left(\frac{F\Delta U_{p}}{2\overline{R}T}\right) = c_{s,p}\exp\left(-\frac{F\Delta U_{p}}{2\overline{R}T}\right) \tag{3.83}$$

$$\left(c_{\max,n} - c_{s,n}\right)c_{e,n}\exp\left(-\frac{F\Delta U_{n}}{2\overline{R}T}\right) = c_{s,n}\exp\left(\frac{F\Delta U_{n}}{2\overline{R}T}\right) \tag{3.84}$$

Then $\Delta U_{p,eq}$ and $\Delta U_{n,eq}$ can be further calculated:

$$\Delta U_{p,eq} = \frac{\overline{R}T}{F}\ln\left[\frac{c_{s,p}}{\left(c_{\max,p} - c_{s,p}\right)c_{e,p}}\right] \tag{3.85}$$

$$\Delta U_{n,eq} = \frac{\overline{R}T}{F}\ln\left[\frac{\left(c_{\max,n} - c_{s,n}\right)c_{e,n}}{c_{s,n}}\right] \tag{3.86}$$

Therefore, the overpotential of charge transfer on the cathode and the anode is calculated as follows:

$$\eta_{p} = \Delta U_{p} - \Delta U_{p,eq} \tag{3.87}$$

$$\eta_{n} = \Delta U_{n} - \Delta U_{n,eq} \tag{3.88}$$

Substituting Eqs. (3.85)–(3.88) into (3.81) and (3.82), we have

$$j_{p} = Fk_{0,p}\sqrt{\left(c_{\max,p} - c_{s,p}\right)c_{s,p}c_{e,p}}\left[\exp\left(\frac{F\eta_{p}}{2\overline{R}T}\right) - \exp\left(\frac{F\eta_{p}}{2\overline{R}T}\right)\right] \tag{3.89}$$

$$j_{n} = Fk_{0,n}\sqrt{\left(c_{\max,n} - c_{s,n}\right)c_{s,n}c_{e,n}}\left[\exp\left(-\frac{F\eta_{n}}{2\overline{R}T}\right) - \exp\left(\frac{F\eta_{n}}{2\overline{R}T}\right)\right] \tag{3.90}$$

According to hyperbolic sine transformation, Eqs. (3.89) and (3.90) can be simplified as:

$$j_{p} = 2i_{0,p}\sinh\left(\frac{F\eta_{p}}{2\overline{R}T}\right) \tag{3.91}$$

$$j_{n} = -2i_{0,n}\sinh\left(\frac{F\eta_{n}}{2\overline{R}T}\right) \tag{3.92}$$

where, $i_{0,p} = Fk_{0,p}\sqrt{\left(c_{\max,p} - c_{s,p}\right)c_{s,p}c_{e,p}}$, $i_{0,n} = Fk_{0,n}\sqrt{\left(c_{\max,n} - c_{s,n}\right)c_{s,n}c_{e,n}}$ is the current density in the internal equilibrium state of the lithium-ion battery, that is, the Exchange Current Density.

3. Mass conservation and concentration polarization effect of solid electrodes

In the porous electrode theory, it is assumed that redox reactions all occur on the electrode surface, that is, the solid–liquid interface. During electrochemical reaction, lithium-ions are intercalated or deintercalated from the surface of porous electrode, which leads to the increase or decrease of local lithium-ion concentration, thus resulting in concentration difference. In order to keep the electrochemical reaction going, more lithium-ions must be embedded or removed from the porous electrode. Therefore, in the process of charge and discharge, the instantaneous power of lithium-ion battery is limited by the time of lithium-ion transfer from the inside to the surface of porous electrode, which is closely related to the diffusion coefficient of solid electrode. Fick's law is widely used to describe the change of particle concentration gradient in solid electrodes;

$$\frac{\partial c_i}{\partial t} = \frac{D_i}{r^2}\frac{\partial}{\partial r}\left(r^2\frac{\partial c_i}{\partial r}\right) \tag{3.93}$$

where, c_i refers to the lithium-ion concentration (mol/m^3) in the ith particle in the solid-phase electrode; t (s) is the time; D_i is the diffusion coefficient of lithium-ions in the solid electrode (m^2/s), which is generally assumed to be a constant; r is the radial coordinate (m) of spherical particles, and $0 < r < R_i$.

According to the law of conservation of mass, we can know the boundary conditions of Eq. (3.93):

$$\left.\frac{\partial c_i}{\partial r}\right|_{r=0} = 0, \quad D_i\left.\frac{\partial c_i}{\partial r}\right|_{r=R_i} = \pm\frac{j_i(x,t)}{F} \tag{3.94}$$

where, the positive sign indicates that the current density on the cathode is positive; the negative sign indicates that the current density on the anode is negative.

Therefore, when a current passes through the battery, the solid–liquid phase will undergo electrochemical reaction, and the surface lithium concentration and volume lithium concentration on the porous electrode will change, resulting in concentration polarization effect, which will affect not only the charge transfer potential, but also the calculation of the open circuit voltage.

4. Ohm's law in solid phase electrodes

The relationship between current density and potential in solid phase electrodes complies with ohm's law:

$$\sigma_{\mathrm{i}}\frac{\partial^2 \Phi_i}{\partial x^2} = a_i j_i(x,t) \tag{3.95}$$

where, Φ_t is the solid phase potential (V); σ is the effective conductivity (Ω/m). It is usually assumed that σ is independent of the x-axis coordinate.

Equation (3.95) is applicable to both cathode and anode, and boundary conditions are set in the current collectors and diaphragm:

$$-\sigma_{\mathrm{n}}\frac{\partial \Phi_{\mathrm{n}}}{\partial x}\bigg|_{x=0} = \sigma_{\mathrm{p}}\frac{\partial \Phi_{\mathrm{p}}}{\partial x}\bigg|_{x=L_{\mathrm{cell}}} = \frac{I(t)}{A} \tag{3.96}$$

where, I is the charge and discharge current (A); A is the effective surface area of current collector (m^2).

5. Mass conservation and concentration polarization effect of liquid phase electrodes

Electrolyte, as a carrier, provides a path for lithium-ion transmission between the cathode and the anode. Where, lithium-ion transmission modes mainly include diffusion, migration and convection. Generally, when lithium-ions leave the solid phase electrodes and enter the electrolyte, the concentration of lithium-ions in the vicinity increases due to oxidation reaction. On the contrary, the lithium-ion concentration decreases due to the reduction reaction. As with solid electrode surfaces, when lithium-ions are transmitted in the electrolyte, there is a concentration difference in the x-axis direction, which will further accelerate the diffusion of lithium-ions from one electrode with high concentration to the other with low concentration. In addition, due to the diffusion of lithium-ions, there will be potential difference between the cathode and the anode, which will accelerate the migration of lithium-ions in the electrolyte. Convection has little influence on the transmission of lithium-ions in the electrolyte, which is usually ignored.

(1) Dilute solution theory

In dilute solution theory, the ion transmission process in the electrolyte can be expressed with Eq. (3.97):

$$N_i = -D_i\frac{\partial c_i}{\partial x} - z_i u_i F c_i\frac{\partial \Phi_e}{\partial x} + c_i v \tag{3.97}$$

where, the first item on the right side of the equal sign represents diffusion phenomenon; The second item represents migration phenomenon; The third item represents convection phenomenon; N_i is the flow density of ions in electrolyte [mol/(m^2 s)]; Di is the diffusion coefficient in electrolyte (m^2/s); z_i is the number of charged particles transported in electrolysis; Φ_e is the electrostatic potential (V) in the electrolyte, $\partial \Phi_e/\partial X = -E$ c_i is the concentration of ions in electrolyte (mol/m^3); v is the transport speed of ions in electrolyte (m/s); u_i is the mobility of ions in electrolyte [m^2 mol/(J s)], which is usually calculated with the Nernst-Einstein equation, $u_i{=}D_i/\left(\overline{RT}\right)$.

According to the law of conservation of mass, lithium-ion concentration in electrolyte can be calculated with Eq. (3.98):

$$\varepsilon_e \frac{\partial c_e}{\partial t} = \frac{a_i j_i(x,t)}{F} - \frac{\partial N}{\partial x} \tag{3.98}$$

where, ε_e is the volume fraction of electrolyte; J_i is the current density during the redox reaction.

(2) Concentrated solution theory

Compared with the dilute solution theory, the concentrated solution theory is more widely used because it can accurately express the interaction between ions in electrolyte, where the Nernst-Einstein equation is no longer applicable (Newman et al. 1965). Different from the dilute solution theory, the concentrated solution theory no longer classifies the modes of lithium-ion transmission in electrolyte, and considers that lithium-ion transmission in the electrolyte is related to electrochemical potential.

$$\overline{\mu}_i = \mu_i + z_i F \Phi \tag{3.99}$$

where, $\overline{\mu}_i$ is the electrochemical potential of charged ions (J/mol), which determines the migration of lithium-ions in electrolyte; μ_i is the chemical potential of charged ions (J/mol), which is equal to the slope of Gibbs free energy at normal temperature and constant pressure, $\mu_i = \partial G/\partial n_i$ and determines the diffusion of lithium-ions in electrolyte; z_i is the number of charged particles transported in electrolyte, uncharged ions ($z_i = 0$) will not be affected by electrostatic potential, that is, for uncharged ions, we have $\overline{\mu}_i = \mu_i$.

In the concentrated solution theory, we have,

$$c_i \frac{\partial \overline{\mu}_i}{\partial x} = \sum_j K_{ij}(v_j - v_i) \tag{3.100}$$

where, K_{ij} is the interaction coefficient between particles i and j (J s/m^5), which is related to the concentration of particles.

K_{ij} can be calculated with Eq. (3.101):

$$K_{ij} = \frac{\overline{R} T c_i c_j}{c_T \overline{D_{ij}}} \tag{3.101}$$

where, $c_T = \sum_i c_i$ is the sum of the concentrations of all ions; $\overline{D}$ is the diffusion coefficient, which is used to describe the interaction between different ions, and is considered as an empirical parameter in most cases.

For example, $LiPF_6$ is the most commonly used electrolyte in lithium-ion batteries, and the equation after decomposition is $LiPF_6 \rightarrow Li^+ + PF_6^-$, including cations ($Li^+$), anions $\left(PF_6^-\right)$ and uncharged solvent ions ($LiPF_6$). Therefore, according to Eq. (3.100), the electrochemical potentials of different ions in the electrolyte are:

$$c_{+}\frac{\partial \overline{\mu}_{+}}{\partial x} = K_{+-}(v_{-} - v_{+}) + K_{+0}(v_{0} - v_{+}) \tag{3.102}$$

$$c_{-}\frac{\partial \overline{\mu}_{-}}{\partial x} = K_{-+}(v_{+} - v_{-}) + K_{-0}(v_{0} - v_{-}) \tag{3.103}$$

$$c_{0}\frac{\partial \overline{\mu}_{0}}{\partial x} = K_{0+}(v_{+} - v_{0}) + K_{0-}(v_{-} - v_{0}) \tag{3.104}$$

But according to Newton's third law of motion, $K_{ji} = K_{ij}, \overline{D}_{ji} = \overline{D}_{ij}$, so the right sides of Eqs. (3.102)–(3.104) are equal to 0 (Eide and Maybeck 1996). According to Gibbs–Duhem relation, the left side of Eqs. (3.102)–(3.104) are also equal to 0. Since the current density in the electrolyte is defined as $i_{\mathrm{e}} = F\sum_{i} z_{i}N_{i} = F\sum_{i} z_{i}c_{i}v_{i}$, Eqs. (3.102)–(3.104) can be simplified to two equations:

$$N_{+} = c_{+}v_{+} = -\frac{v_{+}\overline{D}_{\mathrm{e}}}{v_{\mathrm{e}}RT}\frac{c_{\mathrm{T}}}{c_{0}}c_{\mathrm{e}}\frac{\partial \overline{\mu}_{\mathrm{e}}}{\partial x} + \frac{i_{\mathrm{e}}(t)t_{0}^{+}}{z_{+}F} + c_{+}v_{0} \tag{3.105}$$

$$N_{-} = c_{-}v_{-} = -\frac{v_{-}\overline{D}_{\mathrm{e}}}{v_{\mathrm{e}}RT}\frac{c_{\mathrm{T}}}{c_{0}}c_{\mathrm{e}}\frac{\partial \overline{\mu}_{\mathrm{e}}}{\partial x} + \frac{i_{\mathrm{e}}(t)t_{0}^{-}}{z_{-}F} + c_{-}v_{0} \tag{3.106}$$

where, v_{+} and v_{-} are the number of cations and anions in the electrolyte; The amount of solvent can be defined as $v_{\mathrm{e}} = v_{+} + v_{-}$, $\overline{\mu}_{\mathrm{e}}v_{\mathrm{e}} = v_{+}\overline{\mu}_{+} + v_{-}\overline{\mu}_{-}$; Based on the assumption of electrical neutrality, the concentration of lithium salt is $c_{\mathrm{e}} = \frac{c_{+}}{v_{+}} = \frac{c_{-}}{v_{-}}$; Ion diffusion coefficient in the electrolyte is $\overline{D}_{\mathrm{e}} = \frac{\overline{D}_{0+}\overline{D}_{0-}(z_{+}-z_{-})}{z_{+}\overline{D}_{0+}-z_{-}\overline{D}_{0-}}$, which indicates the effective diffusion coefficient between cation and anion; t_{0}^{i} is the number of ion migration, which represents the current density ratio of ion i in the charge and discharge process, $t_{0}^{+} = 1 - t_{0}^{-} = \frac{z_{+}\overline{D}_{0+}}{z_{+}\overline{D}_{0+}-z_{-}\overline{D}_{0-}}$.

In order to further obtain the relationship between lithium-ions in electrolyte and time, Eqs. (3.97) and (3.105) are combined to obtain the following

$$D_{\mathrm{e}} = \overline{D}_{\mathrm{e}}\frac{c_{\mathrm{T}}}{c_{0}}\left(1 + \frac{\mathrm{d}\ln\gamma_{+-}}{\mathrm{d}\ln m}\right) \tag{3.107}$$

where, γ_{+-} is the average molar activity coefficient; m is the number of moles of lithium salt per kilogram of electrolyte.

In addition, the electrochemical potential gradient can be replaced by the concentration gradient.

$$\frac{\overline{D}_{\mathrm{e}}}{v_{\mathrm{e}}RT}\frac{c_{\mathrm{T}}}{c_{0}}c_{\mathrm{e}}\frac{\partial \overline{\mu}_{\mathrm{e}}}{\partial x} = D_{\mathrm{e}}\left(1 - \frac{\mathrm{d}\ln c_{0}}{\mathrm{d}\ln c_{\mathrm{e}}}\right)\frac{\partial c_{\mathrm{e}}}{\partial x} \tag{3.108}$$

Therefore, the flow rate of lithium-ions in the electrolyte is described as follows:

$$N_+ = -D_e\left(1 - \frac{d\ln c_0}{d\ln c_e}\right)\frac{\partial c_e}{\partial x} + \frac{i_e(t)t_0^+}{z_+ v_+ F} + c_e v_0 \tag{3.109}$$

Assuming that there is no convection in the electrolyte, i.e., $v_0 = 0$, according to Eq. (3.97), the variation of lithium-ion with time can be obtained as follows.

$$\varepsilon_e \frac{\partial c_e}{\partial t} = \frac{a_i j_i(x,t)}{F} - \frac{\partial}{\partial x}\left[-D_e\left(1 - \frac{d\ln c_0}{d\ln c_e}\right)\frac{\partial c_e}{\partial x} + \frac{i_e(t)t_0^+}{z_+ v_+ F}\right] \tag{3.110}$$

In addition, because the change of lithium salt concentration in the electrolyte inside a given lithium-ion battery is very low, that is $d\ln c_0/d\ln c_e \approx 0$, Eq. (3.110) can be further simplified as:

$$\varepsilon_e \frac{\partial c_e}{\partial t} = \frac{a_i j_i(x,t)}{F} + D_e \frac{\partial^2 c_e}{\partial x^2} - \frac{\partial i_e}{\partial x}\frac{t_0^+}{z_+ v_+ F} \tag{3.111}$$

The assumption of electricity is that the total current density inside the battery is constant, that is, the gradient of the total current density is zero. Where, the charge carrier in the solid phase electrode is electrons flowing through the current collector, while the charge carrier in the liquid phase area is lithium-ions, so there are.

$$\nabla \cdot \boldsymbol{i}_s + \nabla \cdot \boldsymbol{i}_e = 0 \tag{3.112}$$

where, only electrons flow through the current collector, so $\boldsymbol{i}_e = 0$; Instead, only lithium-ions cross the membrane, so $\boldsymbol{i}_s = 0$.

During the charge and discharge of the lithium-ion battery, there is current flowing through the solid phase and liquid phase area, and an oxidation–reduction reaction takes place.

$$\nabla \cdot \boldsymbol{i}_s = -\nabla \cdot \boldsymbol{i}_e = auser2j \tag{3.113}$$

According to Eq. (3.113), (3.111) can be further transformed into:

$$\varepsilon_e \frac{\partial c_e}{\partial t} = D_e \frac{\partial^2 c_e}{\partial x^2} \pm \frac{3\varepsilon_i\left(1 - t_0^+\right)}{F R_i} j_i(x,t) \tag{3.114}$$

where the current density is positive at the anode, negative at the cathode and 0 at the separator; D_e is the ion diffusion coefficient in electrolyte, which is generally regarded as a constant.

The boundary conditions are as follows:

$$c_e|_{x=L_n^-} = c_e|_{x=L_n^+}, \qquad c_e|_{x=(L_n+L_{sep})^-} = c_e|_{x=(L_n+L_{sep})^+}$$

$$\left.\frac{\partial c_{\mathrm{e}}}{\partial t}\right|_{x=L_{\mathrm{n}}^{-}}=\left.\frac{\partial c_{\mathrm{e}}}{\partial t}\right|_{x=L_{\mathrm{n}}^{+}},\quad \left.\frac{\partial c_{\mathrm{e}}}{\partial t}\right|_{x=(L_{\mathrm{n}}+L_{\mathrm{sep}})^{-}}=\left.\frac{\partial c_{\mathrm{e}}}{\partial t}\right|_{x=(L_{\mathrm{n}}+L_{\mathrm{sep}})^{+}} \tag{3.115}$$

6. Liquid phase potential

Cations, anions and lithium salts participating in electrochemical reaction in electrolyte have the following relationship:

$$s_{+}N_{+}^{z+}+s_{-}N_{-}^{z-}\leftrightharpoons s_{0}N_{0}+n_{\mathrm{e}}e^{-} \tag{3.116}$$

where, N_i represents the chemical equation of ion i; s_i is the stoichiometric coefficient; n_{e} is the number of electrons generated in the electrochemical reaction.

Therefore, the electrochemical potential gradient between ions in the electrolyte has the following relationship with the electrostatic potential gradient:

$$s_{+}\frac{\partial\overline{\mu}_{+}}{\partial x}+s_{-}\frac{\partial\overline{\mu}_{-}}{\partial x}+s_{0}\frac{\partial\overline{\mu}_{0}}{\partial x}=-n_{\mathrm{e}}F\frac{\partial\Phi_{\mathrm{e}}}{\partial x} \tag{3.117}$$

Because of $s_{+}z_{+}+s_{-}z_{-}=-n$, $\overline{\mu}_{\mathrm{e}}=v_{+}\overline{\mu}_{+}+v_{-}\overline{\mu}_{-}$ and the Gibbs–Duhem equation $c_{0}\mathrm{d}\overline{\mu}_{0}+c_{\mathrm{e}}\mathrm{d}\overline{\mu}_{\mathrm{e}}=0$, in combination with Eqs. (3.100)–(3.106), Eq. (3.117) can be converted into

$$\frac{1}{z_{-}}\frac{\partial\overline{\mu}_{-}}{\partial x}=\frac{F}{\sigma_{\mathrm{e}}}i_{\mathrm{e}}(t)-\frac{\partial\overline{\mu}_{\mathrm{e}}}{\partial x}\frac{t_{0}^{+}}{z_{+}v_{+}} \tag{3.118}$$

where, the conductivity σ_{e} of electrolyte is defined as follows:

$$\frac{1}{\sigma_{\mathrm{e}}}=-\frac{\overline{R}T}{c_{\mathrm{T}}z_{+}z_{-}F^{2}}\left(\frac{1}{\overline{D}_{+-}}+\frac{c_{0}t_{0}^{-}}{c_{+}\overline{D}_{0-}}\right) \tag{3.119}$$

In combination with Eqs. (3.107) and (3.108), the potential gradient in electrolyte has the following relationship with current density:

$$\frac{\partial\Phi_{\mathrm{e}}}{\partial x}=\frac{i_{\mathrm{e}}(t)}{\sigma_{\mathrm{e}}}-\frac{v_{\mathrm{e}}\overline{R}T}{F}\left(\frac{s_{+}}{n_{\mathrm{e}}v_{+}}+\frac{t_{0}^{+}}{z_{+}v_{+}}-\frac{s_{0}c_{\mathrm{e}}}{nc_{0}}\right)\left(1+\frac{\mathrm{d}\ln\gamma_{+-}}{\mathrm{d}\ln m}\right)\frac{1}{c_{\mathrm{e}}}\frac{\partial c_{\mathrm{e}}}{\partial x} \tag{3.120}$$

Because $\mathrm{d}\ln c_{0}/\mathrm{d}\ln c_{\mathrm{e}}\approx 0$, and the electrochemical reaction of lithium-ion in electrolyte is $\mathrm{Li}^{+}+\mathrm{e}^{-}\rightarrow\mathrm{Li}$, we can know, $v_{+}=1$, $v_{\mathrm{e}}=2$, $z_{+}=1$, $s_{+}=1$, $s_{0}=s_{-}=0$ and $n_{\mathrm{e}}=-1$. Therefore, Eq. (3.120) can be simplified as:

$$\frac{\partial\Phi_{\mathrm{e}}}{\partial x}=\frac{i_{\mathrm{e}}(t)}{\sigma_{\mathrm{e}}}-(1+\gamma)\frac{2\overline{R}T\left(1-t_{0}^{+}\right)}{F}\frac{\partial\ln(c_{\mathrm{e}})}{\partial x} \tag{3.121}$$

where, $\gamma=\mathrm{d}\ln\gamma_{+-}/\mathrm{d}\ln m$ represents the effective activity coefficient.

Equation (3.121) represents the potential difference caused by ohmic resistance and concentration polarization in the electrolyte.

3.4.2 Extended Single Particle Electrochemical Model

1. Single particle model

Based on the pseudo-two-dimensional model (P2D), the single-particle electrochemical model is further assumed, and the whole electrode is replaced by a single active material particle, thus eliminating the influence of x-axis dimension on lithium-ion concentration and potential in the P2D model, reducing the number of nonlinear equations in the model and improving the calculation efficiency. The current density of the whole active electrode surface in the lithium-ion battery can be calculated with Eq. (3.122):

$$j_i(x,t) = j_i(t) = \frac{I(t)R_i}{3AL_i\varepsilon_i} \tag{3.122}$$

where, $i = p, n$, respectively representing the cathode and the anode; R is the particle radius; A is the surface area of the active electrode current collector; L is the electrode thickness; ε is the volume fraction of active material in the active electrode, which can be replaced by $a_i = 3\varepsilon_i/R_i$ in the spherical volume.

In addition, on the solid-phase active electrode, it is assumed that the current density is uniformly distributed on the surface of each electrode and in the thickness direction of each electrode. In the electrolyte, when a current passes, it is assumed that the voltage drop in the electrolyte is caused by the ohmic resistance of the liquid phase, and the voltage drop of the liquid phase potential caused by the concentration polarization effect is ignored. It is assumed that the accuracy of the model will not be affected in the case of small current, but it is not suitable for the case of high rate current. Therefore, based on the single particle model, the voltage drop caused by non-ohmic resistance in the liquid phase is considered, and the accuracy of the model under the condition of high rate current is improved, and an extended single particle model is established, as shown in Fig. 3.14.

See Table 3.4 for the main governing equations of the extended single particle model. Where, Eq. (3.124) represents the lithium-ion diffusion phenomenon in the solid-phase active electrode. Equations (3.125) and (3.126) are Butler–Volmer equations, namely, equations of electrochemical reaction on the surface of solid phase active electrode; Eq. (3.127) represents the mass conservation of lithium-ion concentration in electrolyte. Equation (3.128) represents the potential generated in electrolyte, including ohmic resistance potential and concentration difference polarization potential in the liquid phase.

Equation (3.130) represents the output voltage of the extended single particle model. $R_{c,i}$ are contact resistances on the positive and negative solid electrodes,

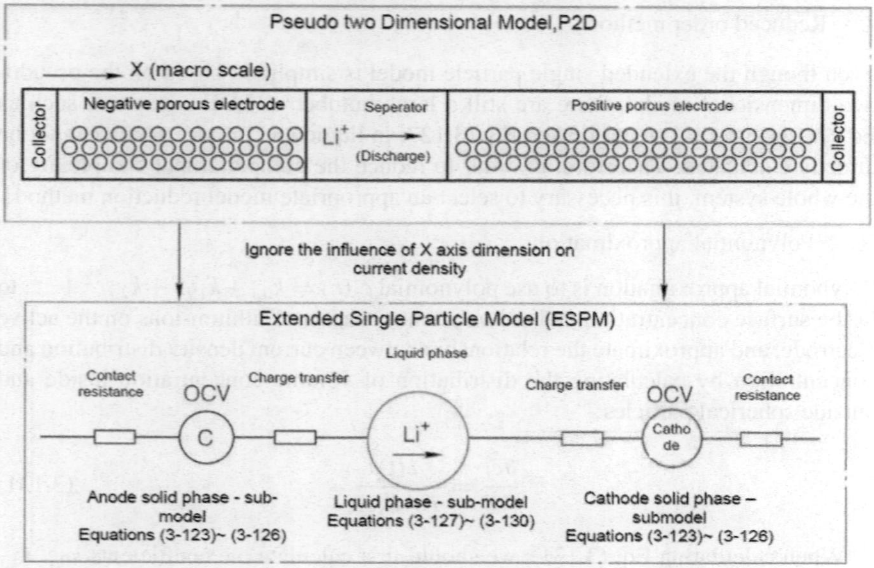

Fig. 3.14 Schematic diagram of an extended single particle electrochemical model

Table 3.4 Control equations of the extended single particle model system

Variable	Governing equation	
η_i	$\eta_i(t) = U_i - \phi_i - R_{c,i} I_i(t)$	(3.123)
c_i	$\frac{\partial c_i}{\partial t} = \frac{D_{s,i}}{r}\left(r^2 \frac{\partial c_i}{\partial t}\right)$, $\frac{\partial c_i}{\partial t} = 0 \; at \; r = 0, \; D\frac{\partial c_i}{\partial t} = \frac{j_i(t)}{F} \; at \; r = R_{p,i}$	(3.124)
j_i	$j_i(t) = 2i_{0,i} \sinh\left[\frac{F}{2RT}\eta_i(t)\right] = \frac{I_i}{a_i A L \varepsilon_i}$	(3.125)
$i_{0,i}$	$i_{0,i} = Fk_i\sqrt{c_i(R_{p,i})c_e[c_{\max,i} - c_i(R_{p,i})]}$	(3.126)
c_e	$\varepsilon_e \frac{\partial c_e}{\partial t} = D_e \frac{\partial^2 c_e}{\partial x^2} + \frac{a_i(1-t_+^0)}{F} j_i(t)$	(3.127)
Φ_e	$\frac{\partial \Phi_e}{\partial x} = -\frac{i_e(t)}{\sigma_e} + (1+\gamma)\frac{\overline{R}T(1-t_+^0)}{F}\frac{\partial \ln(c_e)}{\partial x}$	(3.128)
V_e	$V_e = \phi_e(L_{cell,t}) - \phi_e(0, t)$	(3.129)
U	$U(t) = \Phi_p(t) - \Phi_n(t) - V_e(t) - R_{cell} I(t)$	(3.130)

$R_{c,i}\, I_i(t)$ indicating the ohmic voltage drop caused by the conductivity of the material when electrons pass through the current collector from the surface of the solid electrode to the external positive and anode lugs. R_{cell} is the ohmic resistance of lithium-ion battery.

2. Reduced order method

Even though the extended single particle model is simplified based on the pseudo-two-dimensional model, there are still a large number of PDE equations, such as Eq. (3.124) in solid electrode and Eq. (3.127) in liquid electrolyte, which cannot be directly calculated. Therefore, in order to reduce the computational complexity of the whole system, it is necessary to select an appropriate model reduction method.

(1) Polynomial approximation

Polynomial approximation is to use polynomial $c_i(r) = k_{0,i} + k_{1,i}r + k_{2,i}r^2 + \ldots$ to fit the surface concentration and volume concentration of lithium-ions on the active electrode, and approximate the relationship between current density distribution and concentration by calculating the distribution of volume concentration inside and outside spherical particles:

$$\frac{\partial \overline{c}_i}{\partial t} = \frac{I(t)}{FAL_i\varepsilon_i} \tag{3.131}$$

When calculating Eq. (3.131), we should first calculate the coefficients $k_{0,i}$, $k_{1,i}$ and $k_{2,i}$, etc. in the polynomial, while considering the boundary conditions:

$$\begin{cases} \frac{\partial c_i(r)}{\partial r}, & (r = 0) \\ \frac{\partial c_i(r)}{\partial r} = \frac{j_i(t)}{FD_i} = \frac{R_i I(t)}{3FD_i AL_i \varepsilon_i}, & (r = R_i) \end{cases} \tag{3.132}$$

where, $\frac{\partial c_i(r)}{\partial r} = k_{1,i} + 2k_{2,i}r + \ldots$.

Assuming that the volume concentration distribution inside spherical particles is uniform, then,

$$\overline{c}_i = \int_0^{R_i} c_i(r) \frac{4\pi \mathrm{r}^2}{\frac{4}{3}\pi \mathrm{R}_\mathrm{i}^3} \mathrm{d}r \tag{3.133}$$

Therefore, in combination with Eqs. (3.132) and (3.133), the polynomial coefficient can be calculated as follows:

$$k_{0,i} = \overline{c}_i - \frac{R_i^2 I(t)}{10FD_i AL_i\varepsilon_i}, \quad k_{1,i} = 0, \quad k_{2,i} = \frac{I(t)}{6FD_i AL_i\varepsilon_i}, \tag{3.134}$$

Therefore, on the particle surface, that is, when $r = R_i$, the relationship between the surface concentration and the volume concentration of lithium-ions is as follows:

$$c_{s,i} = \overline{c}_i + \frac{R_i^2 I(t)}{15FD_i AL_i\varepsilon_i} \tag{3.135}$$

It should be noted that the error of volume concentration is directly proportional to the current, and the larger the current is, the greater the estimation error of volume concentration is. Therefore, in the calculation of model reduction, first, the lithium-ion volume concentration is calculated with Eq. (3.131), then the lithium-ion surface concentration is calculated with Eq. (3.135), and finally, the open circuit voltage and charge transfer potential are further calculated according to the obtained surface concentration.

The Laplace transform and re-integration of Eqs. (3.131) and (3.135) are carried out to obtain the relationship between surface concentration and volume concentration and current density:

$$\frac{\overline{c}_i(s)}{I(s)} = \frac{1}{FAL_i\varepsilon_i s} \tag{3.136}$$

$$\frac{c_{s,i}(s)}{I(s)} = \frac{15D_i + R_i^2 s}{15D_i AFL_i\varepsilon_i s} \tag{3.137}$$

(2) Pade approximation

Pade approximation is a model reduction method widely used in frequency domain. Usually, low-order linear equations are used to estimate high-order or nonlinear equations. The differential order of linear approximate equation has a great influence on the accuracy of simulation results, that is, the higher the selected order, the higher the accuracy, but the amount of calculation will be greatly increased, so whether the selected order is reasonable or not directly determines the calculation accuracy and efficiency of the whole reduced-order model.

For example, Laplace transform is applied to Eq. (3.93) and boundary condition Eq. (3.94) in the solid electrode:

$$sc_i(s) = D_i\frac{\mathrm{d}^2c_i(s)}{\mathrm{d}r^2} + \frac{2D_i}{r}\frac{\mathrm{d}c_i(s)}{\mathrm{d}r} \tag{3.138}$$

$$\frac{\mathrm{d}c_i(s)}{\mathrm{d}r} = 0,\quad r = 0;\quad D_i\frac{\mathrm{d}c_i(s)}{\mathrm{d}r} = \pm\frac{R_i I(s)}{3FAL_i\varepsilon_i},\quad r = R_i \tag{3.139}$$

When s is regarded as an independent variable, Eq. (3.139) can be converted into an ordinary differential equation:

$$c_i(s,r) = \frac{c_1}{r}\exp\left(r\sqrt{\frac{s}{D_i}}\right) + \frac{c_2}{r}\exp\left(-r\sqrt{\frac{s}{D_i}}\right) \tag{3.140}$$

In combination with the boundary conditions of Eq. (3.139), the coefficients c_1 and c_2 in Eq. (3.138) can be solved, and the transfer function equation of lithium-ion surface concentration and current density on the particle surface can be obtained:

$$\frac{c_{s,i}(s)}{I(s)} = G_0(s) = \frac{R_i^2}{3D_iFAL_i\varepsilon_i}\frac{\sinh\left(\sqrt{\frac{sR_i}{D_i}}\right)}{R_i\sqrt{\frac{s}{D_i}}\cosh\left(\sqrt{\frac{sR_i}{D_i}}\right) - \sinh\left(R_i\sqrt{\frac{s}{D_i}}\right)} \tag{3.141}$$

In the Pade approximation method, the transfer function equation can be linearized by "Moment-matching", and the specific linearized equation form can be selected according to the specific application (Zhu 2011). For the extended single particle model, Eq. (3.141) can be estimated with Eq. (3.142):

$$\frac{c_{s,i}(s)}{I(s)} \approx G_x(s) = \frac{a_{0,i} + a_{1,i}s + a_{2,i}s^2 + \ldots}{s\left(b_{0,i} + b_{1,i}s + b_{2,i}s^2 + \ldots\right)} \tag{3.142}$$

where, time-moment matching is mainly used to estimate the nonlinear equation under steady-state or low-frequency conditions, that is, it is considered that the limit of transfer function equation, linear approximate equation and its differential equation is equal under the condition of zero pole (Hu and Lin 1989), as shown in Eq. (3.143). In combination with Eq. (3.142), the calculation results of the coefficients in the transfer function are shown in Table 3.5.

$$\begin{aligned} \lim_{s\to 0} &= \lim_{s\to 0} G_x(s) \\ \lim_{s\to 0} \frac{dG_0(s)}{ds} &= \lim_{s\to 0} \frac{dG_x(s)}{ds} \\ \lim_{s\to 0} \frac{d^2G_0(s)}{ds^2} &= \lim_{s\to 0} \frac{d^2G_x(s)}{ds^2} \\ &\vdots \\ \lim_{s\to 0} \frac{d^{m-1}G_0(s)}{ds^{m-1}} &= \lim_{s\to 0} \frac{d^{m-1}G_x(s)}{ds^{m-1}} \end{aligned} \tag{3.143}$$

In order to compare the accuracy of different order reduction methods, the single particle model was discharged for 30 s with the current of 1C rate, and then stood still for 30 s. The simulation results are shown in Figs. 3.15 and 3.16. The approximate results of different order reduction methods under step response and the numerical results of partial differential equations in time domain and frequency domain were compared respectively. The results in Fig. 3.15 show that during discharging or standing, the second-order and third-order Pade reduction methods are more accurate

Table 3.5 Transfer function coefficients calculated with Pade approximation method in solid phase electrodes

$\frac{c_{s,i}(s)}{I(s)}$	First order Pade approximation	Second order Pade approximation	Third order Pade approximation
	$\frac{R_i}{3FAL_i\varepsilon_i}\frac{a_{0,i}}{s}$	$\frac{R_i}{3FAL_i\varepsilon_i}\frac{a_{0,i}+a_{1,i}}{s(1+b_{2,i}s)}$	$\frac{R_i}{3FAL_i\varepsilon_i}\frac{a_{0,i}+a_{1,i}s+a_{2,i}s^2}{s(1+b_{2,i}s+b_{3,i}s^2)}$
$a_{0,i}$	$3/R_i$	$3/R_i$	$3/R_i$
$a_{1,i}$	–	$2R_i/(7D_i)$	$4R_i/(11D_i)$
$a_{2,i}$	–	–	$R_i^3/(165D_i^2)$
$b_{2,i}$	–	$2R_i^2/(35D_i)$	$3R_i^2/(55D_i)$
$b_{3,i}$	–	–	$R_i^4/(3465D_i^2)$

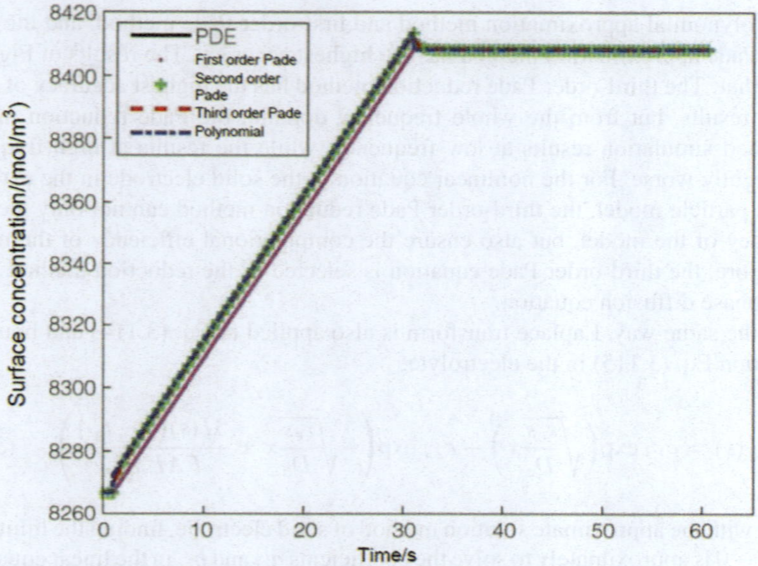

Fig. 3.15 Comparison of approximate simulation results of different order reduction methods in solid electrode in time domain

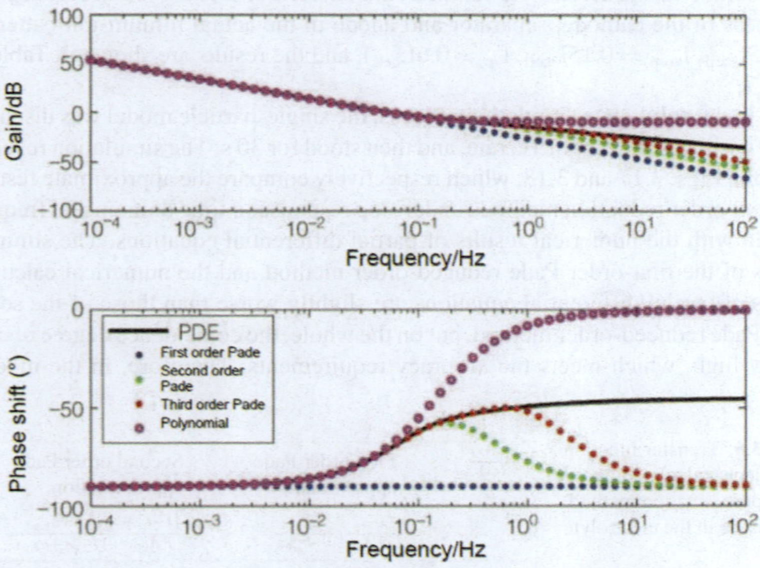

Fig. 3.16 Comparison of approximate simulation results of different order reduction methods in solid electrode in frequency domain

than polynomial approximation method and first-order Pade method, and the third-order Pade approximation method has the highest accuracy. The results of Fig. 3.16 show that: The third-order Pade reduction method has the highest accuracy of simulation results, but from the whole frequency domain, the Pade reduction method has good simulation results at low frequency, while the results at high frequency are slightly worse. For the nonlinear equation in the solid electrode in the extended single particle model, the third-order Pade reduction method can not only meet the accuracy of the model, but also ensure the computational efficiency of the model. Therefore, the third-order Pade equation is selected as the reduction method of the solid phase diffusion equation.

In the same way, Laplace transform is also applied to Eq. (3.114) and boundary condition Eq. (3.115) in the electrolyte:

$$c_{e,i}(s) = c_{1,i}\exp\left(\sqrt{\frac{\varepsilon_e s}{D_e}}x\right) + c_{2,i}\exp\left(-\sqrt{\frac{\varepsilon_e s}{D_e}}x \pm \frac{3I(s)\left(1-t_0^+\right)}{FAL_i\varepsilon_i\varepsilon_e}\right) \tag{3.144}$$

As with the approximate solution method of solid electrode, finding the limit at the pole $s\rightarrow 0$ is approximately to solve the coefficients $a_{j,i}$ and $b_{j,i}$ in the linear equations. However, the difference is that the current density distribution in the solid electrode is the same, while the current density distribution from the cathode and separator to the anode in the electrolyte is different. Therefore, it is solved according to the thickness of the cathode, separator and anode in the actual lithium-ion battery ($L_n = 0.25L_{cell}$, $L_{sep} = 0.15L_{cell}$, $L_p = 0.6L_{cell}$), and the results are shown in Tables 3.6 and 3.7.

As in the solid-state simulation process, the single-particle model was discharged for 30 s with the current of 1C rate, and then stood for 30 s. The simulation results are shown in Figs. 3.17 and 3.18, which respectively compare the approximate results of different order reduction methods under step response in time domain and frequency domain with the numerical results of partial differential equations. The simulation results of the first-order Pade reduced-order method and the numerical calculation results of partial differential equations are slightly worse than those of the second-order Pade reduced-order method, but on the whole, the coincidence degree of curves is very high, which meets the accuracy requirements. Therefore, in the modeling

Table 3.6 Transfer function coefficients calculated by Pade approximation method for cathode in the electrolyte

$\frac{c_{e,p}(s)}{I(s)}$	First order Pade approximation $\frac{(1-t_0^+)}{FA\varepsilon_p}\frac{a_{0,p}}{1+b_{2,p}s}$	Second order Pade approximation $\frac{(1-t_0^+)}{FA\varepsilon_p}\frac{a_{0,p}+a_{1,p}s}{1+b_{2,p}s+b_{3,p}s^2}$
$a_{0,p}$	$-L_{cell}/4D_e$	$-L_{cell}/4D_e$
$a_{1,p}$	–	$-0.0045L_{cell}^3\varepsilon_e/D_e^2$
$b_{2,p}$	$0.11L_{cell}^2\varepsilon_e/D_e$	$0.13L_{cell}^2\varepsilon_e/D_e$
$b_{3,p}$	–	$0.0029L_{cell}^4\varepsilon_e^2/D_e^2$

Table 3.7 Transfer function coefficients calculated by Pade approximation method for anode in the electrolyte

$\frac{c_{e,n}(s)}{I(s)}$	First order Pade approximation	Second order Pade approximation
	$\frac{(1-t_0^+)}{FA\varepsilon_n}\frac{a_{0,n}}{1+b_{2,n}s}$	$\frac{(1-t_0^+)}{FA\varepsilon_n}\frac{a_{0,n}+a_{1,n}s}{1+b_{2,n}s+b_{3,n}s^2}$
$a_{0,n}$	$L_{cell}/3D_e$	$L_{cell}/3D_e$
$a_{1,n}$	–	$0.012L_{cell}^3\varepsilon_e/D_e^2$
$b_{2,n}$	$0.092L_{cell}^2\varepsilon_e/D_e$	$0.13L_{cell}^2\varepsilon_e/D_e$
$b_{3,n}$	–	$0.0027L_{cell}^4\varepsilon_e^2/D_e^2$

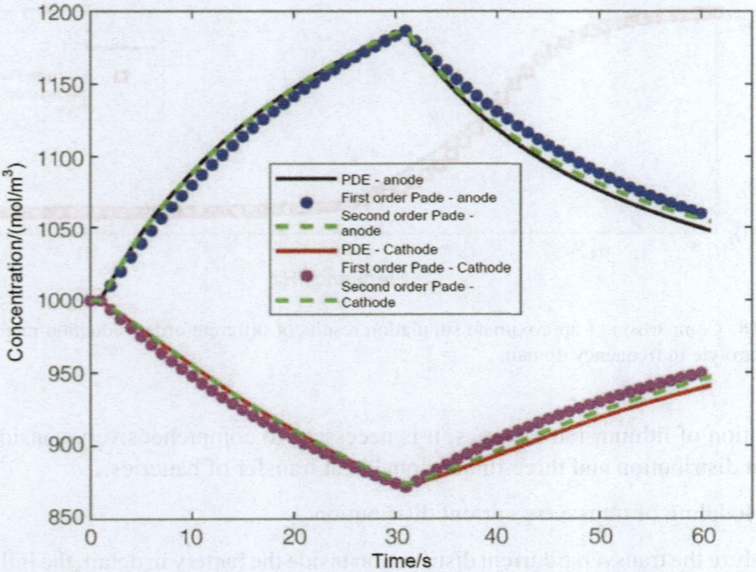

Fig. 3.17 Comparison of approximate simulation results of different order reduction methods in the electrolyte in time domain

process of the extended single particle model, the first-order Pade method is adopted to reduce the order of the nonlinear equation of liquid phase diffusion.

3.4.3 Thermal Model of Lithium-ion Batteries

In the process of engineering application, the temperature distribution on the battery surface and the local transverse current density distribution inside the battery are uneven, especially for the prismatic ALF battery, the unevenness will further increase with the increase of the size. Therefore, when considering the three-dimensional heat

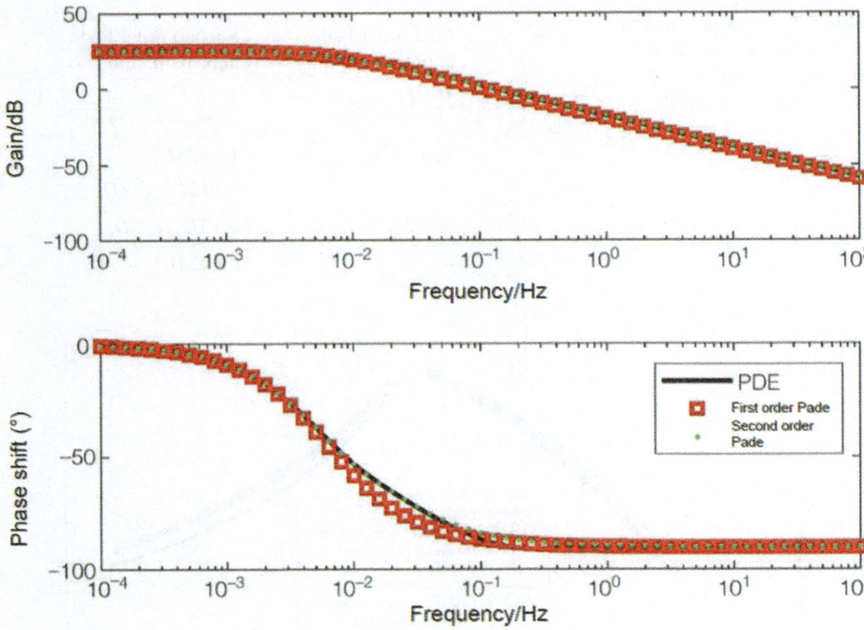

Fig. 3.18 Comparison of approximate simulation results of different order reduction methods in the electrolyte in frequency domain

generation of lithium-ion batteries, it is necessary to comprehensively consider the current distribution and three-dimensional heat transfer of batteries.

1. Modeling of transverse current distribution

To analyze the transverse current distribution inside the battery in detail, the influence of current and voltage in the longitudinal direction (i.e., Plane y–z) is ignored, as shown in Fig. 3.19. The specific assumptions are as follows:

(1) It is assumed that the electrochemical properties and parameters inside the battery on Plane y–z are the same.
(2) It is assumed that the heat generation coefficient of the battery on Plane y–z is consistent with the local temperature distribution.
(3) It is assumed that the local potential of the battery on Plane y–z is consistent.

According to Poisson's equation, the charge conservation equation on the positive current collector of the lithium-ion battery in the working process is:

$$\frac{\partial^2 \Phi_{\mathrm{A}}(y,z)}{\partial y^2}+\frac{\partial^2 \Phi_{\mathrm{A}}(y,z)}{\partial z^2}+\frac{i_{\mathrm{N}}(y,z)}{\sigma_{\mathrm{p}}\delta_{\mathrm{p}}}=0 \tag{3.145}$$

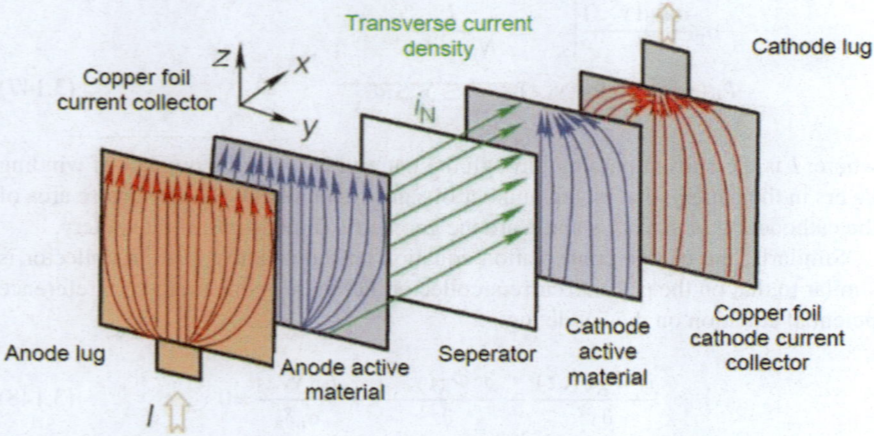

Fig. 3.19 Schematic diagram of transverse current distribution of lithium-ion batteries

where, Φ_A is the potential on the positive current collector; σ_p is the conductivity of the positive current collector; δ_p is the thickness of the cathode current collector; i_N is the transverse current density of the battery, as shown in Fig. 3.19.

On the lug, there is no transverse current density, and the charge conservation equation is:

$$\frac{\partial^2 \Phi_B(y,z)}{\partial y^2} + \frac{\partial^2 \Phi_B(y,z)}{\partial z^2} = 0 \tag{3.146}$$

where, Φ_B is the potential of the cathode lug on the positive current collector. The boundary conditions are as follows:

$$\begin{aligned}
&-\sigma_p \frac{\partial \Phi_A(y,z)}{\partial y}\bigg|_{y=0} = 0 \\
&-\sigma_p \frac{\partial \Phi_A(y,z)}{\partial y}\bigg|_{y=a} = 0 \\
&-\sigma_p \frac{\partial \Phi_A(y,z)}{\partial z}\bigg|_{z=0} = 0 \\
&-\sigma_p \frac{\partial \Phi_A(y,z)}{\partial z}\bigg|_{z=c} = \begin{cases} 0, & (y < d, y > e) \\ -\sigma_p \frac{\partial \Phi_B(y,z)}{\partial z}\Big|_{z=c} & (d \le y \le e) \end{cases} \\
&-\sigma_p \frac{\partial \Phi_B(y,z)}{\partial y}\bigg|_{y=d} = 0 \\
&-\sigma_p \frac{\partial \Phi_B(y,z)}{\partial y}\bigg|_{y=e} = 0
\end{aligned}$$

$$
\begin{aligned}
&-\sigma_{\mathrm{p}}\frac{\partial \Phi_{\mathrm{B}}(y,z)}{\partial z}\bigg|_{z=h}=\frac{I}{N_{\mathrm{layer}}A_{\mathrm{cs}}}\\
&\Phi_{\mathrm{B}}(y,z)=\Phi_{\mathrm{A}}(y,z),\quad d\le y\le e
\end{aligned}
\tag{3.147}
$$

where, I is the current passing through the battery; N_{layer} is the number of winding layers in the battery, that is, the number of sandwich units; A_{cs} is the surface area of the cathode lug; *a, b, c, d, e* and *h* are the geometric dimensions of the battery.

Similarly, the charge conservation equation on the negative current collector is similar to that on the positive current collector, but there is one more zero reference potential equation on the anode lug:

$$\frac{\partial^2 \Phi_{\bar{\mathrm{A}}}(y,z)}{\partial y^2}+\frac{\partial^2 \Phi_{\bar{\mathrm{A}}}(y,z)}{\partial z^2}+\frac{i_{\mathrm{N}}(y,z)}{\sigma_{\mathrm{n}}\delta_{\mathrm{n}}}=0 \tag{3.148}$$

$$\frac{\partial^2 \Phi_{\mathrm{C}}(y,z)}{\partial y^2}+\frac{\partial^2 \Phi_{\mathrm{C}}(y,z)}{\partial z^2}=0 \tag{3.149}$$

$$-\sigma_{\mathrm{n}}\frac{\partial \Phi_{\bar{\mathrm{A}}}(y,z)}{\partial y}\bigg|_{y=0}=0 \tag{3.150}$$

where, Φ_{A} and Φ_{C} are the potentials on the negative current collector and the anode lug, respectively. The boundary conditions are as follows:

$$
\begin{aligned}
&-\sigma_{\mathrm{n}}\frac{\partial \Phi_{\bar{\mathrm{A}}}(y,z)}{\partial y}\bigg|_{y=a}=0\\
&-\sigma_{\mathrm{n}}\frac{\partial \Phi_{\bar{\mathrm{A}}}(y,z)}{\partial z}\bigg|_{z=0}=0\\
&-\sigma_{\mathrm{n}}\frac{\partial \Phi_{\bar{\mathrm{A}}}(y,z)}{\partial z}\bigg|_{z=c}=\begin{cases}0, & (y<d,\ y>e)\\ -\sigma_{\mathrm{n}}\frac{\partial \Phi_{\mathrm{C}}(y,z)}{\partial z}\Big|_{z=c} & (d\le y\le e)\end{cases}\\
&-\sigma_{\mathrm{n}}\frac{\partial \Phi_{\mathrm{C}}(y,z)}{\partial y}\bigg|_{y=d}=0\\
&-\sigma_{\mathrm{n}}\frac{\partial \Phi_{\mathrm{C}}(y,z)}{\partial y}\bigg|_{y=e}=0\\
&-\sigma_{\mathrm{n}}\frac{\partial \Phi_{\mathrm{C}}(y,z)}{\partial z}\bigg|_{z=h}=\frac{I}{N_{\mathrm{layer}}A_{\mathrm{cs}}}\\
&\Phi_{\mathrm{C}}(y,c)=\Phi_{\bar{\mathrm{A}}}(y,c),\quad d\le y\le e\\
&\Phi_{\mathrm{C}}(y,h)=0
\end{aligned}
\tag{3.151}
$$

Therefore, the potential distribution on the positive current collector depends on Φ_{A} and Φ_{B}, and the potential distribution on the negative current collector depends on Φ_A and Φ_C, which can be calculated by Eqs. (3.152) and (3.153):

$$\Phi_p = \{\Phi_A(y, z), \Phi_B(y, z)\} \tag{3.152}$$

$$\Phi_n = \{\Phi_{\overline{A}}(y, z), \Phi_C(y, z)\} \tag{3.153}$$

The current density on Plane y–z can be expressed as:

$$i_{y,k}(y, z) = -\sigma_k \frac{\partial \Phi_k(y, z)}{\partial y} \tag{3.154}$$

$$i_{z,k}(y, z) = -\sigma_k \frac{\partial \Phi_k(y, z)}{\partial z} \tag{3.155}$$

where, when $k = \mathrm{p}$ represents the positive current collector. When $k = \mathrm{n}$ represents the negative current collector.

2. Three dimensional heat transfer modeling

For prismatic ALF lithium-ion batteries, the internal energy conservation equation can be expressed with Eq. (3.156):

$$\begin{aligned} \rho c \frac{\partial T(y, z, t)}{\partial t} =& k_y \frac{\partial^2 T(y, z, t)}{\partial y^2} + k_z \frac{\partial^2 T(y, z, t)}{\partial z^2} \\ &+ Q_{\mathrm{Gen}}(y, z, t) - Q_{\mathrm{Diss}}(y, z, t) \end{aligned} \tag{3.156}$$

where, ρ is the density; c is specific heat capacity; T is the temperature; k_y and k_z are thermal conductivity coefficients; Q_{Gen} is the total heat generation rate inside the battery; Q_{Diss} is the heat dissipation rate of the battery.

For a prismatic ALF battery, the battery thickness is much smaller than the other two dimensions, so we can only consider the temperature change on Plane y–z and ignore the temperature distribution change on the x axis (Kim et al. 2011; Guo and White 2013; Xu et al. 2014). The heat generated inside the battery mainly includes the electrochemical reaction heat on the active electrode and separator and Joule heat on the current collector (Guo and White 2013), and the equation is as follows:

$$\begin{aligned} Q_{\mathrm{Gen}}(y, z, t) =& \varepsilon_{\mathrm{psn}} Q_{\mathrm{psn}}(y, z, t) + \varepsilon_{\mathrm{cc,p}} Q_{\mathrm{cc,p}}(y, z, t) \\ &+ \varepsilon_{\mathrm{cc,n}} Q_{\mathrm{cc,n}}(y, z, t) + Q_{\mathrm{abuse}} \end{aligned} \tag{3.157}$$

where, Q_{psn} is the heat generation rate on the cathode, anode and separator; $Q_{\mathrm{cc,p}}$ and $Q_{\mathrm{cc,n}}$ are Joule heat generated inside the battery and on the positive and anode lugs. Q_{abuse} is the heat generation rate when each side reaction occurs in the battery; $\varepsilon_{\mathrm{psn}}$, $\varepsilon_{\mathrm{cc,p}}$ and $\varepsilon_{\mathrm{cc,n}}$ are the volume ratios of heat generated between the positive

Table 3.8 Volume fraction and thickness parameter distribution on cathode and anode

Parameters	Cathode lug	Anode lug	Between the positive and anode lugs
ε_{psn}	0	0	$\frac{L}{\delta_\text{n}+L+\delta_\text{p}}$
$\varepsilon_{\text{cc,p}}$	1	0	$\frac{\delta_\text{p}}{\delta_\text{n}+L+\delta_\text{p}}$
$\varepsilon_{\text{cc,n}}$	0	1	$\frac{\delta_\text{n}}{\delta_\text{n}+L+\delta_\text{p}}$
d	δ_p	δ_n	$\delta_\text{n}+L+\delta_\text{p}$

and negative current collectors, on the positive current collector and on the negative current collector respectively (see Table 3.8).

The equation of Joule heat generation rate is as follows:

$$Q_{\text{cc},k}(y,z,t)=\frac{1}{\sigma_k}\left[i_{y,k}^2(y,z,t)+i_{z,k}^2(y,z,t)\right] \tag{3.158}$$

where, $i_{y,k}$ and $i_{z,k}$ are the current density on Plane y–z, as shown in Eqs. (3.146) and (3.147); $k=\text{p, n}$ representing cathode and anode respectively.

The equation for reversible and irreversible heat generation rates is as follows (Newman et al. 1985; Gu and Wang 2000):

$$Q_{\text{psn}}(y,z,t)=\frac{1}{L}\int_0^L[q_{\text{rev}}(y,z,t)+q_{\text{irrev}}(y,z,t)]\text{d}x \tag{3.159}$$

where, L is the thickness of the battery, as shown in Fig. 3.20; q_{rev} and q_{irrev} are reversible heat and irreversible heat (Newman et al. 1985) generated during chemical reaction inside the battery, and the specific calculation equation is as follows:

$$q_{\text{rev}}=\sum_{k=\text{p,n}}a_kJ_kT\frac{\partial U_k}{\partial T} \tag{3.160}$$

$$q_{\text{irrev}}=\sum_{k=\text{p,n}}a_kJ_k\eta_k+\sum_{k=\text{p,n}}\sigma_k^{\text{eff}}\nabla\phi_{s,k}\cdot\nabla\phi_{s,k}+\kappa^{\text{eff}}\nabla\phi_e\cdot\nabla\phi_e+\kappa_D^{\text{eff}}\cdot\nabla(\ln c_\text{e})\cdot\nabla\phi_\text{e} \tag{3.161}$$

When there are side reactions in lithium-ion batteries, reversible heat accounts for a low proportion. Therefore, for the convenience of calculation, reversible heat is usually ignored in thermal runaway model, but it cannot be ignored during normal charge and discharge of the battery (Ren et al. 2017).

The heat dissipation rate of the battery is related to the ambient temperature and cooling mode. Under the adiabatic condition, $Q_{\text{Diss}}=0$; Under non-adiabatic conditions, according to Newton's cooling law, we have:

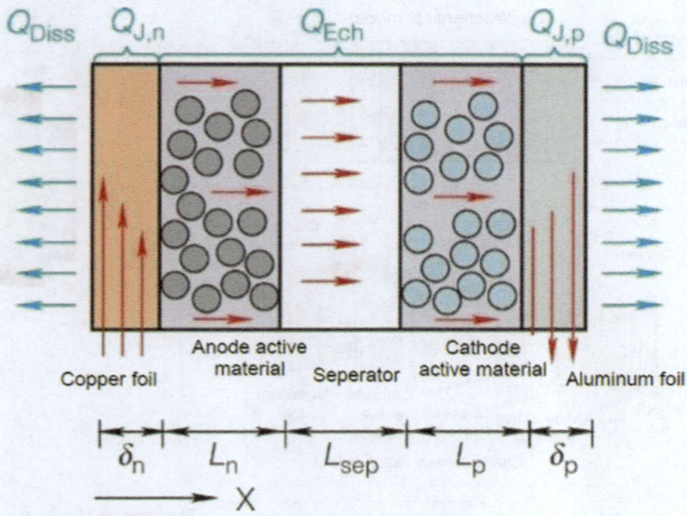

Fig. 3.20 Lithium-ion battery heat generation

$$Q_{\text{Diss}}(y, z, t) = 2\frac{h[T(y, z, t) - T_{\text{amb}}]}{d} \tag{3.162}$$

where, h is the heat transfer coefficient; d is the thickness of the battery heat dissipation surface; T_{amb} is ambient temperature.

3.4.4 Electrothermal Coupling Model

The electro-thermal coupling model of lithium-ion batteries is mainly the coupling calculation of the above electrochemical model and thermal model. The schematic diagram of the electro-thermal coupling model is shown in Fig. 3.21. During the solution, the main difficulty lies in how to solve the local transverse current density inside the battery and the potential on the positive and negative current collectors, that is, the problem of solving the model is transformed into the problem of solving i_N, Φ_p and Φ_n.

Firstly, assuming that the whole y–z plane of the battery is divided into M nodes, the electrothermal coupling model of the whole battery is transformed into an electrothermal coupling model with M nodes. Based on the electrochemical model and charge conservation model established above, the potentials on the positive and negative current collectors are calculated respectively.

The potential difference between the positive and negative current collectors of the battery is:

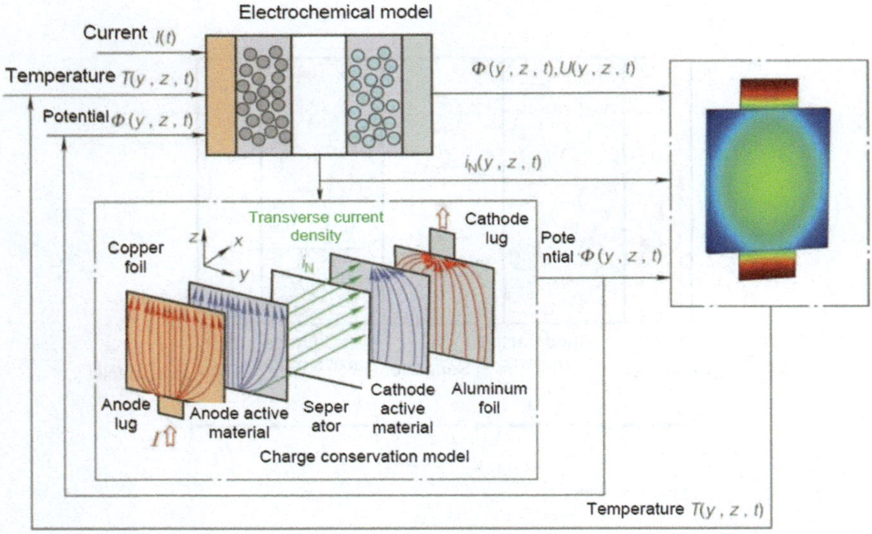

Fig. 3.21 Electro-thermal coupling model of lithium-ion batteries

$$\begin{aligned} V(t) &= \Phi_{\mathrm{p}}(t) - \Phi_{\mathrm{n}}(t) - V_{\mathrm{e}}(t) \\ &= \left\{U_{\mathrm{p}}\left[i_{\mathrm{N},j}(t)\right] + \eta_{\mathrm{p},j}\left[i_{\mathrm{N},j}(t)\right] + \phi_{\mathrm{e},j}\left[i_{\mathrm{N},j}(t)\right]\big|_{x=0}\right\} \\ &\quad - \left\{U_{\mathrm{n}}\left[i_{\mathrm{N},j}(t)\right] + \eta_{\mathrm{n},j}\left[i_{\mathrm{N},j}(t)\right] + \phi_{\mathrm{e},j}\left[i_{\mathrm{N},j}(t)\right]\big|_{x=L}\right\} \end{aligned} \tag{3.163}$$

where, $j = 1, 2, \ldots$, and M.

According to the charge conservation equation, the potential difference between the positive and negative current collectors can be calculated with Eq. (3.164):

$$\begin{aligned} \overline{V}_j(t) &= \Phi_{\mathrm{p},j}(t) - \Phi_{\mathrm{n},j}(t) \\ &= \overline{\Phi}_{\mathrm{p},j}(t) + \Phi_{\mathrm{p,ref}}(t) - \overline{\Phi}_{\mathrm{n},j}(t) - \Phi_{\mathrm{n,ref}}(t) \\ &= \overline{\Phi}_{\mathrm{p},j}(t) + \Phi_{\mathrm{p,ref}}(t) - \overline{\Phi}_{\mathrm{n},j}(t) \end{aligned} \tag{3.164}$$

where, $\Phi_{\mathrm{n,ref}}(t)$ is the anode zero reference electrode.

According to Eqs. (3.139) and (3.143), we can see that the boundary conditions in the charge conservation equation are Neumann boundary conditions. Therefore, the positive and negative potentials on the current collector can be expressed with Eqs. (3.165) and (3.166):

$$\Phi_{\mathrm{p},j}(t)\big|_{x=0} = \Phi_{\mathrm{p},j}(t) \tag{3.165}$$

$$\Phi_{\mathrm{n},j}(t)\big|_{x=L} = \Phi_{\mathrm{n},j}(t) \tag{3.166}$$

During calculation and solution, the potential difference between the positive and negative current collectors is constant, and the total transverse current density is equal to the externally applied current:

$$\overline{V}_j(t) = V(t) \quad (j = 1, 2, \ldots, \mathrm{M}) \tag{3.167}$$

$$\iint\limits_{\Omega \Phi_{\mathrm{p}}} i_N(y, z, t)\mathrm{d}y\mathrm{d}z = I(t) \tag{3.168}$$

The initial conditions of the above calculation process are as follows:

$$\begin{aligned} i_{\mathrm{N},j}(0) &= \frac{I_0}{A} \\ V_j(0) &= \overline{V}_j(0) = V_0 \\ T_j(0) &= T_0 \end{aligned} \tag{3.169}$$

where, A is the effective surface area of the current collector; I_0 is the current when $t = 0$; T_0 is the initial temperature of the lithium-ion battery before charge and discharge.

3.4.5 Electrothermal Coupling Model Verification

In order to verify the accuracy of the electro-thermal coupling model, a prismatic ALF battery is taken as an example to compare and analyze the experimental results of charge and discharge. The curves of battery voltage with time in simulation and experiment under 1C charge are shown in Fig. 3.22, and that under 1C discharge is

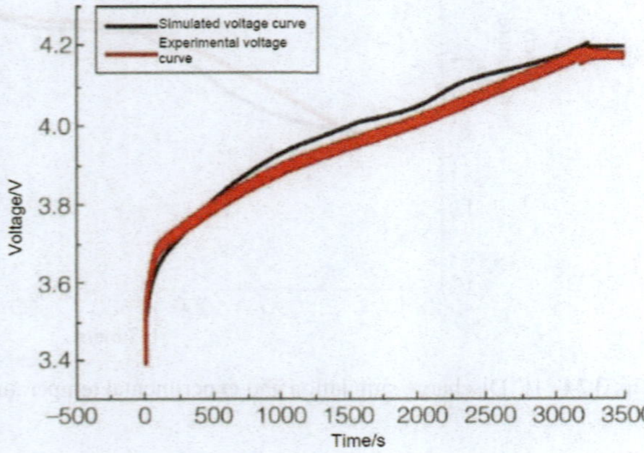

Fig. 3.22 1C charging simulation and experimental voltage curve

shown in Fig. 3.23. Besides, the curves of battery temperature with time in simulation and experiment under 1C discharge is shown in Fig. 3.24.

It can be clearly seen from Figs. 3.22 and 3.23 that there is a very high similarity between the simulated voltage of the model and the voltage measured by the experiment in the overall trend. Although the overall simulation results of the model are about 0.05 V higher, compared with the overall voltage platform, the error is

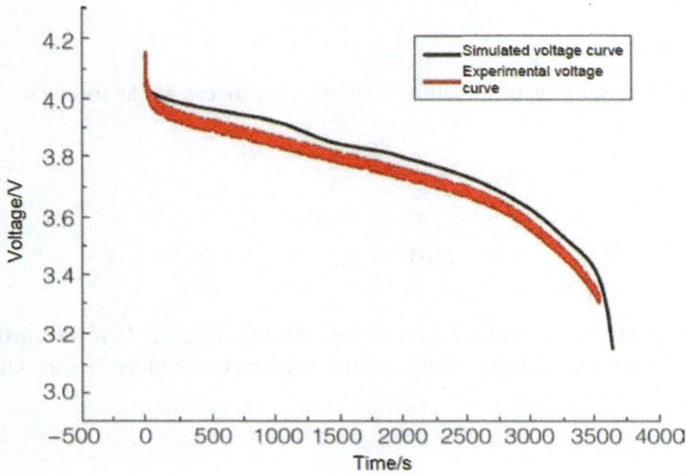

Fig. 3.23 1C discharging simulation and experimental voltage curve

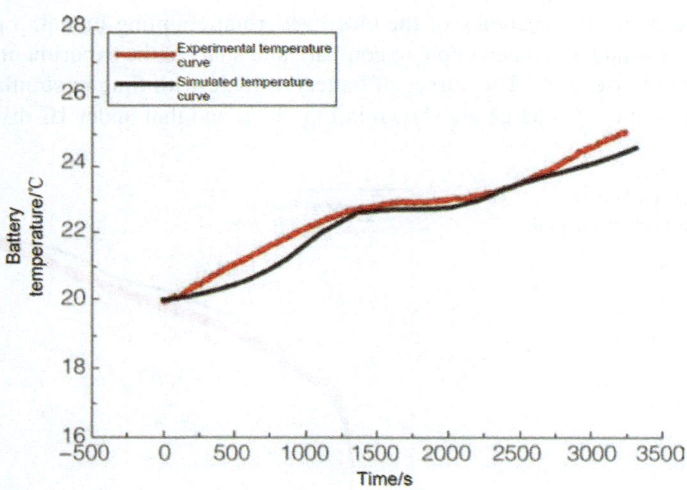

Fig. 3.24 1C Discharge simulation and experimental temperature curve

only 1.3%, so it can be considered that the simulation results of the electrothermal coupling model have high accuracy.

Figure 3.24 shows the experimental and simulation curves of battery temperature during 1C discharging. It can be seen that the overall trend of the simulated temperature of the model is in good agreement with the measured temperature, and there is only some difference in temperature rise rate between the initial 0–1000 s, which may be caused by the calculation error of the model at the initial stage. The temperature curves of the model and experiment tend to be flat in 1500–2000 s, which may be caused by the negative electrochemical reaction heat and the decrease of the total heat generation rate of the battery. After 2000 s, the model and experimental temperature curves began to rise almost simultaneously. Figure 3.24 shows a good agreement between the simulated temperature and the experimental temperature, which shows that the model has high accuracy in simulating the battery temperature field.

3.5 Radial Layered Electrothermal Coupling Model of Cylindrical Battery

3.5.1 Radial Layered Electrothermal Coupling Modeling

For cylindrical batteries, the heat balance equation of each micro-cell can be established in cylindrical coordinates:

$$\rho c \frac{\partial t}{\partial \tau} = \frac{1}{r}\frac{\partial}{\partial r}\left(\lambda_r r \frac{\partial t}{\partial r}\right) + \frac{1}{r^2}\frac{\partial}{\partial \varphi}\left(\lambda_\varphi \frac{\partial t}{\partial \varphi}\right) + \frac{\partial}{\partial z}\left(\lambda_z \frac{\partial t}{\partial z}\right) + \dot{\Phi}_V \tag{3.170}$$

where, ρ is the density; c is specific heat capacity; t is temperature; τ is time; r, φ and z are three coordinate axes of cylindrical coordinates; r also indicates the value of the micro-cell on the r axis of the cylindrical coordinate axis; λ_r, λ_φ and λ_z are the values of the thermal conductivity coefficient of the battery in three directions of cylindrical coordinates; Φ_V is the internal heat source of the micro-cell.

Equation (3.170) indicates the temperature change rate of the micro-cell $\rho c\frac{\partial t}{\partial \tau}$, the first three items on the right are the heat transfer rate obtained by calculating the heat conduction with other micro-cells, and the last item on the right is the heat generation rate generated by internal heat generation. Equation (3.170) is the heat balance equation of a micro-cell and the differential equation of heat conduction of a cylindrical battery.

According to the principle of Eq. (3.170) and related boundary conditions, time conditions, geometric conditions and physical conditions, the temperature field distribution inside and on the surface of the battery cell can be finally determined, and then the calculated surface temperature changes are compared with those in actual experiments to verify the correctness of the model. If the deviation of the calculated results is too high, it is necessary to modify the relevant parameters, and then recalculate and compare them until they are consistent with the experimental results.

The above-mentioned method can be realized by related finite element software, by dividing grids, setting boundary conditions, etc., and then iteratively solving, finally, the temperature field of the whole surface can be simulated and compared with the surface temperature variation obtained by experiments.

The above method is more suitable for modeling the thermal characteristics of batteries in positive thinking mode. When modeling, the thermalphysical parameters of batteries must be obtained through other ways before the model can be solved smoothly. If the exact thermophysical parameters of the battery are not known before modeling, the approximate values of these thermophysical parameters are generally determined by referring to other literatures or by experience, and these parameters are determined through continuous simulation, debugging and comparison with experimental results. Therefore, the radial layered modeling method of cylindrical battery has been proposed. This method can directly calculate the temperature of the battery cell through software programming without the help of finite element thermal simulation software, and can also identify the thermophysical parameters and self-defined parameters in the model through genetic algorithm, which can be applied to the battery heat generation modeling in reverse thinking mode. When the battery thermophysical parameters have not been determined, the best thermophysical parameters can be identified directly according to the experimental data. After the identified parameters are substituted into the model, the model can reflect the actual situation well.

The radial stratification model of cylindrical batteries only considers the heat conduction and convection of a cylindrical battery in the radial direction, ignoring the heat exchange with the external environment through the upper and lower cylindrical bottom surfaces in the axial direction. There are two main reasons for neglecting the heat exchange in the axial direction:

① From the point of view of simplifying the model, the more factors considered in the model, the greater the cumulative error in the calculation process, and the more factors affecting the final result.

② Cylindrical battery cells are usually connected by welding the top and bottom surfaces of the battery cells in the axial direction.

To analyze the heat exchange in the axial direction, we need to consider the heat conduction between the top and bottom surfaces of the battery cells and welding materials, the heat conduction between welding materials and wires, and the heat exchange between the non-welded parts of the top and bottom surfaces of the battery and the external environment. When considering the heat exchange in the axial direction, if too many factors are considered, it will easily lead to major errors. Therefore, when the radial stratification model is established, the boundary conditions of the top and bottom surfaces in the axial direction are regarded as adiabatic boundary conditions.

Because the heat exchange in the axial direction is neglected, the thermalphysical parameters of the battery calculated according to the above radial layered heat balance

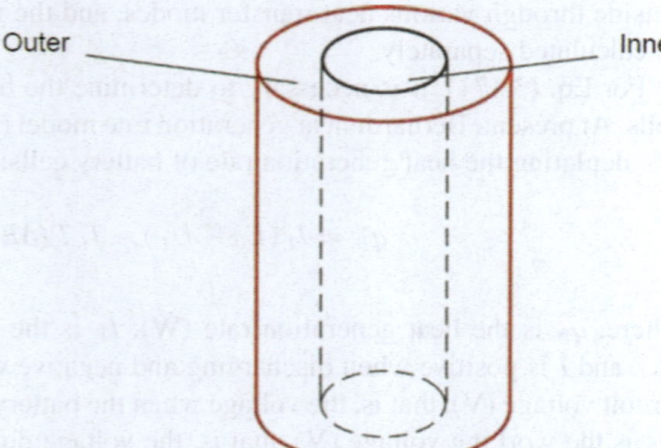

Fig. 3.25 Schematic diagram of a radial two-layer model of cylindrical battery

model will be subject to certain errors. However, the reliability and practicability of the model can be fully verified by comparing with the experimental results according to the simulated temperature rise curve of the battery surface.

In the radial stratification model, the simulation results will be different if the number of concrete stratification is different. In this paper, the calculation results of the two-layer model (with fewer layers) and the nine-layer model (with more layers) will be compared and analyzed. Firstly, the two-layer model is taken as an example, and the modeling principle of the nine-layer model is the same as that of the two-layer model.

The two-layer model can also be called inner and outer layer model, that is, the battery cell is divided into inner layer and outer layer in radial direction, as shown in Fig. 3.25.

In Fig. 3.25, the inner layer is a small cylinder inside, and the outer layer is the remaining part of the whole battery cell after removing the inner layer. The radius of the inner layer can be used as a user-defined parameter of the model and determined by parameter optimization.

Whether it is the inner layer or the outer layer, the heat balance equation can be established according to Eq. (3.171):

$$cm(\mathrm{d}T/\mathrm{d}t) = q_\mathrm{s} + q_\mathrm{e} \tag{3.171}$$

In the equation, c is the specific heat capacity [J(kg K)], and the specific heat capacity of the inner and outer layers is regarded as the same value; m is the mass (kg), and the mass of the inner and outer layers is calculated separately; t is the time; $\mathrm{d}T/\mathrm{d}t$ is the temperature change rate, which can be regarded as the temperature change value in a short time, and the inner and outer layers need to be calculated separately; q_s is the heat generation rate (W). The heat generation rate of the battery cell as a whole can be calculated at first, and then that of the inner and outer layers can be determined separately according to the volume ratio of the inner and outer layers. q_e is the heat transfer rate (W), which is used to calculate the heat rate obtained from outside through various heat transfer modes, and the inner and outer layers need to be calculated separately.

For Eq. (3.171), it is necessary to determine the heat generation rate of battery cells. At present, Bernardi heat generation rate model (Bernardi 1985) is widely used in calculating the heat generation rate of battery cells:

$$q_B = I_L(E_0 - U_L) - I_L T(\mathrm{d}E_0/\mathrm{d}T) \tag{3.172}$$

where, q_B is the heat generation rate (W); I_L is the charge and discharge current (A), and I is positive when discharging and negative when charging; E_0 is the open circuit voltage (V), that is, the voltage when the battery is not charged or discharged; U_L is the working voltage (V), that is, the voltage during the charge and discharge process of the battery; T is the temperature (K), which is the Kelvin temperature. When calculating, the average temperature of the battery cell can be taken. $\mathrm{d}E_0/\mathrm{d}T$ is the rate of change of battery open circuit voltage with temperature.

Equation (3.172) mainly considers Joule heat and reaction heat, where $I_\mathrm{L}(E_0 - U_\mathrm{L})$ represents Joule heat and $I_\mathrm{L}T(\mathrm{d}E_0/\mathrm{d}T)$ represents reaction heat; $\mathrm{d}E_0/\mathrm{d}T$ in reaction heat can be obtained by analyzing the change of open circuit voltage of battery cells at different temperatures.

Equation (3.172) calculates the total heat generation rate of battery cells, and it is necessary to determine the heat generation rate for the inner and outer layers respectively. Assuming that the inside of the battery cell is uniform and the heat generation rate per unit volume is equal everywhere, so as long as the volume of the inner and outer layers is determined, the total heat generation rate of the inner and outer layers can be determined respectively. Let the cell volume be v_z, the inner layer volume be v_n and the outer layer volume be v_w, we have

$$q_\mathrm{sn} = q_\mathrm{B}(v_\mathrm{n}/v_\mathrm{z}) \tag{3.173}$$

$$q_\mathrm{sw} = q_B(v_\mathrm{w}/v_\mathrm{z}) \tag{3.174}$$

In Eqs. (3.173) and (3.174), q_sn is the inner heat generation rate; q_sw is the outer heat generation rate.

According to the calculation of heat transfer rate in Eq. (3.171), it is necessary to analyze the heat transfer of the inner and outer layers respectively. Since the heat exchange between the inner and outer layers and the outside in axial direction is not considered, the inner layer only exchanges heat with the outer layer through heat conduction in radial direction, so the heat flux density of heat transfer can be calculated according to Fourier heat conduction equation on the contact surface of the inner and outer layers:

$$q = -\lambda(\mathrm{d}t/\mathrm{d}x) \tag{3.175}$$

where, q is heat flux density (W/m^2); λ is the thermal conductivity coefficient [W/(m K)]; (dt/dx) is the temperature change rate in one-dimensional direction of x axis; The negative sign indicates that the heat flux density always flows from high temperature to low temperature.

By simplifying Eq. (3.175), the heat transfer rate at the contact surface between inner and outer layers can be obtained:

$$q_{\mathrm{nw}} = -\lambda S_{\mathrm{nw}} \cdot \frac{(T_{\mathrm{n}} - T_{\mathrm{w}})}{\frac{1}{2}r_{\mathrm{n}} + \frac{1}{2}(r - r_{\mathrm{n}})} \tag{3.176}$$

where q_{nw} is the heat transfer rate (W) on the contact surface between the inner and outer layers; S_{nw} is the area of the inner and outer contact surfaces (m^2); T_{n} is the average temperature of inner layer (K); T_{w} is the average temperature of outer layer (K); r_{n} is the inner radius (m).

In the Eq. (3.176), S_{nw} can be obtained by the radius of the inner layer, and let the battery length be l, we have

$$S_{\mathrm{nw}} = 2\pi r n l \tag{3.177}$$

For the outer layer, besides the heat conduction at the contact surface between the inner and outer layers, it also conducts convection heat transfer with air at the outer side. According to the heat transfer principle, the equation of convection heat transfer is:

$$q = h(T - T_{\mathrm{a}}) \tag{3.178}$$

where, q is the heat flux density of convective heat transfer (W/m^2); h is the surface heat transfer coefficient [W/(m^2 K)]; T is the temperature (K) at the surface where convection heat transfer is performed for the battery cell; T_{a} is the temperature (K) of the fluid for convection heat transfer, which can generally be set as the ambient temperature.

According to Eq. (3.178), the heat transfer rate at the outer side of the outer layer can be obtained as follows:

$$q_{\mathrm{w}} = hS_{\mathrm{w}}(T_{\mathrm{w}} - T_{\mathrm{a}}) \tag{3.179}$$

where, q_{w} is the heat transfer rate (W) of the outer side of the outer layer; S_{w} is the area of the outer side of the outer layer (m^2); T_{w} is the average temperature of outer layer (K). In this equation, the average temperature of the outer layer is approximately regarded as the average temperature of the outer side surface of the outer layer.

According to the above deduction, the temperature rise of inner and outer layers within Δt can be obtained.

$$\Delta T_{\mathrm{n}} = \frac{(q_{\mathrm{sn}} + q_{\mathrm{nv}}) \cdot \Delta t}{c\rho v_{\mathrm{n}}} \tag{3.180}$$

$$\Delta T_{\mathrm{w}} = \frac{(q_{\mathrm{sw}} - q_{\mathrm{mv}} - q_{\mathrm{w}})\Delta t}{c\rho v_{\mathrm{w}}} \tag{3.181}$$

where, ΔT_{n} and ΔT_{w} are the temperature rise of inner layer temperature and outer layer temperature within Δt respectively; ρ is the density, and the average density of battery cells is taken for both the inner and outer layers.

With the above equation of temperature rise of the inner and outer layers of the battery within Δt, the iterative calculation can be started, and the initial value needs to be set before the iteration:

$$T_{\mathrm{w}}(1) = T_{\mathrm{a}} \tag{3.182}$$

$$T_{\mathrm{n}}(1) = T_{\mathrm{a}} \tag{3.183}$$

$$T_{\mathrm{nw}}(1) = [T_{\mathrm{n}}(1) + T_{\mathrm{w}}(1)]/2 \tag{3.184}$$

where, T_{nw} is the average temperature (K) of the inner and outer layers of the battery.

Equations (3.182)–(3.184) assign initial values to the inner and outer temperature and the average temperature of the battery, and set the initial values of these temperatures as the ambient temperature.

After setting the initial value, the iterative calculation can be started. First, calculate the heat generation of the battery according to Eq. (3.172):

$$Q_{\mathrm{B}}(i) = \frac{1}{g_{\mathrm{s}}}\{I(i)[E_0 - E(i)] - I(i)T_{\mathrm{nw}}(i)(\mathrm{d}E_0/\mathrm{d}T)\}\Delta t \tag{3.185}$$

where i is the specific value of each parameter in the ith iteration; g_{s} is the number of battery cells, depending on the number of battery cells in the battery pack; $I(i)$ is the current value at the ith Δt, which can be directly set as a definite and constant value if it is a constant current discharge; E_0 is the open circuit voltage of the battery, that is, the voltage of the battery before charge and discharge begins; $E(i)$ is the working voltage of the battery at the ith Δt; $\mathrm{d}E_0/\mathrm{d}T$ is the rate of change of the open circuit voltage with temperature.

Because of different volumes, the inner and outer layers generate different volumes of heat. According to the principles of Eqs. (3.178) and (3.179), we have,

$$Q_{\mathrm{sn}}(i) = Q_{\mathrm{B}}(i)(v_{\mathrm{n}}/v_{\mathrm{z}}) \tag{3.186}$$

$$Q_{\mathrm{sw}}(i) = Q_{\mathrm{B}}(i)(v_{\mathrm{w}}/v_{\mathrm{z}}) \tag{3.187}$$

After calculating the heat generation rate, the heat transfer at the inner and outer contact surfaces can be calculated:

$$Q_{\mathrm{nw}}(i) = -\lambda S_{\mathrm{nw}} \frac{T_{\mathrm{n}}(i) - T_{\mathrm{w}}(i)}{\frac{1}{2}r_{\mathrm{n}} + \frac{1}{2}(r - r_{\mathrm{n}})} \Delta t \tag{3.188}$$

Equation (3.188) is based on Eq. (3.176) and multiplied by Δt, so as to calculate the heat transfer amount passing through the contact surfaces of the inner and outer layers within Δt, and then the temperature rise of the inner layer in Δt can be calculated:

$$\Delta T_{\mathrm{n}}(i) = \frac{Q_{\mathrm{sn}}(i) + Q_{\mathrm{nw}}(i)}{c\rho v_{\mathrm{n}}} \tag{3.189}$$

According to Eq. (3.189), the average temperature of the inner layer at the next moment can be obtained:

$$T_{\mathrm{n}}(i+1) = T_{\mathrm{n}}(i) + \Delta T_{\mathrm{n}}(i) \tag{3.190}$$

After calculating the temperature of the inner layer, you can calculate the temperature of the outer layer. First, the convection heat transfer between the outer layer and the environment should be considered. According to Eq. (3.179), we have,

$$Q_{\mathrm{w}}(i) = hS_{\mathrm{w}}[T_{\mathrm{w}}(i) - T_{\mathrm{a}}]\Delta t \tag{3.191}$$

Equation (3.191) is used to calculate the heat transfer between the outer layer of the battery and the environment. With this equation, the temperature rise of the outer layer within Δt can be calculated:

$$\Delta T_{\mathrm{w}}(i) = \frac{Q_{\mathrm{sw}}(i) - Q_{\mathrm{nw}}(i) - Q_{\mathrm{w}}(i)}{c\rho v_{\mathrm{w}}} \tag{3.192}$$

According to Eq. (3.192), the average temperature of the outer layer at the next moment can be calculated:

$$T_{\mathrm{w}}(i+1) = T_{\mathrm{w}}(i) + \Delta T_{\mathrm{w}}(i) \tag{3.193}$$

Combined with Eq. (3.190), calculate the average temperature of inner and outer layers at the next moment:

$$T_{\mathrm{nw}}(i+1) = [T_{\mathrm{n}}(i+1) + T_{\mathrm{w}}(i+1)]/2 \tag{3.194}$$

Thus, an iterative process is completed, and the temperatures of the inner and outer layers of the battery at any time during charge and discharge can be calculated as long as the above process is repeated continuously.

Through the above modeling analysis, it can be seen that the heat source of the inner layer in the two-layer model has two aspects: first, the heat generated by the inner layer itself; second, the heat conduction between the inner and outer contact surfaces.

There are three sources of heat in the outer layer: first, the heat generated by the outer layer itself; second, the heat conduction between the inner and outer contact surfaces; third, the convection heat transfer on the outer side of the outer layer.

Based on the two-layer model, a nine-layer model can be established. The heat sources of the innermost layer and outermost layer of the nine-layer model are the same as those of the two-layer model. The heat of the middle seven layers has three sources: first, self-generated heat; second, heat conduction with the contact surface of the previous layer; third, the heat conduction with the contact surface of the next layer.

The radius of the inner layer is not determined when the two-layer model is established, but the radius of the inner layer is taken as a user-defined parameter of the model for parameter optimization, which is to make the calculation result approach the experimental result better. For the nine-layer model, because there are too many layers, if the radius of each layer is also optimized, too many parameters will also affect the correctness of the calculation results. Therefore, the radius of each layer in the nine-layer model should be determined in advance.

Taking a 18650 cylindrical battery cell as an example, the radius in the radial direction is 9 mm, so the first layer of the nine-layer model is a cylinder with a radius of 1 mm, the second layer is a cylinder with a radius of 2 mm after removing the first layer, and the third layer is a cylinder with a radius of 3 mm after removing the first and second layers. By analogy, the ninth layer is the remaining part of the whole cylindrical battery cell after removing the first eight layers. Of the nine layers, only the first layer (i.e., the innermost layer) is a cylinder, and the other eight layers are circular cylinders.

3.5.2 Identification of Battery Thermophysical Parameters Based on Genetic Algorithm

When building a battery thermal model, it is necessary to determine the relevant parameters in the model, mainly including the thermalphysical parameters of the battery and the user-defined parameters in the model. Take the two-layer model of cylindrical battery as an example, it includes,

① Specific heat capacity c.
② Heat transfer coefficient h of convective heat transfer surface.
③ Thermal conductivity coefficient λ.
④ Inner radius r_n.

In order to identify the above parameters, the genetic algorithm will be briefly introduced below.

The concept of genetic algorithm was put forward by Professor J. Holland of Michigan University in 1975. The basic principle of the algorithm is based on the evolutionary theory of biology (Wang et al. 1999; Liu et al. 2001; Zhuo et al. 2014). When identifying the best parameters, a group of parameters, which is the first generation of population in evolution, can be randomly generated. This group of parameters are substituted into the fitness function for calculation. According to the calculation results, the parameters with low fitness have a greater probability of being eliminated, and the remaining parameters are "propagated" into the second generation population by means of "mating" and "gene mutation", and then substituted into the fitness function for calculation, and the above process is repeated continuously. When the number of iterations is increasing, the probability of obtaining the best parameters will become higher and higher. The flow chart of genetic algorithm is shown in Fig. 3.26.

Genetic algorithm is a random full-range search algorithm. Because of the randomness of "mating" and "gene mutation" in the process of finding the best parameters, all individuals in the whole range may participate in this evolution process, thus ensuring that the parameters found must be the optimal solution or approximate optimal solution in the global range.

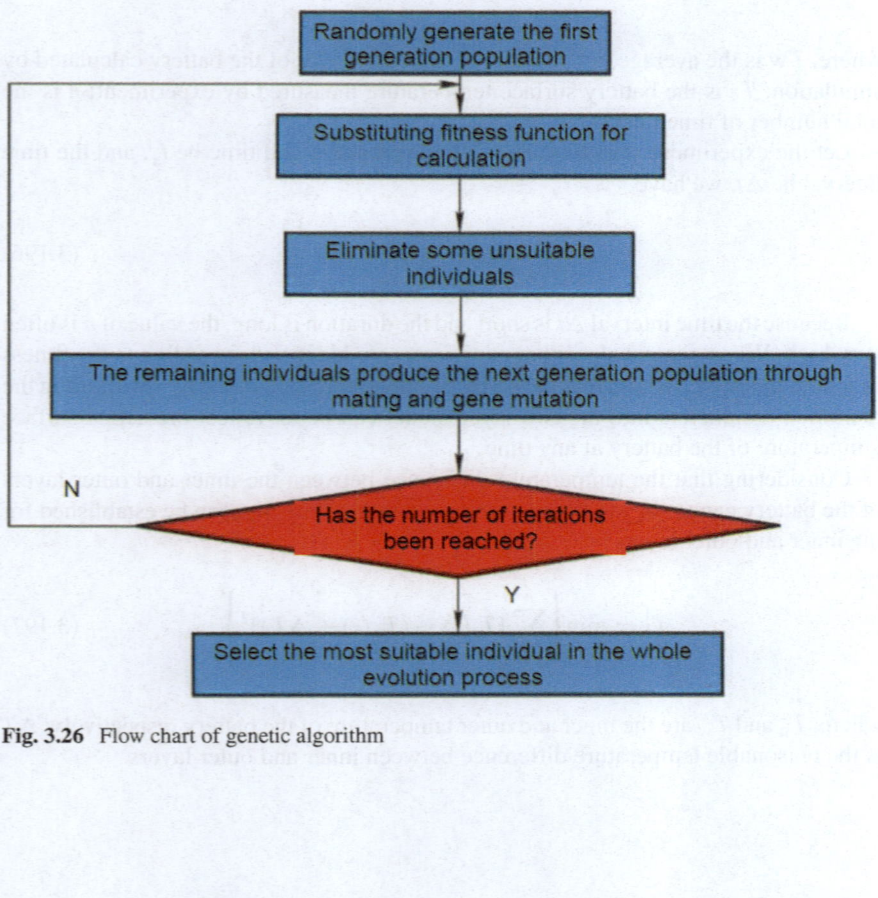

Fig. 3.26 Flow chart of genetic algorithm

The genetic algorithm is used to identify parameters according to the following steps:

① Determining the fitness function.
② Coding the parameters.
③ Realizing "mating", "gene mutation" and "reproduction" of the population.

(1) Determining the fitness function.

The fitness function in genetic algorithm is the weather vane to guide evolution. All parameters need to pass the test of fitness function, and unsuitable parameters will be eliminated with great probability.

For parameter identification of a battery heat generation model, considering that the simulated temperature should be as close as possible to the temperature obtained by experiment, the fitness function of cylindrical battery two-layer model can be established according to the principle of least prismatic method, i.e. the principle of the minimum sum of quadratics:

$$F = \min\left\{\sum_{i=1}^{n}[T_{w}(i) - T_{s}(i)]^{2}\right\} \tag{3.195}$$

where, Tw is the average temperature of the outer layer of the battery calculated by simulation; T_s is the battery surface temperature measured by experiment; n is the total number of time intervals.

Let the experiment start time be t_0, the experiment end time be t_1, and the time interval be Δt, we have,

$$n = \frac{t_1 - t_0}{\Delta t} \tag{3.196}$$

Because the time interval Δt is short and the duration is long, the value of n is often very high. When genetic algorithm parameters are identified according to the fitness function, it can be ensured that the surface temperature calculated by substituting the identified parameters into the two-layer model can better reflect the actual surface temperature of the battery at any time.

Considering that the temperature difference between the inner and outer layers of the battery cannot be too high or too low, a fitness function can be established for the inner and outer layers:

$$G = \min\left\{\sum_{i=1}^{n}[T_{n}(i) - (T_{w}(i) + \Delta T)]^{2}\right\} \tag{3.197}$$

where, T_n and T_w are the inner and outer temperature of the battery respectively; ΔT is the reasonable temperature difference between inner and outer layers.

Equation (3.195) can be combined with Eq. (3.197) to determine an overall fitness function:

$$H = \eta_1 F + \eta_2 G \tag{3.198}$$

where, η_1 and η_2 are the weight values, and the corresponding weight values can be determined according to their importance and specific order of magnitude.

For the cylindrical nine-layer model, the sum of the quadratic variances of the outermost temperature of the battery and the actual surface temperature of the battery at all times can also be taken as the fitness function. In order to prevent the temperature difference between layers from being too high, it is also possible to set fitness functions between layers, and then set different weight values for the above fitness functions. Because there are many layers in the nine-layer model, it is easy to cause major errors when setting each weight value, so the error of the final simulation result may be great, which will be analyzed later.

For prismatic battery cells, since there is no stratification, it is only necessary to set the fitness function directly for the average temperature of prismatic battery cells:

$$J = \min\left\{\sum_{i=1}^{n} [T(i) - T_{\mathrm{s}}(i)]^2\right\} \tag{3.199}$$

where, T is the average temperature of prismatic battery cells; T_{s} is the surface temperature of prismatic battery measured by experiment.

(2) Parameter coding

In genetic algorithm, the parameters to be identified are generally encoded by binary number strings. Taking the two-layer model of cylindrical battery as an example, 32-bit binary number strings can be used for coding.

Generally, before coding, it is necessary to determine the value range of each parameter. For the two-layer model of cylindrical battery, the value range of each parameter can be determined according to (Shi 2015) and the actual situation. Take a cylindrical 18650 battery cell as an example, and set it as follows:

① The range of specific heat capacity c is [800, 1200], and the unit is J/(kg K).
② The value range of thermal conductivity coefficient λ is [1, 116], and the unit is W/(m K).
③ The range of heat transfer coefficient h of convective heat transfer surface can be set as [1, 100], and the unit is W/(m^2 K).
④ The value range of the inner layer radius r_{n} in the inner and outer layer model can be set as [0.005, 0.009], and the unit is m.

For the above four parameters, a 32-bit binary number string will be used, in which the 1st–10th bits represent the specific heat capacity c, the 11–18th bits represent the heat transfer coefficient h of convective heat transfer surface, the 19–26th bits represent the thermal conductivity coefficient λ, and the last 6 bits represent the inner layer radius r_n.

In order to convert binary numbers into decimal parameters for operation, the conversion can be carried out according to the following methods. Taking the specific heat capacity c as an example, take the 1st–10th bits of a 32-bit binary number string to form a 10-bit binary number, and then convert this binary number into a decimal number. Let the converted decimal number be c_{10}, and the value of c_{10} will not exceed 2^{10}–1, so the final value of specific heat capacity c can be determined with Eq. (3.200):

$$c = 800 + (1200 - 800) \times \frac{c_{10}}{2^{10} - 1} \tag{3.200}$$

With Eq. (3.200), it can be ensured that the value range of c is within a given interval, and the value precision of c is also very high within a given interval. Other parameters are converted into decimal specific parameter values in the same way.

(3) "mating", "gene mutation" and "reproduction" of the population

Mating operation means that after one or more mating positions are determined, two individuals exchange partial codes at the mating positions to form two new sub-individuals. For example, after $X_1 = 001110001$ and $X_2 = 1011000111$ exchange the last two bits, new $X_1{}' = 0011100011$ and $X_2{}' = 1011000101$ will be formed. Mutation is to avoid premature convergence of population in the later period, and to carry out gene mutation with low probability on binary coding, so as to expand the scope of optimization. For example, if the third position of $X_3 = 1000101101$ is mutated, a new offspring code $X_3{}' = 10001001$ will be obtained.

For the four parameters in the two-layer model, the 32-bit binary number string composed of these four parameters can be called an individual, and a certain number of individuals can be generated by random function, and these individuals will constitute the first generation population. Each individual in the population is substituted into the two-layer model for calculation, and then the fitness of each individual is determined according to the fitness function. Individuals with high fitness will have a higher probability of staying, while individuals with low fitness will have a lower probability of staying. Then, the remaining individuals will be "mated" and "mutated" to "reproduce" the second generation population. Because each individual contains four parameters, different parameters can be recombined by mating without changing the parameter values. According to the bit distribution of the four parameters in the 32 bits, the mating bit can be fixed among the 10, 18 and 26th bits, which can be determined randomly, and the individuals participating in mating can also be determined randomly. In general, not all individuals are involved in mating to ensure that there are still outstanding individuals of the previous generation in the new population.

After mating, the values of parameters do not change, but different parameters are recombined. In order to avoid limiting the parameter values in the previous generation population, it is necessary to carry out "gene mutation" on the parameters. Because each individual is a 32-bit binary number string, "gene mutation" can be achieved by changing the value of any number of 32 bits. For example, if a certain bit of an individual is 1, change this bit to 0. "Gene mutation" is carried out on the whole population, and multiple individuals can be randomly determined to carry out gene mutation. Through gene mutation, the value of parameters will be changed, thus avoiding that the value of parameters is limited to the parameter range of the previous generation population.

Through the above analysis, it can be seen that the previous generation of population experienced the "survival of the fittest" of fitness function, and then "mating" and "gene mutation". Therefore, the new generation of population not only has the outstanding individuals of the previous generation, but also has new parameter combinations formed by splicing different individuals, and also has new parameter values produced by "gene mutation". Then, the new population is substituted into the operation, and the optimal or near-optimal parameters can be found through multiple cycles.

Because of the randomness of genetic algorithm, all values in the range of values may appear in the operation process, and the optimal value can be searched in the whole range. However, genetic algorithm is different from the random search algorithm featuring "looking for a needle in a haystack", because the generation of new population in genetic algorithm is always based on the outstanding individuals in the previous generation, and the optimal solution is always near the approximate optimal solution for the general parameter optimization process, so the method of generating new population in genetic algorithm is in line with this principle.

Through the above genetic algorithm, in combination with the specific temperature rise during the battery experiment, we can identify the thermophysical parameters of the battery and the user-defined parameters of the model.

3.5.3 *Verification of Radial Layered Model*

By programming in software and using genetic algorithm, the parameters of radial layered heat generation model of cylindrical batteries are optimized. For the two-layer model, taking the 1C discharging of a 18650 battery module at 0 °C as an example, the four parameter values obtained from the identification results are:

① The specific heat capacity is 995.50 J/(kg K).
② The heat transfer coefficient of convective heat transfer surface is 76.70 W/(m^2 K).
③ The radius of the inner layer is 1.26×10^{-3} m.
④ The average radial thermal conductivity coefficient of the inner radius is 27.27 W/(m K).

Because the blower in the thermostat is always turned on in the process of 1C discharging at 0 °C, the value of heat transfer coefficient h on the convective heat transfer surface is too large, which shows that the identified value of h is reasonable.

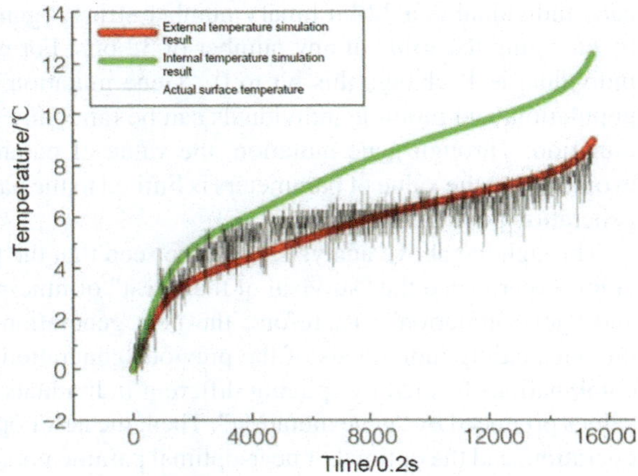

Fig. 3.27 Comparison between simulation and practice of two-layer model for 1C discharge temperature rise at 0 °C

The value of thermal conductivity coefficient does not represent the average thermal conductivity of the whole battery cell, but only represents the average radial thermal conductivity of the side at the radius of the inner layer, so the value is also reasonable.

The above parameters are substituted into the two-layer model for calculation, and the average temperature of the inner and outer layers is obtained by combining the data of voltage variation with time obtained by simulation of Thevenin model of the 18650 battery module, as shown in Fig. 3.27.

It can be seen from Fig. 3.27 that in the two-layer model, the simulated outer layer temperature has a high degree of fitting with the actual surface temperature, with the highest temperature difference of 2.49 °C and the average temperature difference of 0.56 °C. The simulated inner layer temperature will be higher than the actual surface temperature, with the max. temperature difference of 5.64 °C and the average temperature difference of 2.07 °C. On the whole, the simulation results of the above two-layer model reflect the actual situation well.

According to the same principle, the parameters of the nine-layer model are identified by using the principle of genetic algorithm. Taking the 1C discharging of a 18650 battery module at 0 °C as an example, the identified three parameters are:

① The specific heat capacity is 816.03 J/(kg K).

② The heat transfer coefficient of convective heat transfer surface is 76.19 W/(m^2 K).

③ The total average radial thermal conductivity coefficient of each side in the nine-layer model is 95.25 W/(m K).

The above parameters are substituted into the nine-layer model for calculation, and combined with the data of voltage variation with time obtained by the simulation of Thevenin model of battery, the final calculated results are shown in Fig. 3.28.

It can be seen from Fig. 3.28 that when a nine-layer model is adopted, although the temperature of the outermost layer is close to the actual surface temperature, the average simulated temperature of the nine layers and the simulated temperatures of

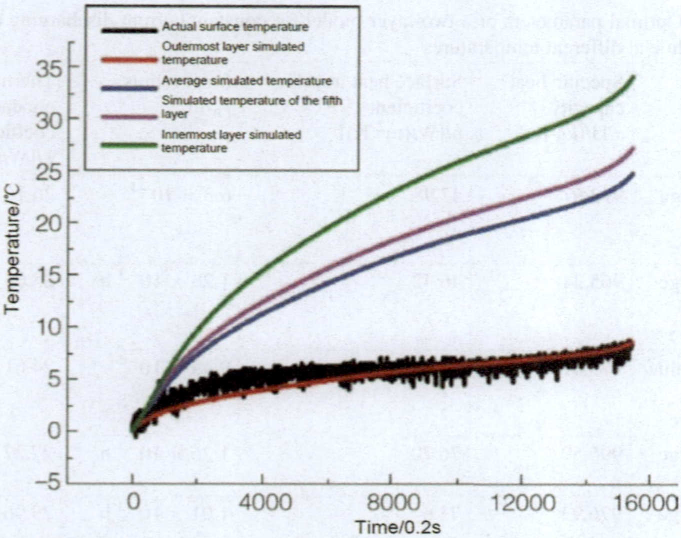

Fig. 3.28 Comparison between simulation and practice of nine-layer model for 1C discharge temperature rise at 0 °C

the fifth and innermost layers are much higher than the actual surface temperature, which is mainly due to the fact that only the radial heat transfer is considered in the simulation process, and the inner layer can only rely on the contact with the outer layer for heat dissipation, so the heat accumulation causes the increase of the inner layer and the total average temperature. Compared with the two-layer model, the temperature of the inner layers of the nine-layer model is too high. This may be inconsistent with the actual situation.

Theoretically speaking, the nine-layer model can better reflect the temperature distribution inside the battery cell because of more layers. However, due to the lack of understanding of the composition materials, thermophysical properties and material distribution in the battery, if all layers are simply considered to be uniform, it is easy to cause large errors after iterative calculation of nine layers. Therefore, the nine-layer model is not suitable for the situation that the composition and distribution of materials in the battery are unknown.

In contrast, the two-layer model has higher flexibility, and because there are only two layers, there will be fewer iterative errors in the calculation. In addition, the inner layer radius in the two-layer model is an adjustable parameter. In the process of parameter identification, the constant adjustment of the inner layer radius can ensure that the outer layer temperature of the battery is close to the actual surface temperature, while the inner and outer layer temperatures do not differ much. This situation is more in line with the actual situation, so the two-layer model has strong practicability.

Table 3.9 Optimal parameters of a two-layer model for constant current discharging of a 18650 battery module at different temperatures

	Specific heat capacity c/[J/(kg K)]	Surface heat transfer coefficient h/[W/(m^2 K)]	Inner radius r_n/m	Thermal conductivity coefficient λ/[W/(m K)]
1C discharge at normal temperature	992.67	47.98	6.3×10^{-4}	26.16
2C discharge at normal temperature	965.34	46.42	1.26×10^{-3} h	23.06
Average value at normal temperature	979.01	47.2	9.45×10^{-4}	24.61
1C discharge at 0 °C	995.50	76.70	1.26×10^{-3} h	27.27
1C discharge at −10 °C	976.93	73.6	1.01×10^{-3} h	29.96
1C discharge at −20 °C	995.41	75.93	1.01×10^{-3} h	27.22
Average value at low temperature	989.28	75.41	1.09×10^{-3} h	28.15

According to the principle of two-layer model, the optimal parameters of a 18650 battery module during constant current discharge at different temperatures can be determined through parameter optimization of genetic algorithm, as shown in Table 3.9.

It can be seen from Table 3.9 that the specific heat capacity of the battery cell is between 950 and 1000 J/(kg K), and the average radial thermal conductivity coefficient of the side where the radius of the battery cell is about 1 mm is between 20 and 30 W/(m K). At normal temperature, because the blower is not turned on in the thermostat, the surface heat transfer coefficient is low, while at low temperature, in order to speed up the cooling speed of the battery, the blower is turned on throughout the experiment, so the surface heat transfer coefficient is high. By substituting the parameters corresponding to the row of the average values at room temperature in Table 3.9 into the two-layer model for calculation, the temperature rise curves of the inner and outer layers during the discharge experiment at room temperature can be obtained, as shown in Figs. 3.29 and 3.30. By substituting the low-temperature average parameters into the two-layer model for calculation, the temperature rise curves of the inner and outer layers during the discharge experiment at low temperature can be obtained, as shown in Figs. 3.31 and 3.32.

It can be seen from Figs. 3.29, 3.30, 3.31 and 3.32 that the simulation results well reflect the actual situation, the outer layer temperature curve basically coincides

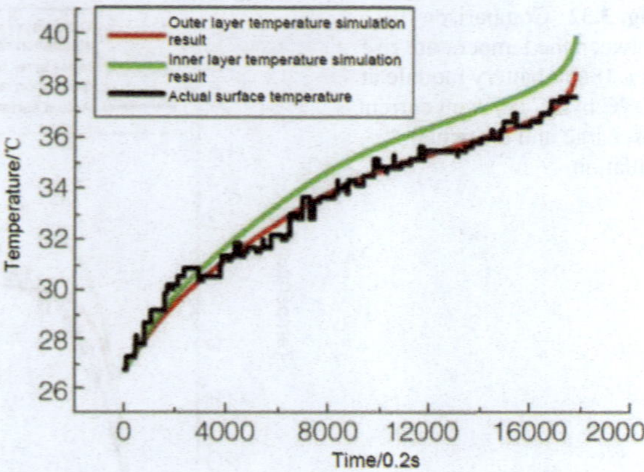

Fig. 3.29 Comparison between simulation and practice of temperature rise of a 18650 battery module under constant current discharging at room temperature of 1C

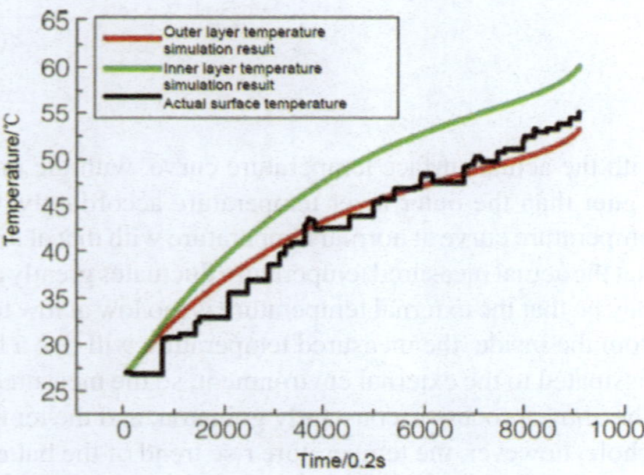

Fig. 3.30 Comparison between the temperature rise of a 18650 battery module under constant current discharge at room temperature of 2C and the actual situation

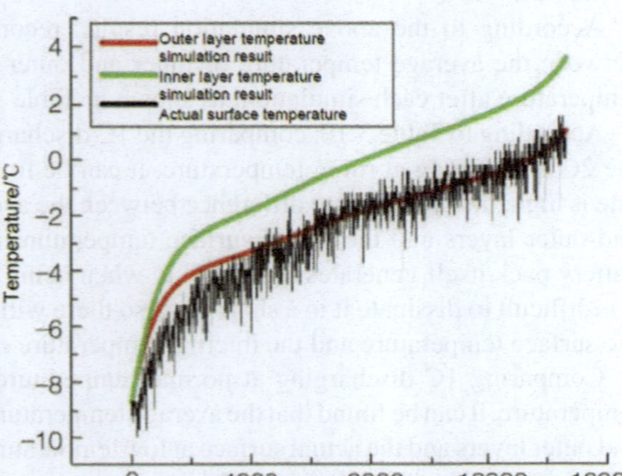

Fig. 3.31 Comparison between simulation and practice of temperature rise of 18650 battery module during 1C constant current discharge at −10 °C

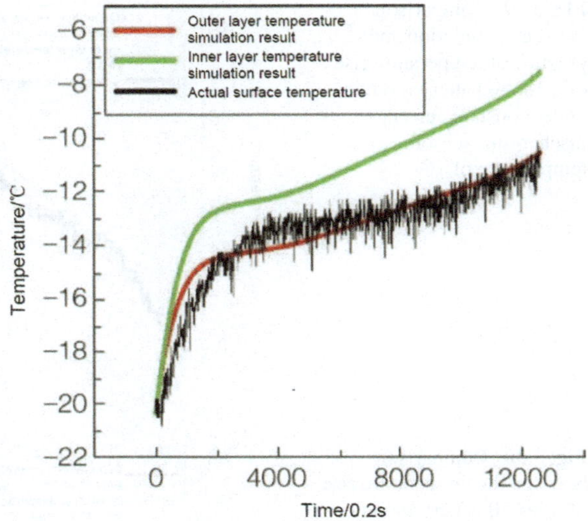

Fig. 3.32 Comparison between the temperature rise of a 18650 battery module at 10 °C by 2C constant current discharge and the actual situation

with the actual surface temperature curve, with the inner layer temperature slightly higher than the outer layer temperature accordingly. Comparing the actual surface temperature curve at normal temperature with that at low temperature, it can be found that the actual measured temperature fluctuates greatly at low temperature. The reason may be that the external temperature is too low at low temperature. When heat comes from the inside, the measured temperature will rise a little, but the heat will soon be dissipated to the external environment, so the measured temperature will drop again. The above process is constantly going on, and the temperature will fluctuate. On the whole, however, the temperature rise trend of the battery surface during discharging is obvious.

According to the above simulation results, record the temperature difference between the average temperature of inner and outer layers and the actual surface temperature after each simulation, as shown in Table 3.10.

According to Table 3.10, comparing the 1C discharging at room temperature with the 2C discharging at room temperature, it can be found that when the discharging rate is high, the temperature difference between the average temperature of the inner and outer layers and the actual surface temperature is high, which is because the battery pack itself generates heat quickly when being discharged at a high rate, and it is difficult to dissipate it in a short time, so there will be a major deviation between the surface temperature and the internal temperature of the battery.

Comparing 1C discharging at normal temperature with 1C discharging at low temperature, it can be found that the average temperature difference between the inner and outer layers and the actual surface at low temperature is higher than that at normal temperature, which is also caused by faster heat generation at low temperature. Because the internal resistance of the battery is high at low temperature, when the battery is discharged at the same rate, the internal temperature of the battery rises

Table 3.10 Temperature difference between inner and outer layers in two-layer model and actual surface temperature of a 18650 battery

	Max. difference between inner layer temp. and real temp./ °C	Average difference between inner layer temp. and real temp./ °C	Max. difference between outer layer temp. and real temp./ °C	Average difference between outer layer temp. and real temp./ °C
1C discharge at normal temperature	2.36	0.96	1.05	0.26
2C discharge at normal temperature	7.00	4.98	4.66	1.54
1C discharge at 0 °C	5.64	2.07	2.49	0.56
1C discharge at −10 °C	4.52	2.16	3.25	0.63
1C discharge at −20 °C	4.68	2.29	2.78	0.55

faster, and the heat is difficult to dissipate in a short time, so there will be a major deviation between the internal temperature and the surface temperature of the battery.

Summary

In this chapter, the modeling method of lithium-ion battery heat generation is introduced in detail. Based on the method of obtaining thermophysical parameters of lithium-ion batteries, the battery heat generation model based on Bernardi heat generation rate, the electro-thermal coupling model based on electrochemical model and the radial layered heat generation model of cylindrical battery are introduced. The main conclusions are as follows:

(1) The thermophysical parameters needed in battery modeling and the methods for obtaining the parameters are introduced. Based on the layered structure of the prismatic battery, the thermal conductivity coefficient, density and specific heat capacity of the batteries in all directions are obtained.
(2) Based on Bernardi heat generation rate and Bernardi heat generation rate with introduced current density, an electrothermal coupling model of charge and discharge of lithium-ion battery is established.
(3) The P2D electrochemical modeling theory based on porous electrode theory and concentrated solution theory is introduced. On this basis, the whole active electrode is regarded as a single active material particle, and the extended single particle model is introduced. Furthermore, in combination with the three-dimensional heat transfer model, the electro-thermal coupling modeling methods based on electrochemical theory are expounded.

(4) The radial layered heat generation characteristics of cylindrical batteries are modeled. Taking the two-layer model as an example, heat balance equations are established for the inner layer and the outer layer respectively, and an iterative calculation method is developed to calculate the temperature rise of each layer. At the same time, the modeling methods and calculation methods of the nine-layer model are briefly described.

References

Bernardi D, Pawiikowski E, Newman J (1985) A general energy balance for battery system. J Electrochem Soc 5:132

Chen K (2010) Electric field theory and thermal characteristics analysis of EV power batteries. Beijing Institute of Technology, Beijing

Christopher D, Wang CY (2013) Battery systems engineering (RA·hn/battery). Batter Manag Syst 10:191–229

Eide P, Maybeck P (1996) An MMAE failure detection system for the F-16. IEEE Trans Aerosp Electron Syst 32(3):1125–1136

Forman JC, Bashash S, Stein J et al (2011) Reduction of an electrochemistry-based lithium-ion battery model via quasi-linearization and Pade approximation. J Electrochem Soc 158(2):A93

Gu WB, Wang CY (2000) Thermal-electrochemical modeling of battery systems. J Electrochem Soc 147(8):2910

Guo M, White RE (2013) A distributed thermal model for a lithium-ion electrode plate pair. J Power Sources 221:334–344

Hu SS, Lin DY (1989) Review on the reduction methods of classical models. J Nanjing Univ Aeronaut Astronaut 4:106–110

Kim GH, Smith K, Lee K-J, et al (2011) Multi-domain modeling of lithium-ion batteries encompassing multi-physics in varied length scales. J Electrochem Soc 158(8):A955

Liu GH, Bao H, Li WC et al (2001) Realization of genetic algorithm program with MATLAB. Appl Res Comput 18(8):80–82

Newman J, Tiedemann W (1974) Porous-electrode theory with battery applications. Aiche J 21(1):25–41

Newman J, Bernardi D, Pawlikowski E (1985) A general energy-balance for battery systems. J Electrochem Soc 132(1):5

Newman J, Bennion D, Tobias Charles W (1965) Mass transfer in concentrated binary electrolytes. Berichte der Bunsengesellschaft für physikalische Chemie 69(7):608–612

Ren DS, Feng XN, Lu LG, et al (2017) An electrochemical-thermal coupled overcharge-to-thermal-runaway model for lithium-ion battery. J Power Sources 364:328–340

Shi N (2015) Research on thermal models of cylindrical lithium-ion batteries for electric vehicles. Beijing Institute of Technology, Beijing

Tao WQ (2006) Heat transfer. Northwestern Polytechnical University Press, Xi'an

Verbrugge MW (1996) Lithium intercalation of carbon-fiber microelectrodes. J Electrochem Soc 143(1):24

Wang BW, Fan Z, Kang XH et al (1999) Realization of genetic algorithm in MATLAB environment. J Wuhan Univ Technol 21(6):25–28

Xia YY, Kumada N, Yoshio M (2000) Enhancing the elevated temperature performance of Li/LiMn2O4 cells by reducing LiMn2O4 surface area. J Power Sources 90(2):135–138

Xu M, Zhang ZQ, Wang X, et al (2014) Two-dimensional electrochemical–thermal coupled modeling of cylindrical LiFePO4 batteries. J Power Sources 256:233–243

Yang SM, Tao WQ (2006) Heat transfer, 4th edn. Higher Education Press, Beijing

Zhao ZN (2008) Heat transfer. Higher Education Press, Beijing

Zhu YL, Yang ZH, Chen XH (2011) Research on model reduction methods. Microcomput Inf 27(6):22–25

Zhuo JW et al (2014) Application of MATLAB in mathematical modeling. Beihang University Press, Beijing

Chapter 4
Modeling and Optimization of Air Cooling Heat Dissipation of Lithium-ion Battery Packs

In this chapter, battery packs are taken as the research objects. Based on the theory of fluid mechanics and heat transfer, the coupling model of thermal field and flow field of battery packs is established, and the structure of aluminum cooling plate and battery boxes is optimized to solve the heat dissipation problem of lithium-ion battery packs, which provides theoretical basis and effective research methods for the design of heat dissipation systems of lithium-ion battery packs.

4.1 Classification of Air Cooling Heat Dissipation of Lithium-ion Batteries

Air cooling is divided into serial type and parallel type according to different air duct structures of cooling systems. According to the presence of fans, it is also divided into natural cooling and forced cooling.

1. Serial and parallel cooling modes

In 1999, Ahmad A. Pesaran of the National Renewable Energy Laboratory of the United States put forward serial and parallel cooling methods, as shown in Fig. 4.1. Figure 4.1a shows serial cooling, in which air blows in from one side of the battery pack and takes heat away from the battery box from the other side. As the air passes through the left side first, it is easy to cause uneven heat dissipation of the battery pack, and the temperature of the battery on the right side is higher than that on the left side. Figure 4.1b shows parallel cooling. The air blows in from the bottom of the battery pack and blows out from the top. Almost the same amount of air flows over the surface of each battery cell, which can ensure even heat dissipation of the battery pack. Ahmad et al. (1999) established a two-dimensional model to simulate the effects of serial and parallel cooling, as shown in Fig. 4.2. With other conditions being the same, the parallel cooling is relatively uniform, and the max. temperature difference

J. Li, *Modeling and Simulation of Lithium-ion Power Battery Thermal Management*,
Key Technologies on New Energy Vehicles,
https://doi.org/10.1007/978-981-19-0844-6_4

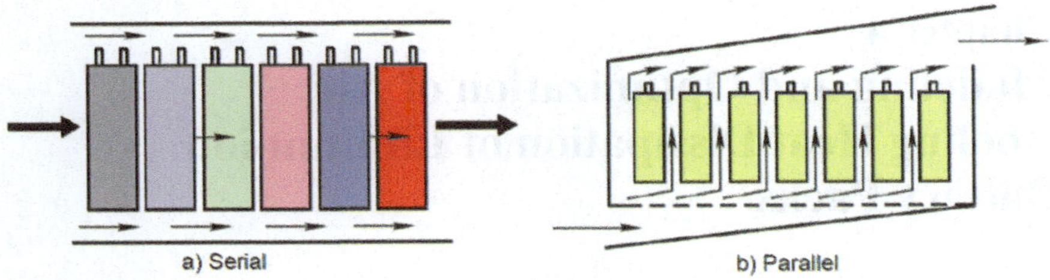

Fig. 4.1 Serial and parallel ventilation modes

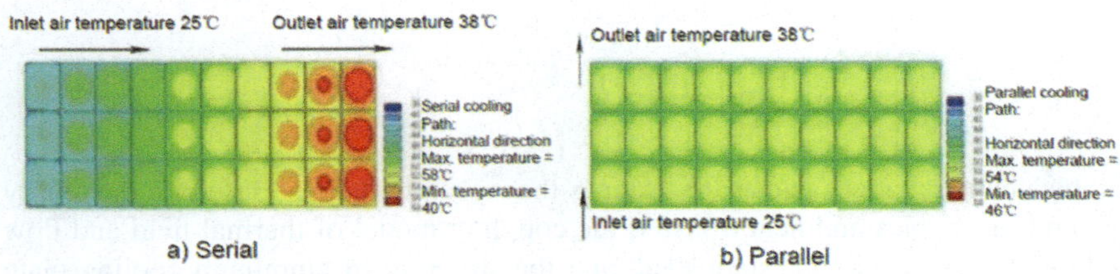

Fig. 4.2 Two-dimensional simulation of serial and parallel cooling effects

in the battery pack is 8 °C. When the serial cooling is adopted, although the min. temperature of the battery pack decreases, the temperature difference in the battery pack is as high as 18 °C, so the parallel cooling method has obvious advantages in reducing the max. temperature and reducing the temperature difference of the battery pack.

2. Natural and forced cooling modes

Natural cooling means that no cooling fan is used for cooling, and its cooling effect is relatively poor; besides, it has higher requirements for the locations of the battery boxes. For series hybrid electric vehicles and battery electric vehicles, natural cooling can no longer meet the heat dissipation requirements of battery packs. Forced cooling refers to cooling with cooling fans, which is currently used by most air-cooled electric vehicles, such as Toyota Prius and Honda Insight.

In 2002, Kenneth J. Kelly of the National Renewable Energy Laboratory of the United States tested the thermal management system of lithium-ion battery packs of Prius in 2001 and Honda Insight in 2000 (Kenneth et al. 2002). The test results showed that both HEVs achieved good results in controlling the temperature of battery packs. The researchers have also tested the cooling fans of Prius, which has four working modes: stop, low speed, medium speed and high speed. When a fan is in different working modes, the energy consumption is different, which is 4–5 W at low speed and 17 W at medium speed. The thermal management system makes the fans work in different modes according to the different battery pack temperatures, so as to reduce the energy consumption of the cooling system.

The effect of forced cooling has been tested and simulated (Rami et al. 2008). 18,650 lithium-ion batteries were used in the experiment, and the battery pack was composed of 68 battery cells. Through the experiment and simulation calculation of different environmental temperatures and different discharge rates, the following conclusions were drawn: when the environmental temperature was 45 °C and the discharge rate was 6.67C, no matter how high the air flow rate was, the temperature of the battery pack could not be controlled below the set 55 °C. When the air flow rate increased, the surface temperature difference of the battery cell would also increase, resulting in uneven temperature distribution of the battery pack.

Forced cooling is a mature and widely used cooling method of thermal management systems of lithium-ion battery packs today. However, when the ambient temperature is high, forced cooling cannot control the max. temperature of the battery pack within a safe range. To solve this problem, the active thermal management system proposed (Pesaran 2001) can be adopted. Before the air is filled into the battery box, the air is cooled by a cooling device, and the cooled air can effectively control the max. temperature of the battery pack.

4.2 Heat Dissipation Flow Field Theory of Battery Packs

Heat conduction, heat convection and heat radiation are the three basic ways of heat transfer, and the heat dissipation model of battery boxes mainly involves two aspects: heat conduction and heat convection. There is no relative displacement between parts of an object, and the heat transfer generated by the thermal movement of microscopic particles such as molecules, atoms and free electrons belongs to heat conduction. Due to the macroscopic movement of the fluid, the heat transfer caused by the relative displacement between various parts of the fluid and the mixing of cold and hot fluids belongs to thermal convection (Tao 2006). For lithium-ion batteries, the heat transfer inside and between battery cells belongs to heat conduction, and the heat transfer between battery cells and the air in the box belongs to heat convection.

Flow field analysis mainly includes theoretical fluid mechanics and computational fluid mechanics. Theoretical fluid mechanics was founded in the eighteenth century, and its development was earlier than computational fluid mechanics. Because fluid flow is nonlinear, many problems can not be solved accurately, so computational fluid dynamics (CFD) is generally used.

The basic governing equation of fluid is the core of numerical analysis. The continuity equation, momentum equation and energy equation of fluid can be derived from the conservation laws of mass, momentum and energy (Liu 2005; Kays 2007; Li et al. 2009).

1. Mass conservation equation

Any flow problem must satisfy the mass conservation equation, that is, the continuity equation, and its integral equation is as follows,

$$\frac{\partial}{\partial t}\iiint\limits_{\text{Vol}} \rho \mathrm{d}x\mathrm{d}y\mathrm{d}z + \oiint \rho v \mathrm{d}A = 0 \tag{4.1}$$

where, Vol is the control body; $\boldsymbol{v}$ is fluid velocity; $\boldsymbol{A}$ is the control plane; The first term indicates the increment of the internal mass of the control body; The second term represents the net flux through the control surface.

The differential equation of Eq. (4.1) in rectangular coordinate system is:

$$\frac{\partial \rho}{\partial t} + \frac{\partial(\rho u)}{\partial x} + \frac{\partial(\rho v)}{\partial y} + \frac{\partial(\rho w)}{\partial z} = 0 \tag{4.2}$$

where, u, v and w are components of fluid velocity; ρ is the fluid density.

Equation (4.2) is a continuity equation, suitable for compressible or incompressible fluids, viscous or inviscid fluids, steady or unsteady flowing fluids.

When the object of study is a steady fluid and the density ρ does not change with time, the Eq. (4.2) can be expressed as:

$$\frac{\partial(\rho u)}{\partial x} + \frac{\partial(\rho v)}{\partial y} + \frac{\partial(\rho w)}{\partial z} = 0 \tag{4.3}$$

When the research object is a steady incompressible fluid and the density is constant, the Eq. (4.2) can be expressed as:

$$\frac{\partial u}{\partial x} + \frac{\partial v}{\partial y} + \frac{\partial w}{\partial z} = 0 \tag{4.4}$$

2. Momentum conservation equation

The law of momentum conservation can be expressed as that the rate of change of fluid momentum to time in any control element is equal to the external action.

The sum of the forces on the infinitesimal can be expressed with Eq. (4.5).

$$\delta_{\mathrm{F}} = \delta_{\mathrm{m}}\frac{\mathrm{d}v}{\mathrm{d}t} \tag{4.5}$$

When the object of study is incompressible viscous fluid with constant physical properties, the momentum equation is as follows:

$$\begin{cases} \rho\left(\frac{\partial u}{\partial t} + u\frac{\partial u}{\partial x} + v\frac{\partial u}{\partial y} + w\frac{\partial u}{\partial z}\right) = \rho F_x - \frac{\partial p}{\partial x} + \mu\left(\frac{\partial^2 u}{\partial x^2} + \frac{\partial^2 u}{\partial y^2} + \frac{\partial^2 u}{\partial z^2}\right) \\ \rho\left(\frac{\partial v}{\partial t} + u\frac{\partial v}{\partial x} + v\frac{\partial v}{\partial y} + w\frac{\partial v}{\partial z}\right) = \rho F_y - \frac{\partial p}{\partial y} + \mu\left(\frac{\partial^2 v}{\partial x^2} + \frac{\partial^2 v}{\partial y^2} + \frac{\partial^2 v}{\partial z^2}\right) \\ \rho\left(\frac{\partial w}{\partial t} + u\frac{\partial w}{\partial x} + v\frac{\partial w}{\partial y} + w\frac{\partial w}{\partial z}\right) = \rho F_z - \frac{\partial p}{\partial z} + \mu\left(\frac{\partial^2 w}{\partial x^2} + \frac{\partial^2 w}{\partial y^2} + \frac{\partial^2 w}{\partial z^2}\right) \end{cases} \tag{4.6}$$

where, F_x, F_y and F_z are components of volume force in x, y and z directions; μ is the viscosity coefficient of fluid.

3. Energy conservation equation

The energy equation can be expressed with Eq. (4.7):

$$\frac{\partial(\rho H)}{\partial t}+\frac{\partial(\rho u H)}{\partial x}+\frac{\partial(\rho v H)}{\partial y}+\frac{\partial(\rho w H)}{\partial z} = -P\mathrm{div}(U)+\mathrm{div}[\lambda\cdot\mathrm{grad}(T)]+\Phi+\dot{q} \tag{4.7}$$

where λ is the thermal conductivity coefficient of the fluid; H is the enthalpy of the fluid; q is the internal heat source of fluid; Φ is the dissipative energy function, which indicates the part of mechanical energy converted into thermal energy due to viscous action, and can be calculated with Eq. (4.8):

$$\Phi=\eta\left\{\begin{array}{l}\left[2\left(\frac{\partial u}{\partial x}\right)^2+\left(\frac{\partial v}{\partial y}\right)^2+\left(\frac{\partial w}{\partial z}\right)^2\right]+\left(\frac{\partial u}{\partial y}+\frac{\partial v}{\partial x}\right)^2\\+\left(\frac{\partial u}{\partial z}+\frac{\partial w}{\partial x}\right)^2+\left(\frac{\partial v}{\partial z}+\frac{\partial w}{\partial y}\right)^2\end{array}\right\}+\lambda\mathrm{div}(U) \tag{4.8}$$

For incompressible fluids, the energy equation can be simplified as.

$$\frac{\partial T}{\partial t}+\mathrm{div}(UT)=\mathrm{div}\left[\frac{\lambda}{\rho C_p}\cdot\mathrm{grad}(T)\right]+\frac{\dot{q}+\Phi}{\rho} \tag{4.9}$$

For solid media, the fluid velocity component $u=v=w=0$, and the energy equation is the heat conduction equation for solving the temperature field inside the solid.

For air, when the flow velocity is less than 1/3 sound velocity, it can be considered as incompressible gas. In this paper, the flow rate of the cooling air is far less than this value, so the cooling air can be considered as an incompressible gas.

4.3 Finite Element Simulation Modeling of Air Cooling Heat Dissipation of Lithium-ion Battery Packs

4.3.1 Finite Element Simulation Process

The battery thermal model describes the laws of heat generation, heat transfer and heat dissipation of a battery, and can calculate the temperature change of the battery in real time. The calculation of battery temperature field based on battery thermal

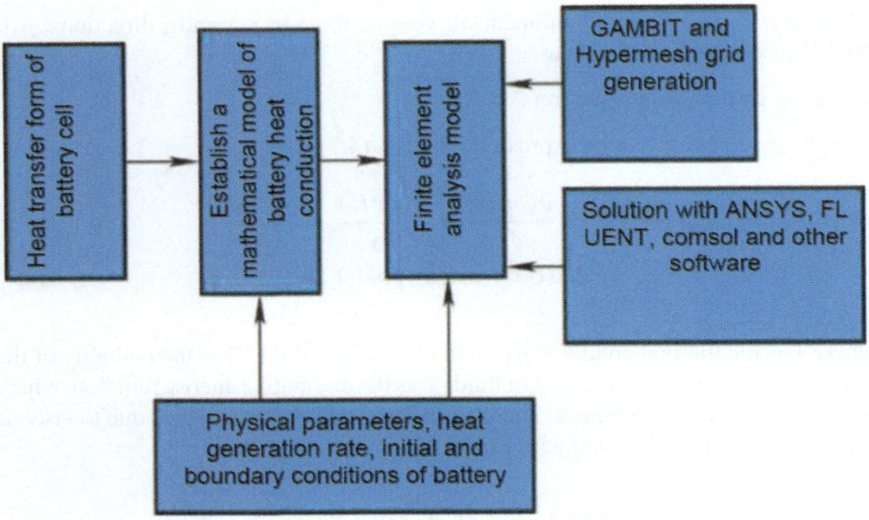

Fig. 4.3 Research method of battery thermal model

model can not only provide guidance for the design and optimization of battery thermal management system, but also provide quantitative basis for the optimization of battery heat dissipation performance.

Based on the principle of heat transfer, the cell heat transfer model can be simplified as follows: under different boundary conditions, the cell body, copper and aluminum poles generate heat at different heat generating rates: part of the heat is transferred to the surrounding air through the cell shell, and the heat transferred to the air shows the heat transfer coefficient of the cell surface; And the other part is used for heating the battery cell itself. The research method is shown in Fig. 4.3.

As for battery heat dissipation, software like FLUENT or ANSYS is used to simulate the fluid flow and heat transfer. GAMBIT or Hypermesh is used to construct the geometric shape of the flow area, generate boundary types and grids, and output the format for calculation by the software solver. The flow area is solved and calculated by using a solver, and the calculation results are post-processed. The solving steps are as follows:

(1) Determining the geometric shape and generating a computational grid (GAMBIT or hypermesh);
(2) Inputting and checking grids, selecting solvers and solving equations: laminar flow or turbulent flow (or inviscid flow), chemical components or chemical reactions, heat transfer models, etc.
(3) Determining the material properties, boundary types and boundary conditions of the fluid;
(4) Flow field initialization, solution and calculation;
(5) Saving the results and post-process.

4.3.2 Geometric Models of Battery Packs

This section focuses on battery packs. As the heating function of wide-wire metal film is integrated in the target battery pack, the battery cells are required to be stacked in structure. Therefore, it is proposed to install slotted aluminum plates between the battery cells to improve the heat dissipation effect of the battery pack. The simplified geometric model of the battery box is shown in Fig. 4.4. The whole battery box includes 48 battery cells, which are arranged into two rows, 24 battery cells in each row, the interval between the battery pack and the surrounding box walls is 15 mm, and the interval between the two rows of cells is 30 mm. The battery box has four air inlets, which are respectively located on the left and right sides of the box body. Two circular openings at the top of the battery box are air outlets, and centrifugal exhaust fans are installed at the air outlets. The power of the exhaust fans is 12 W. Because the battery box is symmetrical, in order to reduce the calculation amount of the model, the quarter battery box is used for modeling when building the battery pack heat dissipation model, as shown in Fig. 4.5. The quarter model includes 12 battery cells, 1 air inlet and 1/2 air outlet. As shown in Fig. 4.6, the slotted aluminum plate has a thickness of 5 mm, a groove width of 15 mm, 5 grooves and a groove

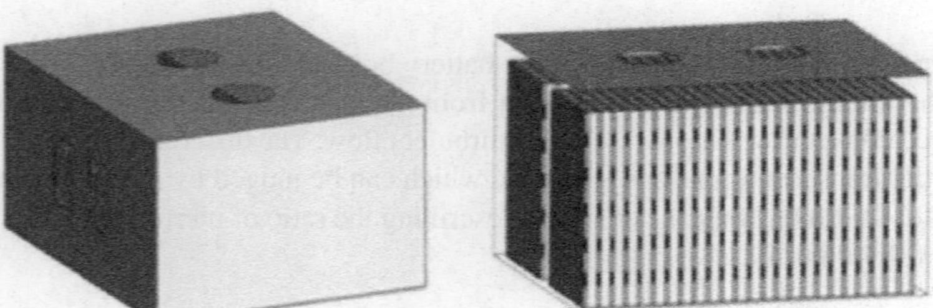

a) b)

Fig. 4.4 Simplified geometric model of battery boxes

Fig. 4.5 Quarter battery box model

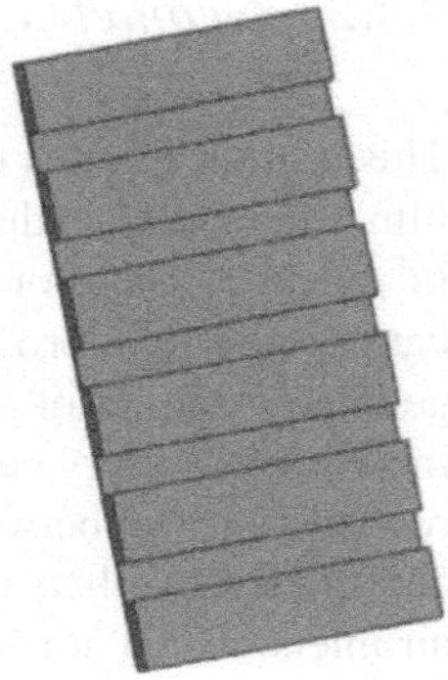

Fig. 4.6 Slotted aluminum plate

depth of 4 mm. The air inlet and outlet positions of the battery box and the related parameters of the slotted aluminum plate are only preliminarily determined here, and will be optimized subsequently.

4.3.3 Battery Flow Field Selection

A centrifugal exhaust fan is used in the battery box, and the cooling air is sucked in from the inlet to forcibly dissipate heat from the battery pack. There are two types of forced convection, laminar flow and turbulent flow. The difference lies in the ratio of inertial transport to viscous transport, which can be judged by Reynolds numbers (Re). Re is a dimensionless parameter describing the ratio of inertial force to viscous force of fluid, i.e.

$$\mathrm{Re}=\frac{\rho L V_0}{\mu} \tag{4.10}$$

where μ is the dynamic viscosity of the fluid; ρ is the fluid density; L is the characteristic scale; V_0 is the fluid flow velocity.

If Re < 2000, the fluid is laminar; If Re > 4000, the fluid is in turbulent state. If 2000 < Re < 4000, the fluid is in a transitional state. With other conditions determined, the greater Re is, the better heat transfer performance is. At 20 °C, the viscosity of the air is 1.808×10^{-5} Pa s, the density is 1.17 kg/m^3, and the wind speed is 1 m/s, and the Re of the cooling air in the battery box is much higher than 4000, so the convection type of the cooling model is turbulence. In this paper, the turbulence standard two-equation $k - \varepsilon$ model will be adopted, where k represents the turbulent motion energy and ε represents the diffusion speed of the turbulent motion energy.

The turbulent viscosity can be calculated with Eq. (4.11):

$$\mu_t=C_{\mu}\frac{k^2}{\varepsilon} \tag{4.11}$$

Equation of turbulence kinetic energy *k*:

$$\frac{\partial(\rho \mathrm{k})}{\partial t}+\frac{\partial(\rho U_i k)}{\partial x_i}-\frac{\partial}{\partial x_i}\left[\left(\mu+\frac{\mu_{\mathrm{t}}}{\sigma_{\mathrm{k}}}\right)\frac{\partial k}{\partial x_i}\right]=G_{\mathrm{k}}-\rho\varepsilon \tag{4.12}$$

Equation of turbulence kinetic energy dissipation rate ε:

$$\frac{\partial(\rho\varepsilon)}{\partial t}+\frac{\partial(\rho U_i\varepsilon)}{\partial x_i}-\frac{\partial}{\partial x_i}\left[\left(\mu+\frac{\mu_{\mathrm{t}}}{\sigma_{\varepsilon}}\right)\frac{\partial \varepsilon}{\partial x_i}\right]=\frac{C_{1\varepsilon}\varepsilon}{k}G_{\mathrm{k}}-C_{2\varepsilon}\rho\frac{\varepsilon^2}{k} \tag{4.13}$$

where, i of U_i is 1, 2 and 3, which respectively represent the velocities u, y and z in the x, y and z directions; *xi* represents three coordinate directions of x, y and z; σ_k and σ_ε respectively represent the effective Prandtl numbers of turbulence kinetic energy and kinetic energy dissipation rate, which are usually 1 and 1.3; C_μ, $C_{1\varepsilon}$ and $C_{2\varepsilon}$ are empirical constants, which are 0.09, 1.44 and 1.92 respectively.

4.3.4 Simulation Calculation of Steady-State Heat Dissipation of Battery Packs

Steady-state heat dissipation simulation of a battery pack means that under certain initial conditions and boundary conditions, the battery pack is continuously discharged, and the steady-state temperature after the heat generated by the battery pack and the heat dissipation of cooling air reach equilibrium is solved. The simulation results can reflect the final trend of the internal temperature distribution of the battery pack and provide a basis for the optimization of the battery box structure.

1. Determination of initial conditions and boundary conditions

(1) Initial conditions

When calculating the steady state heat dissipation, it is assumed that the initial temperature of the battery pack is 20 °C, the ambient temperature is kept at 20 °C, and the inlet cooling air temperature is 20 °C.

(2) Boundary conditions

As to the flow field, the battery box has four air inlets with the dimensions of 100 × 60 mm and two air outlets with a diameter of 90 mm. Each air outlet is equipped with a DC 12 W centrifugal exhaust fan with a voltage of 12 V, a working current of 1.2 A and an air volume of 25.43 CFM (CFM: cubic feet/minute, 1 CFM ≈ 1.7 m^3/h). Set the inlet boundary condition as velocity inlet and the outlet boundary condition as pressure outlet.

2. Calculation results and analysis

Figure 4.7 shows the temperature distribution of the battery pack after discharging at 2C rate given by the steady-state simulation calculation of the heat dissipation model of the battery box. The temperature range of the whole battery box is 294–320 K (21–47 °C), and the temperature range of the lithium-ion battery pack is 304–320 K (31–47 °C). Although the lithium-ion battery can work safely at this temperature, it can be seen from the figure that the temperature distribution of the battery pack is uneven. The temperature of the battery facing the air inlet is lower, while the temperature of the battery away from the air inlet is higher. The sectional temperature distribution of the battery pack is shown in Fig. 4.8, from which it can be clearly seen that the temperature in the middle of the battery pack is on the high

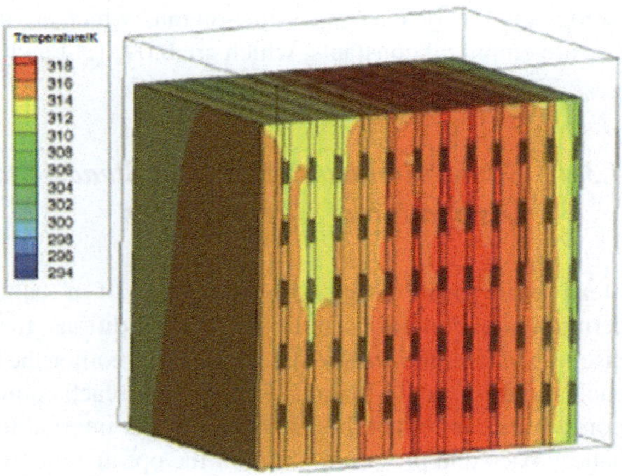

Fig. 4.7 Surface temperature distribution of a battery packs

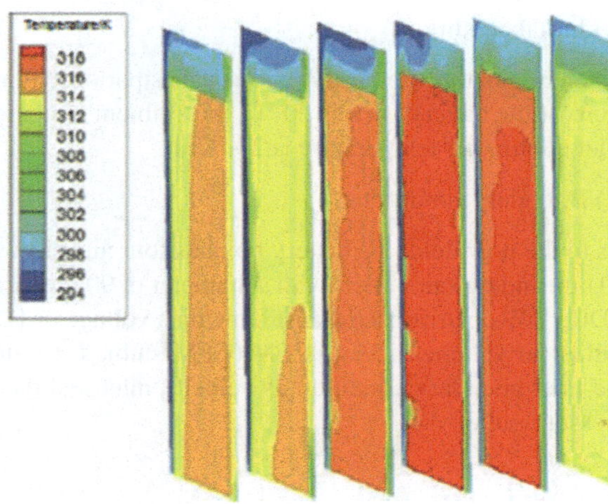

Fig. 4.8 Sectional temperature distribution of a battery pack

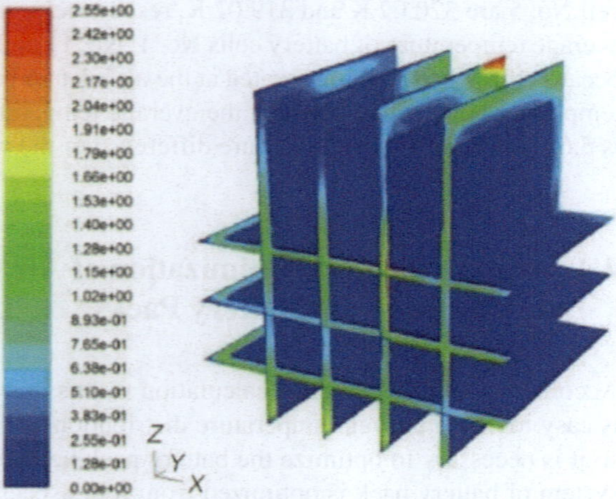

Fig. 4.9 Sectional view of flow field inside the battery box

side. See Fig. 4.9 for the sectional view of the cooling air velocity inside the battery box.

The average temperature, min. temperature and max. temperature curves of 12 battery cells are shown in Fig. 4.10. The batteries are numbered in order from the wall of the battery box to the symmetrical side, i.e. from the right side of Fig. 4.7. Therefore, batteries 1–12 are arranged in sequence. And, subsequent batteries are numbered in the same way. The max. temperature and average temperature of battery

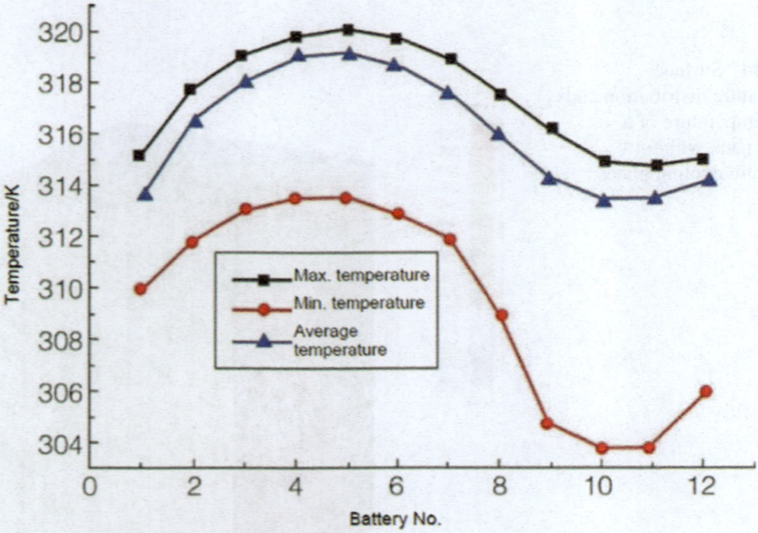

Fig. 4.10 Temperature analysis curves of battery cells

cell No. 5 are 320.02 K and 319.02 K, respectively, while the max. temperature and average temperature of battery cells No. 1, No. 11 and No. 12 are the lowest, mainly because these batteries are located at the air inlet and air duct respectively. The max. temperature difference between the average temperature values of the battery cells is 5.6 K, and the max. temperature difference of the whole battery pack is 16.23 K.

4.4 Simulation and Optimization of Air Cooling Schemes for Lithium-ion Battery Packs

According to the steady-state calculation results of battery pack heat dissipation, it is easy to cause uneven temperature distribution in the process of heat dissipation, so it is necessary to optimize the battery pack heat dissipation system. The cooling system of battery pack is optimized from four aspects: aluminum cooling plate, air inlet, air outlet, height of battery box and wind speed.

4.4.1 Structural Optimization of Heat Conductive Aluminum Plates

Figure 4.11 shows the steady-state calculated temperature distribution of the battery pack without aluminum cooling plates during discharging at 2C rate. Figure 4.12 shows the curves of average temperature and max. temperature of battery cells

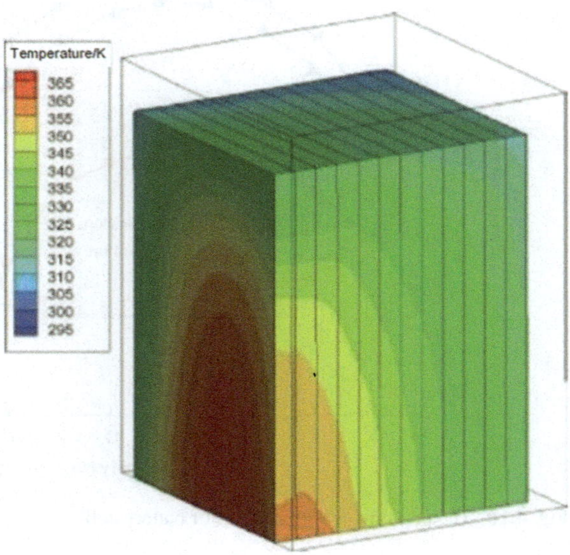

Fig. 4.11 Surface temperature distribution and max. temperature of a battery pack without aluminum cooling plates

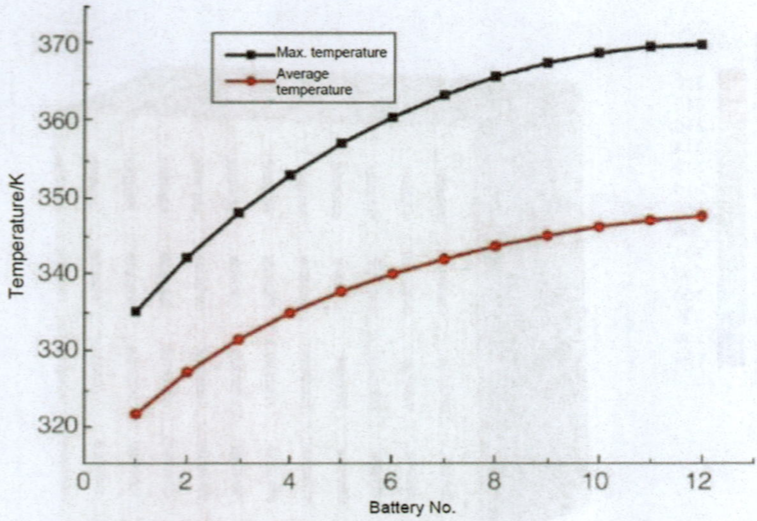

Fig. 4.12 Average temperature and max. temperature of each battery cell without aluminum cooling plates

without aluminum cooling plates. It can be seen from the figure that the temperature of the battery pack increases greatly without installing aluminum cooling plates, and the max. temperature rises to 369 K; The max. average temperature of the battery reaches 347 K, and the max. temperature difference of the battery pack is 67 K. Therefore, aluminum cooling plates play an important role in heat dissipation of battery packs. In order to further study the influence of aluminum cooling plates on the heat dissipation performance of battery packs, this section will study the influence of slot width and slot number of aluminum cooling plates on the heat dissipation performance.

1. The width of each slot of the aluminum cooling plate is 30 mm, and the number of slots is 4

The slot width of the aluminum cooling plate is increased from 15 to 30 mm, and the number of slots is changed to 4. Figure 4.13 shows the surface temperature distribution of a battery pack calculated by steady-state simulation of heat dissipation model. Compared with Fig. 4.7, the temperature distribution of the battery cells near the air inlet is uniform, and the temperature drops significantly, but the temperature of the battery cells far away from the air inlet does not decrease, and the local position is even higher. The average temperature comparison curves of battery cells with the slot width of 15 and 30 mm for aluminum cooling plates is shown in Fig. 4.14. It can be seen from the figure that the average temperature of cells No. 7–12 decreases with the increase of the slot width, especially for cells No. 10–12, but the average temperature of cells No. 2–5 increases slightly.

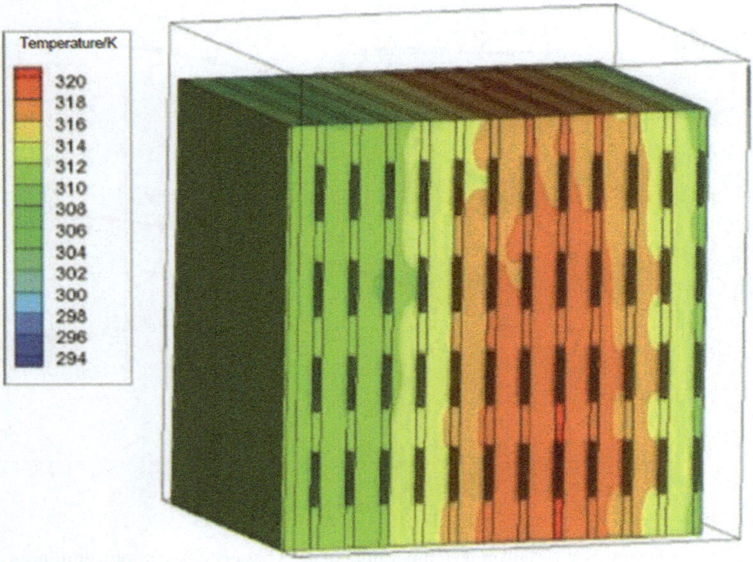

Fig. 4.13 Surface temperature distribution of battery cells with aluminum cooling plates with a slot width of 30 mm

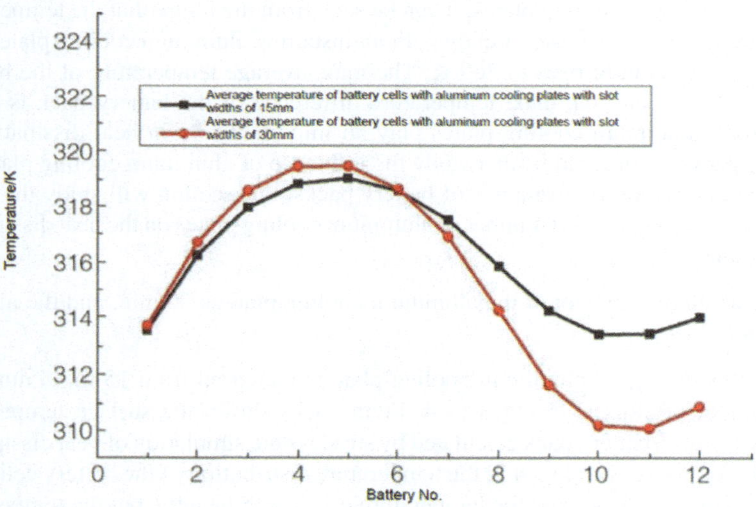

Fig. 4.14 Average temperature of battery cells with aluminum cooling plates with slot widths of 15 and 30 mm

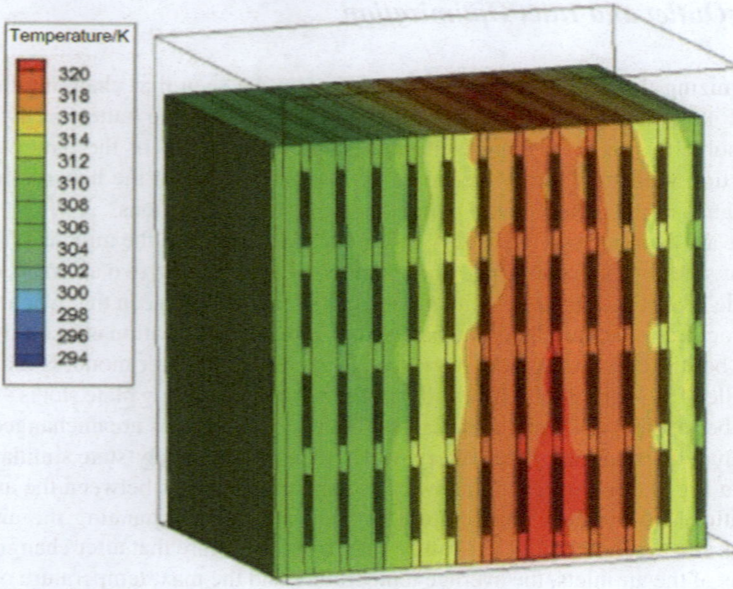

Fig. 4.15 Surface temperature distribution of battery pack with aluminum plate slot width of 40 mm

2. The width of each slot of the aluminum cooling plate is 40 mm, and the number of slots is 4

With the aluminum plate slot width increased from 30 to 40 mm, the number of slots remaining 4, and other conditions unchanged, the temperature distribution of the battery pack calculated by steady-state simulation is shown in Fig. 4.15. It can be seen from the figure that the temperature of battery cells near the air inlet decreases further, but that of cells far away from the air inlet does not decrease, and the uneven temperature distribution of the battery pack has not been improved. This is mainly because after the slot width of the aluminum cooling plate is increased, a large amount of cooling air is discharged from the battery box through the slots near the air inlet, and the cooling air passing through the rear part is reduced, resulting in an increase in the temperature of some battery cells. Although increasing the slot width cannot solve the problem of uneven temperature distribution of the battery pack, it can reduce the temperature of battery cells near the air inlet, and the temperature distribution of these battery cells is relatively uniform. Therefore, if the air intake of cooling air is increased or the position of air inlet is changed, the use of wide-slot aluminum plates will help improve the heat dissipation performance of the battery pack.

4.4.2 Outlet and Inlet Optimization

By optimizing the aluminum cooling plates, it can be seen that changing the slot width of aluminum plates can reduce the temperature of some battery cells, but it cannot solve the problem of uneven temperature distribution of the battery pack. This section will study the influence of air inlet and outlet on the heat dissipation performance of the battery box. Optimization of air inlet locations.

There are 4 air inlets in the battery box, which are located in the middle of the left and right sides of the battery box, and the interval between the two air inlets on the same side is 60 mm. Through simulation calculation, it can be seen that when the air inlet is located in the middle of the battery box, the heat dissipation effect of battery cells on both sides is weakened. Now, the air inlet in the quarter model is moved to the middle of 12 battery cells. The width of the aluminum cooling plate slot is 40 mm, the number of slots is 4, and the initial and boundary conditions are unchanged. The temperature distribution of the battery pack calculated by steady-state simulation is shown in Fig. 4.16. Figure 4.17 shows the comparison curves between the average temperature and the max. temperature of the battery after changing the air inlet locations and the initial model. It can be seen from the figure that after changing the locations of the air inlets, the average temperature and the max. temperature of most battery cells have greatly decreased, and only the temperature of cells No. 11 and 12 has slightly increased. Although the temperature difference of the battery pack has decreased, it is still as high as 15.17 K. This is mainly because the temperature

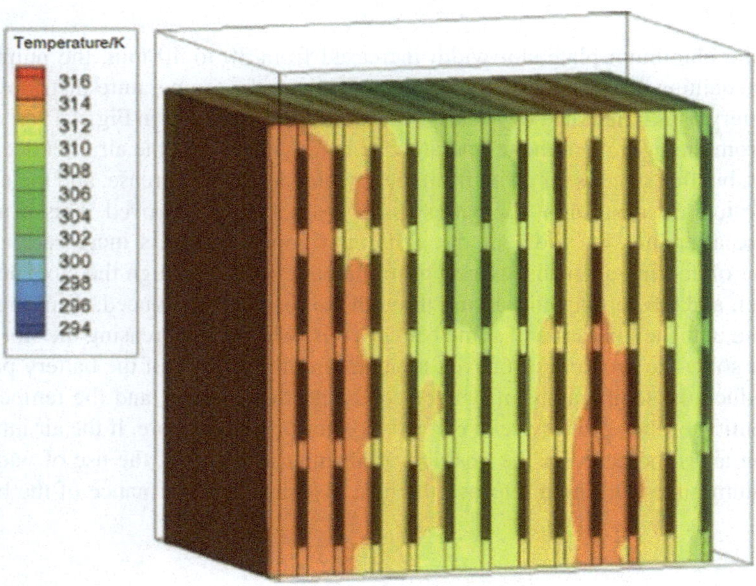

Fig. 4.16 Surface temperature distribution of the battery pack after changing air inlet locations

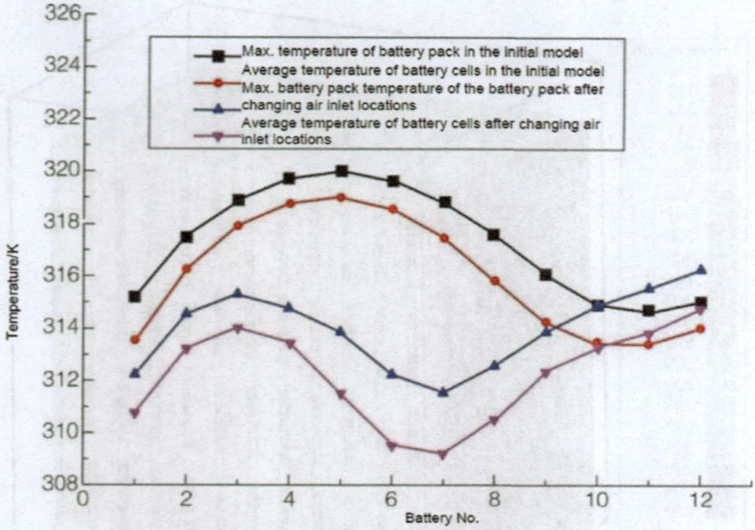

Fig. 4.17 Average temperature and max. temperature of battery cells in two models

of the outer battery cells is greatly reduced by moving the air inlet position, but the temperature of the cells in the center is increased.

1. Adding 1 air inlet

Increase the number of air inlets in the quarter model to 2, and keep the number of air outlets and exhaust fans unchanged, so the wind speed at the air inlet will become 0.5 m/s, and other conditions will remain unchanged. The battery pack temperature distribution obtained through steady-state simulation calculation is shown in Fig. 4.18. As shown in Fig. 4.19, compare the average temperature and max. temperature of the battery with two air inlets with the model with one air inlet. It can be seen from the figure that, among the 12 battery cells, the max. temperature and average temperature of the battery cells at both ends decreases greatly, but the max. temperature of the battery cells at the center decreases slightly, and the average temperature increases slightly. This is mainly because although one air inlet is added, the air outlet and fan remain unchanged, the speed of the air inlet is reduced to 1/2 of the original, the temperature of the battery cells near the two air inlets decreases, and the middle battery cells are far away from the two air inlets, so the temperature decreases little. But overall, the temperature distribution of the battery pack is more uniform, and the temperature difference of the battery pack has been reduced to 11.21 °C.

2. Adding 1 air outlet

On the basis of two air inlets and one air outlet, one air outlet is added, and two inlets and two air outlets are adopted. Figure 4.20 shows the temperature distribution of the battery pack calculated by steady-state simulation. As shown in Fig. 4.21,

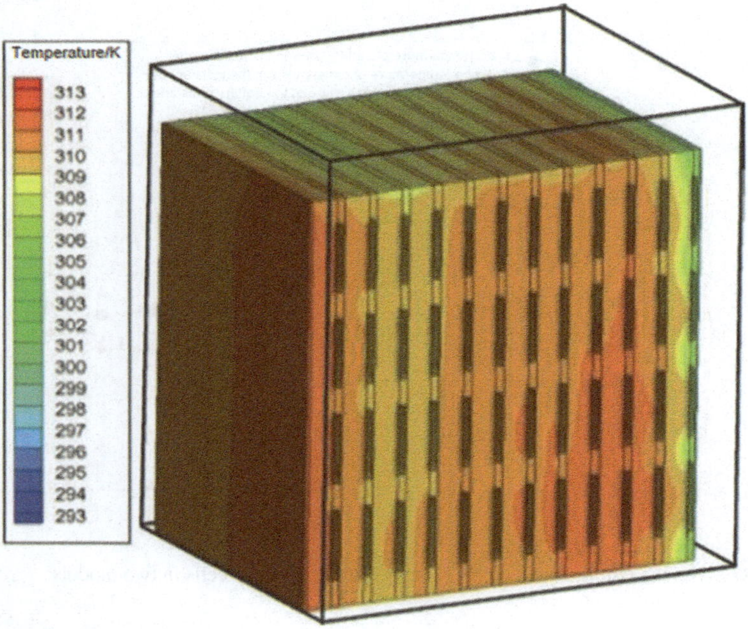

Fig. 4.18 Surface temperature distribution of the battery pack after adding an air inlet

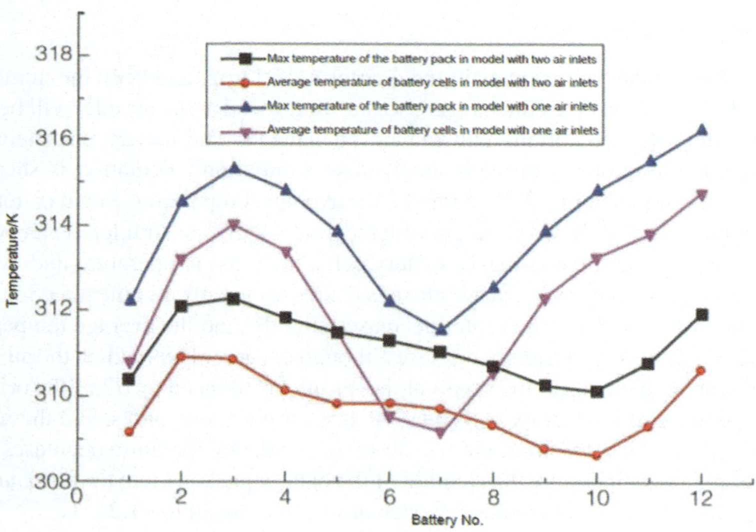

Fig. 4.19 Comparison of max. and average temperatures of the battery pack in models with different number of air inlets

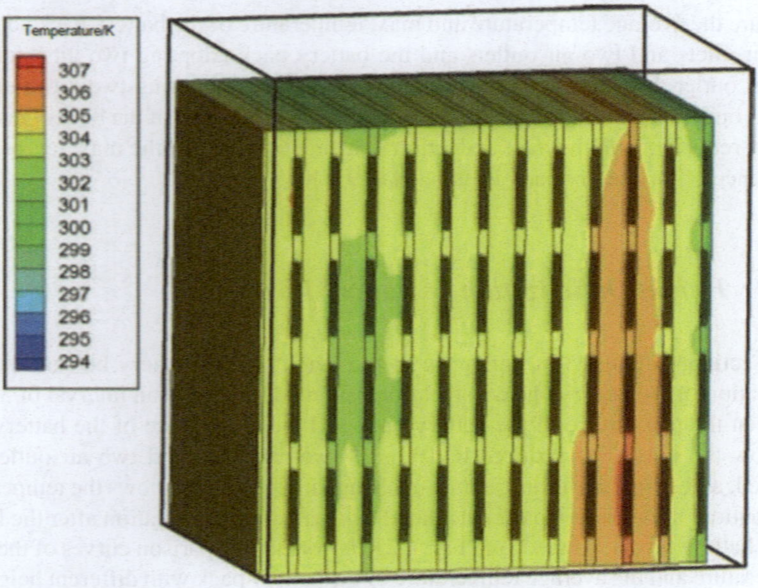

Fig. 4.20 Surface temperature distribution of the battery pack with 2 air inlets and 2 air outlets

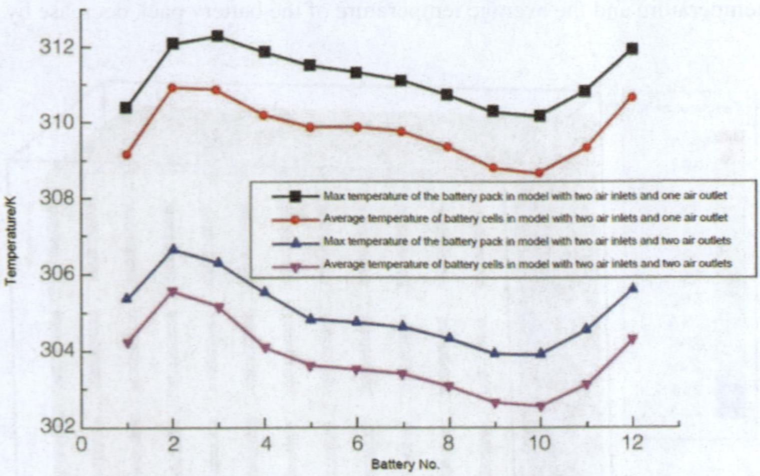

Fig. 4.21 Comparison of max. and average temperatures of battery packs with different outlet models

compare the average temperature and max. temperature of the battery pack adopting two air inlets and two air outlets and the battery pack adopting two air inlets and one air outlet. It can be seen from the figure that after adopting two air inlets and two air outlets, the max. temperature and average temperature of the battery pack are greatly reduced, with the max. reduction reaching 6.71 K, and the max. temperature difference of the battery pack is reduced to 9.3 K.

4.4.3 Height Optimization of Battery Box

This section will study the influence of the height of the battery box on the heat dissipation of the battery pack. In the aforesaid model, there is an interval of 30 mm between the top surface of the battery pack and the top surface of the battery box, but now the interval is reduced to 10 mm. Two air inlets and two air outlets are adopted, and other conditions remain unchanged. Figure 4.22 shows the temperature distribution of the battery pack calculated by steady-state simulation after the height of the battery box is reduced, and Fig. 4.23 shows the comparison curves of the max. temperature and the average temperature of the battery pack with different heights of the battery box. It can be seen from the figure that after the interval between the top surface of the battery pack and the top surface of the battery box is reduced, both the max. temperature and the average temperature of the battery pack decrease by about

Fig. 4.22 Surface temperature distribution of the battery pack with reduced height of battery box

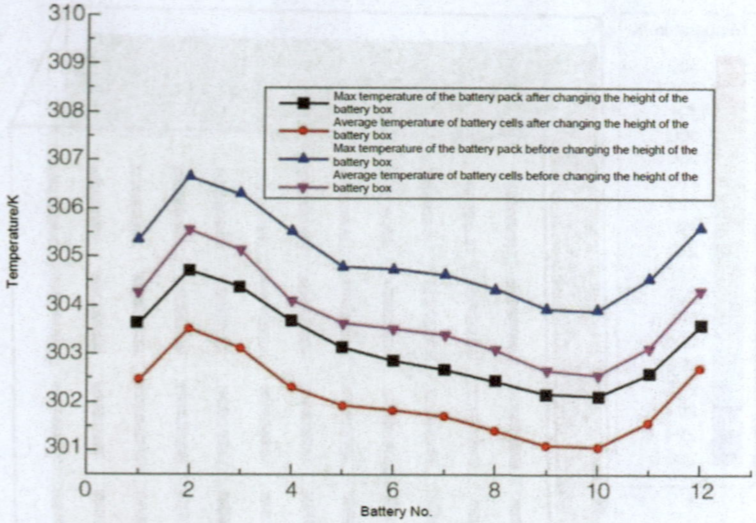

Fig. 4.23 Comparison of max. and average temperatures of battery packs with different battery box heights

2 K. This is mainly because after the interval is reduced, the path of cooling air from the air inlet to the air outlet is shortened, and more cooling air passes through the aluminum cooling plate in the same time, thus enhancing the heat dissipation effect. Although reducing the interval can improve the heat dissipation effect, the space of adjustment is limited due to the limitation of wiring and structure inside the battery box.

4.4.4 Influence of Inlet Air Velocity

In order to keep the temperature difference of the battery pack within a reasonable range, on the basis of reducing the height of the battery box and increasing the air inlet and outlet, the air inlet speed is increased to 2 m/s, and other conditions remain unchanged. The simulated surface temperature distribution of battery pack is shown in Fig. 4.24, and the comparison curves of the max. temperature and the average temperature of battery pack at different air inlet speeds are shown in Fig. 4.25. It can be seen from the figure that the temperature of the battery pack is further decreased, the temperature distribution is relatively uniform, and the max. temperature difference has dropped to 5.73 K.

By optimizing the cooling system, the temperature of the battery pack has been greatly reduced and the non-uniformity of temperature distribution has been greatly improved. As shown in Fig. 4.26, comparing the max. temperature and average temperature of the battery pack between the initial heat dissipation model and the

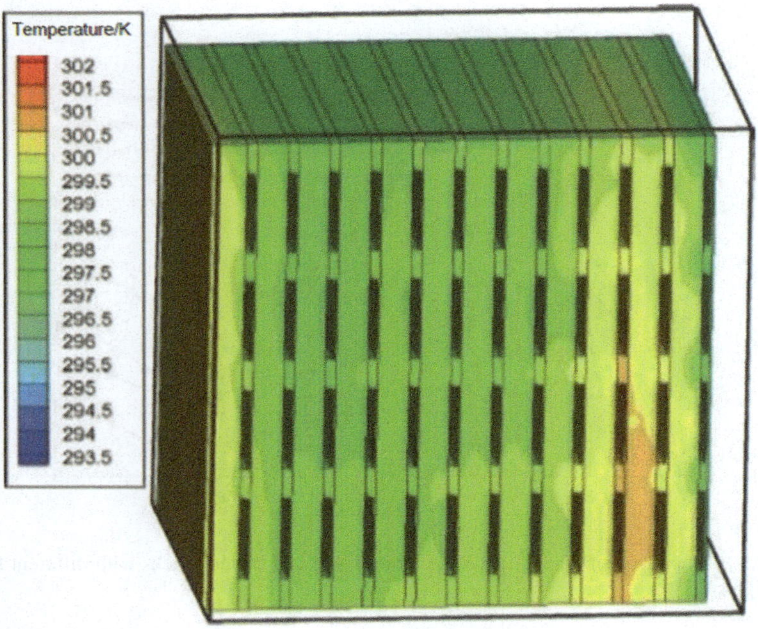

Fig. 4.24 Surface temperature distribution of the battery pack after increasing inlet speed

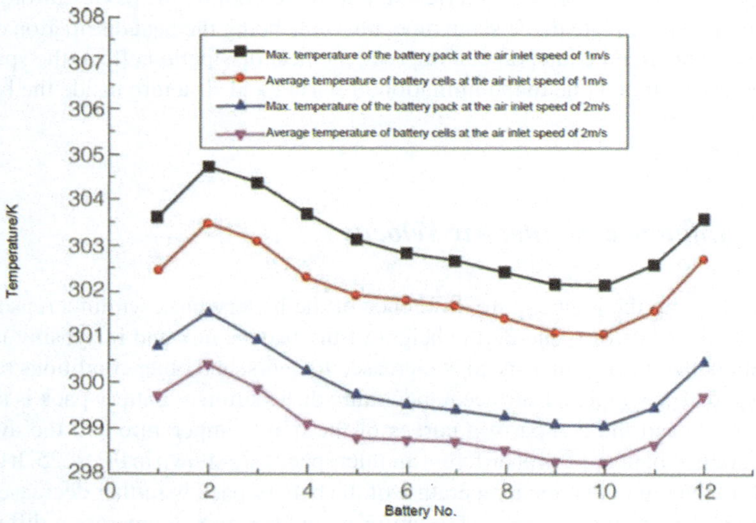

Fig. 4.25 Max. temperature and average temperature of the battery pack with different air inlet speeds

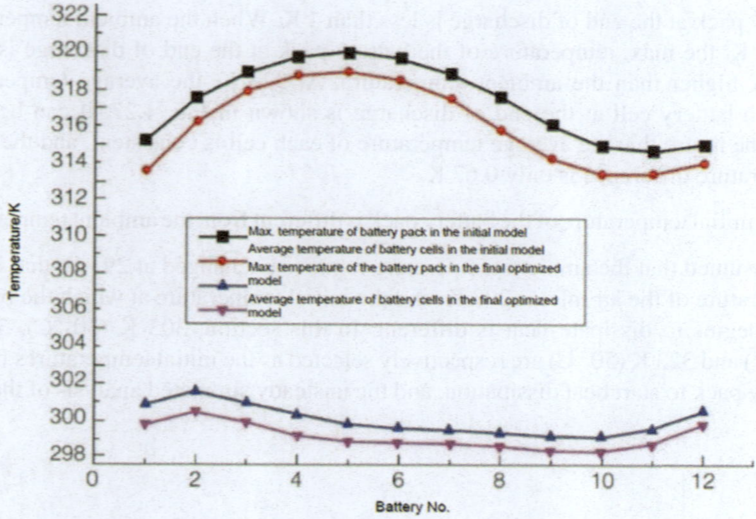

Fig. 4.26 Max. and average temperature of the battery pack in the initial model and the optimized model

final optimized heat dissipation model, the max. drop of the max. temperature of the battery pack and the average temperature of the battery cells after optimization is 20 K, and the optimization effect of the heat dissipation system is remarkable, which provides a theoretical basis for the subsequent design of the battery pack heat dissipation system.

4.4.5 *Simulated Analysis of Heat Dissipation Temperature Consistency of Battery Packs*

In this section, the optimized cooling system will be used for the unsteady simulated analysis of battery packs. Unsteady simulation will be used to study the cooling effect of the cooling system on the battery pack in a limited time, focusing on the analysis of battery temperature consistency. In the unsteady state study, the battery pack still discharges at constant current rate of 2C, and the simulation time is 1800 s.

1. The initial temperature of the battery pack is the same as the ambient temperature

The battery pack works at different temperatures. Assuming that the temperature of the battery pack is the same as the ambient temperature at the beginning of heat dissipation, the unsteady simulated analysis of heat dissipation of the battery pack at ambient temperatures of 303 K (30 °C), 313 K (40 °C) and 323 K (50 °C) is carried out. With the increase of discharge time, the max. temperature of the battery pack is gradually increased. At three temperatures, the max. temperature difference of the battery pack at the end of discharge is less than 4 K. When the ambient temperature is 323 K, the max. temperature of the battery pack at the end of discharge is only 5.64 K higher than the ambient temperature. At 323 K, the average temperature of each battery cell at the end of discharge is shown in Fig. 4.27. It can be seen from the figure that the average temperature of each cell is consistent, and the max. temperature difference is only 0.67 K.

2. The initial temperature of the battery pack is different from the ambient temperature

It is assumed that the ambient temperature remains unchanged at 293 K, that is, the temperature of the air inlet is 293 K, and the initial temperature at which the battery pack begins to dissipate heat is different. In this section, 303 K (30 °C), 313 K (40 °C) and 323 K (50 °C) are respectively selected as the initial temperatures for the battery pack to start heat dissipation, and the unsteady simulated analysis of the heat

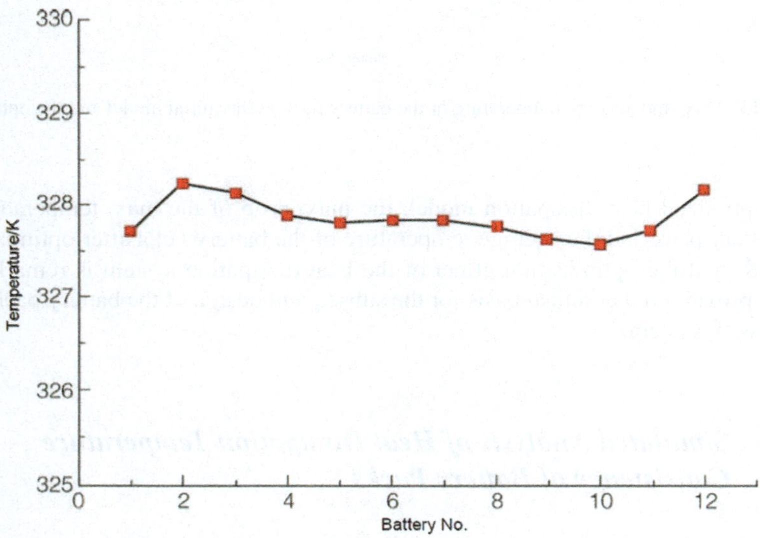

Fig. 4.27 Average temperature of each battery cell after heat dissipation at 323 K for 30 min

Table 4.1 Max. temperature of the battery pack at three different initial temperatures and different moments

Simulation time/s	Initial temperature of battery pack/K		
	303	313	323
300	304.05	313.86	323.62
900	304.72	313.04	321.38
1200	304.82	312.35	319.90
1500	304.86	311.64	318.42
1800	304.87	310.96	317.03

dissipation of the battery pack is carried out. See Table 4.1 for the max. temperatures of the battery pack at different times at three temperatures. Compared with the case where the temperature of the battery pack is the same as the ambient temperature, the max. temperature of the battery pack is greatly reduced, which is because the temperature of the air inlet remains unchanged at 293 K, which enhances the heat dissipation effect.

When the environment is kept at 293 K and the initial temperature of the battery pack is 323 K, the max. temperature difference curves at different times are shown in Fig. 4.28. It can be seen from the figure that the max. temperature difference of the battery pack has reached 18 °C, which is 400% higher than that of the case where the ambient temperature is the same as the initial temperature of the battery pack. It can be seen that the low-temperature cooling air can effectively reduce the temperature of the battery pack, but it is easy to cause the temperature difference of the battery pack to increase, which will adversely affect the consistency and life of the battery pack.

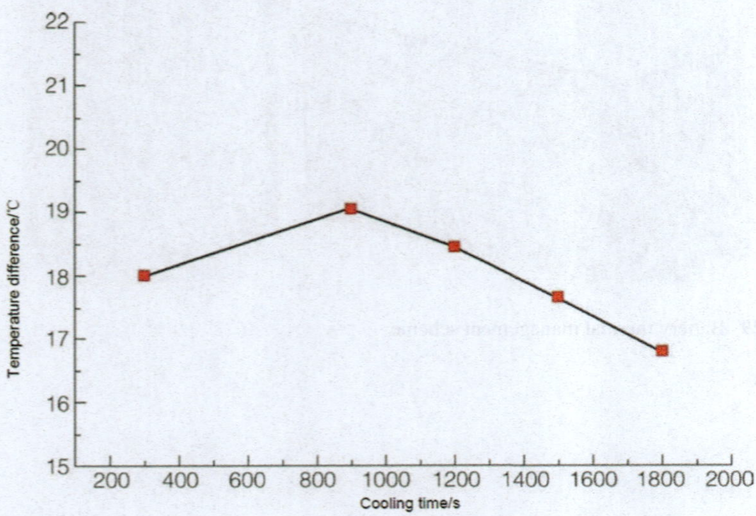

Fig. 4.28 Max. temperature difference at different heat dissipation moments when the initial temperature of battery pack is 323 K

Therefore, the cooling system should reasonably control the temperature difference between the cooling air and the battery pack, and it is not appropriate to set the temperature of the battery starting to dissipate heat too high, nor to charge cooling air with too low a temperature to dissipate heat.

4.5 Case Analysis of Air Cooling Battery Packs

4.5.1 Heat Dissipation Schemes of Battery Packs

The battery box is composed of three battery modules, each module adopts 2P15S (P means parallel; S means series), and the whole battery box adopts 2P45S, with a total of 90 battery cells. The battery box has 3 cooling and exhaust fans and 8 air inlets. After entering the battery box, the wind enters the sides of the three battery modules through the grid in the partition, and exchanges heat with the rectangular sheet heat conductors in the modules; A heat conductive aluminum plate with a thickness of 0.4 mm is embedded between every two battery cells, and extends out of the side of the battery to form the rectangular sheet heat conductor; There are many holes in the rectangular heat conductor to facilitate the formation of an air duct. A PTC resistor tape, which is wound on the rectangular sheet heat conductor, is adopted for heating the battery. The battery thermal management scheme is shown in Fig. 4.29.

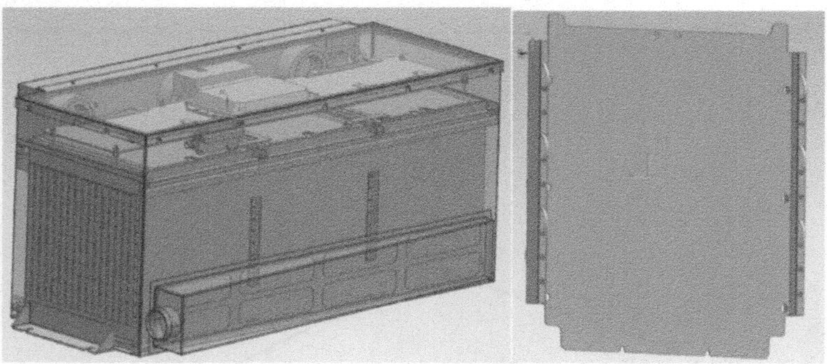

Fig. 4.29 Battery thermal management scheme

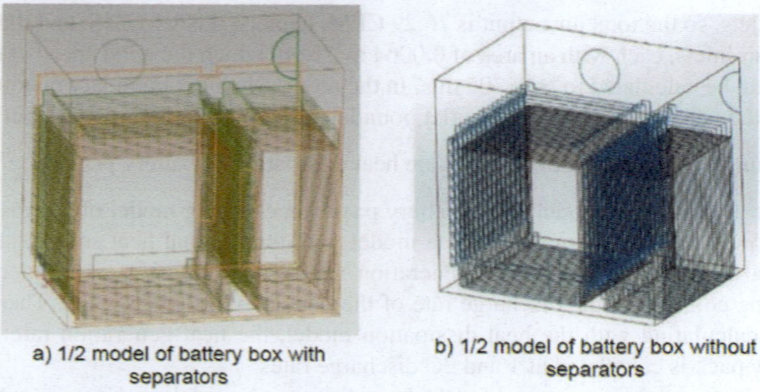

Fig. 4.30 Geometric models of a platform unit battery box

4.5.2 *Simulated Analysis of Battery Pack Heat Dissipation*

1. Geometric model of battery boxes

Each module adopts the battery structure of 2P15S, and the whole battery box adopts the structure of 2P45S, with a total of 90 battery cells. The battery box has 3 heat dissipation and exhaust fans and 8 air inlets. In order to simplify the calculation, the battery box model is simplified, and some components which have little influence on thermal analysis are omitted. Moreover, the battery box is a geometrically symmetrical model, so it is enough to build half of the battery box model. Because there is a partition with grid between the air inlet/outlet and the battery, in order to test the influence of the partition on the heat dissipation of the battery, the model is simplified into two cases. The simplified models are shown in Fig. 4.30.

2. Battery box heat dissipation models

A centrifugal exhaust fan is used in the battery box, and the cooling air is sucked in from the inlet to forcibly dissipate heat from the battery pack. At 20 °C, the viscosity of the air is 1.808×10^{-5} Pa s, the density is 1.17 kg/m^3, and the wind speed is 0.7 m/s, and the Re of the cooling air in the battery box is much higher than 4000, so the convection type of the cooling model is turbulence. In this case, the turbulence standard two-equation $k - \varepsilon$ model will be adopted, where k represents the turbulent motion energy and ε represents the diffusion speed of the turbulent motion energy.

3. Determination of initial conditions and boundary conditions

Initial conditions: the initial temperature of the battery pack is 20 °C, and the inlet cooling air temperature is 20 °C.

Boundary conditions: for the flow field, the battery box has 8 air inlets and 3 air outlets, and the air outlets are equipped with centrifugal exhaust fans. There are three air outlets, so the total air output is 76.29 CFM, namely, 0.0361 CMS, and there are eight air inlets, each with an area of 0.0064 m^2, from which the wind speed at the air inlet can be calculated to be 0.705 m/s. In the model, the inlet boundary condition is set as the velocity inlet, and the outlet boundary condition as the pressure outlet.

4. Simulated calculation of steady-state heat dissipation of battery packs

The heat dissipation model of the battery pack is a coupling model of thermal field and flow field. The thermal field is a model with the internal heat source, and the internal heat source is the heat generation rate of the battery pack. Under most working conditions, the discharge rate of the battery pack is below 2C. Therefore, when calculating with the heat dissipation model, the heat generation rate of the battery pack is calculated at 1 and 2C discharge rates.

The software Fluent is used for the heat dissipation modeling. First, a simplified model of the battery box is established. Then, the geometric model is imported into the software Gambit for mesh generation. Because the structure of the aluminum cooling plate is complicated, the mesh is mainly composed of tetrahedral grids, but hexahedrons, cones and wedge grids can be included in appropriate positions. Finally, the model processed by Gambit is imported into Fluent for calculation.

(1) Battery box without separators

Through the steady-state simulation calculation of the heat dissipation model of the battery pack, the final temperature distribution of the battery pack discharged at 2C rate is obtained (see Fig. 4.31), and the temperature range of the whole lithium-ion battery pack is from 302 to 339 K. The sectional temperature distribution in the height direction of the battery pack is shown in Fig. 4.31b. It can be seen from the figure that the high temperature part of the battery pack is located far away from the air inlet, and the temperature distribution of the battery pack is uneven. The average temperature of each battery cell after steady heat dissipation is shown in Fig. 4.31c. It can be seen from the figure that the battery temperature distribution is uneven. The battery cells are numbered according to the outer and middle rows, each row having 1–30 numbers, starting from the outlet end and ending at the inlet section. Subsequent battery cells are also numbered according to this method. The temperature distribution of the battery

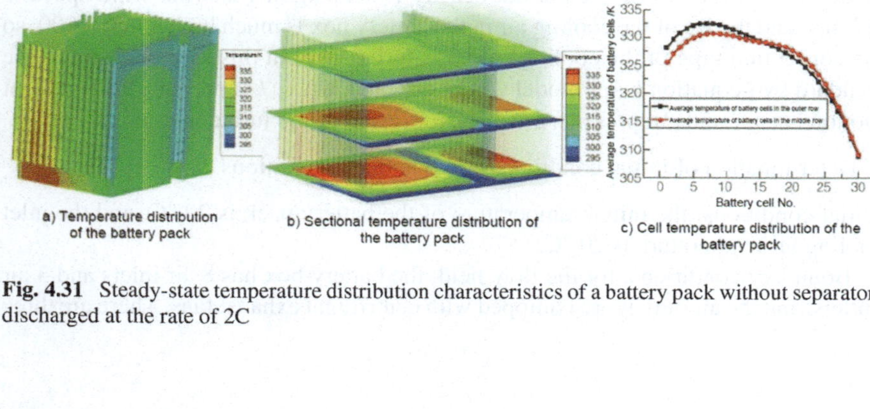

a) Temperature distribution of the battery pack

b) Sectional temperature distribution of the battery pack

c) Cell temperature distribution of the battery pack

Fig. 4.31 Steady-state temperature distribution characteristics of a battery pack without separators discharged at the rate of 2C

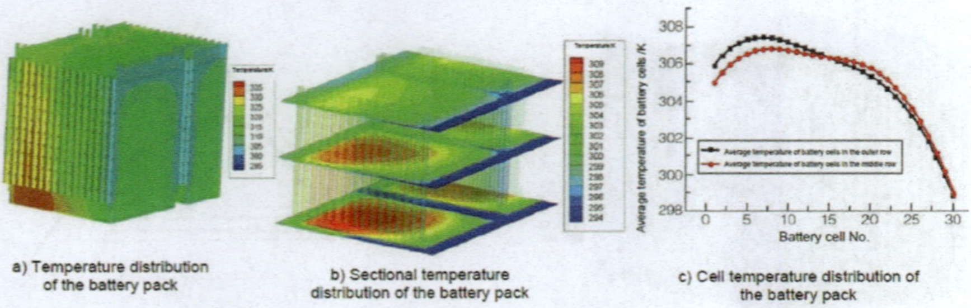

Fig. 4.32 Steady-state temperature distribution characteristics of a battery pack without separators discharged at the rate of 1C

pack discharged at 1C rate is shown in Fig. 4.32. From the steady-state analysis, it can be seen that the structure of the battery box is unreasonable, and long-time operation of the battery may easily cause uneven temperature distribution.

(2) Battery box with separators

Except that two separators with grids are added to the battery box, other conditions of the model are the same, and the calculation results are shown in Figs. 4.33 and 4.34. It can be clearly seen from the figure that the temperature of the battery pack increases significantly and the non-uniformity of temperature distribution increases after separators are added. Therefore, separators increase the structural irrationality of the battery box. Subsequent optimization of the battery box may focus on the locations and sizes of grids the separators, the areas of air inlets and outlets, the power of fans, and the increase of intervals at locations with higher temperature in the simulation model analyzed before.

5. Simulated calculation of unsteady heat dissipation of battery packs

Unsteady simulation will be used to study the cooling effect of the cooling system on the battery pack in limited time. In the unsteady research, it is assumed that

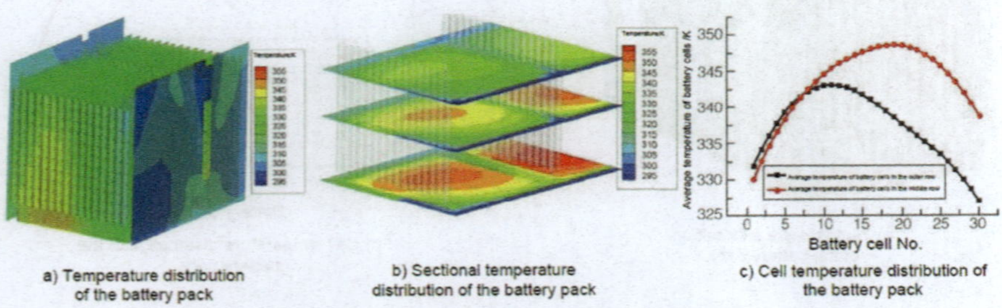

Fig. 4.33 Steady-state temperature distribution characteristics of a battery pack with separators discharged at the rate of 2C

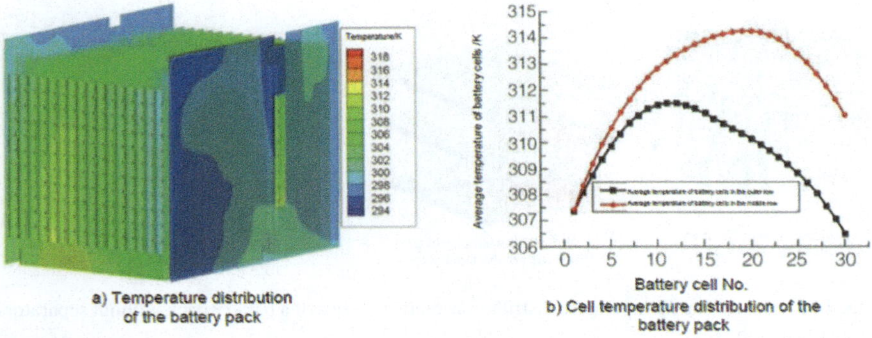

Fig. 4.34 Steady-state temperature distribution characteristics of a battery pack with separators discharged at the rate of 1C

the temperature of the battery pack is the same as the ambient temperature at the beginning of heat dissipation, which is 293 K (20 °C).

(1) Battery box without separators

The battery pack is discharged at a constant current rate of 2C, and the simulation time is 1800 s. The temperature distribution of the battery pack and the average temperature distribution of each monomer are shown in Fig. 4.35. The max. average temperature of the battery cells is 304 K, the minimum average temperature is 299 K, and the temperature difference is 5 K. Figure 4.36 shows the calculation results of the battery pack discharged at 1C rate and the simulation time of 3600 s.

(2) Battery box with separators

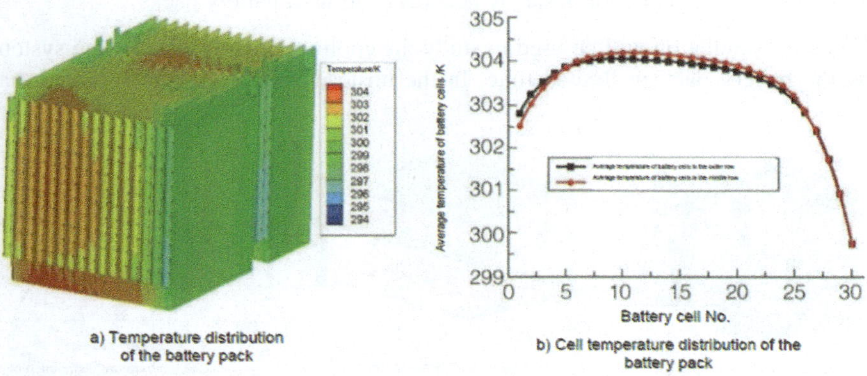

Fig. 4.35 Unsteady temperature distribution characteristics of the battery pack without separators discharged at 2C rate

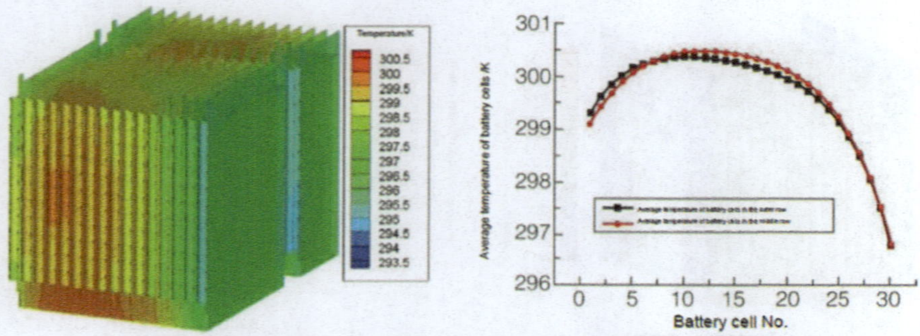

a) Temperature distribution of the battery pack

b) Cell temperature distribution of the battery pack

Fig. 4.36 Unsteady temperature distribution characteristics of the battery pack without separators discharged at 1C rate

Under the same simulation conditions as those without separator, the simulation of 2 and 1C rate discharging of the battery pack is carried out respectively, and the calculation results are shown in Figs. 4.37 and 4.38. Similar to the steady-state situation, the temperature of the middle row of cells is higher than that of the outer cells.

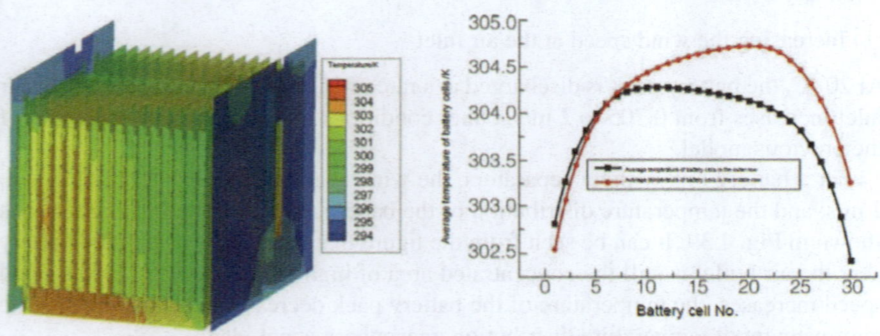

a) Temperature distribution of the battery pack

b) Cell temperature distribution of the battery pack

Fig. 4.37 Unsteady temperature distribution characteristics of the battery pack with separators discharged at 2C rate

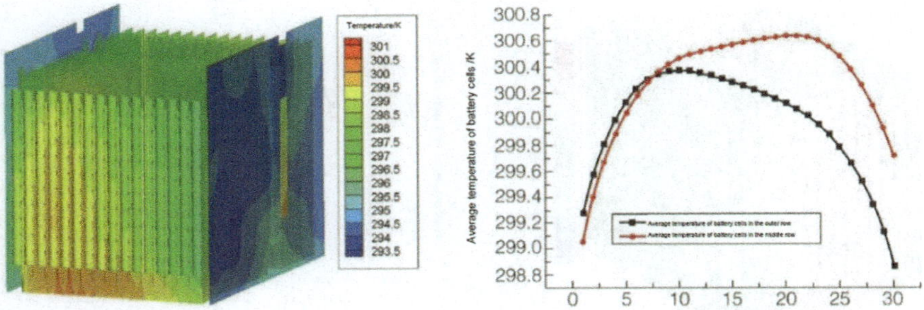

a) Temperature distribution of the battery pack

b) Cell temperature distribution of the battery pack

Fig. 4.38 Unsteady temperature distribution characteristics of the battery pack with separators discharged at 1C rate

6. Optimized design battery boxes

In order to reduce the working temperature of a battery pack and improve the non-uniformity of the temperature distribution of the battery pack, the cooling system of the battery box is optimized by increasing wind speed and the grid area of the separators, which provide a reference for further research on the structural optimization of battery boxes.

(1) Increasing the wind speed at the air inlet

At 20 °C, the battery pack is discharged at a rate of 2C, and the wind speed at the air inlet increases from 0.705 to 2 m/s. Other conditions maintain the same as those of the previous model.

For a battery box without separators, the wind speed at the air inlet increases to 2 m/s, and the temperature distribution of the battery pack discharging at 2C rate is shown in Fig. 4.39. It can be seen from the figure that the middle part of the battery near the air outlet is still the concentrated area of high temperature. After the wind speed increases, the temperature of the battery pack decreases significantly, but the improvement of temperature distribution unevenness is not ideal.

For a battery box with separators, the wind speed at the air inlet increases to 2 m/s, and the temperature distribution of the battery pack discharging at 2C rate is shown in Fig. 4.40, and the average temperatures of the battery pack at two wind speeds are compared. It can be seen from the figure that after the wind speed increases, the temperature of the battery pack decreases significantly, and near the air outlet, although the temperature of the two rows of batteries decreases, the effect is not satisfactory.

(2) Increasing the grid area of the separator

There are two grids with an area of 23 × 147.5 mm^2 in the middle of the battery separator. Now, the area of each grid is increased to 233 × 227.5 mm^2, which is more than 15 times of the original area. After the grid area is increased, the battery

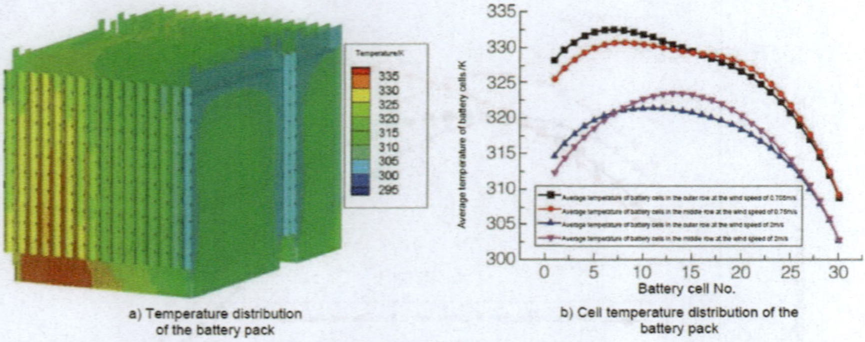

Fig. 4.39 Steady-state temperature distribution characteristics of the battery pack without separators at the wind speed of 2 m/s and discharged at 2C rate

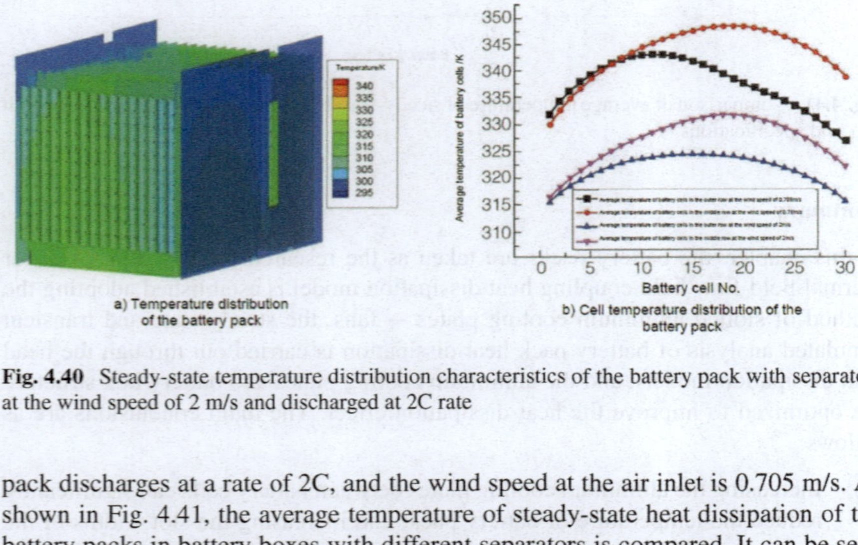

Fig. 4.40 Steady-state temperature distribution characteristics of the battery pack with separators at the wind speed of 2 m/s and discharged at 2C rate

pack discharges at a rate of 2C, and the wind speed at the air inlet is 0.705 m/s. As shown in Fig. 4.41, the average temperature of steady-state heat dissipation of the battery packs in battery boxes with different separators is compared. It can be seen from the figure that the closer the grid area is to the air inlet, the greater the drop of the average temperature of the battery after the grid area has increased. However, near the air outlet, the average temperature of battery cells does not drop, on the contrary, it increases slightly. Although the temperature difference between the two rows of battery cells has been greatly reduced, the overall temperature unevenness of the battery pack has increased.

Through the above simulated analysis, it can be seen that simply increasing the wind speed at the air inlet or increasing the grid area cannot solve the heat dissipation problem of the battery box, and the battery box should be further optimized by adjusting the air inlet locations, grid locations and sizes, battery arrangement and so on.

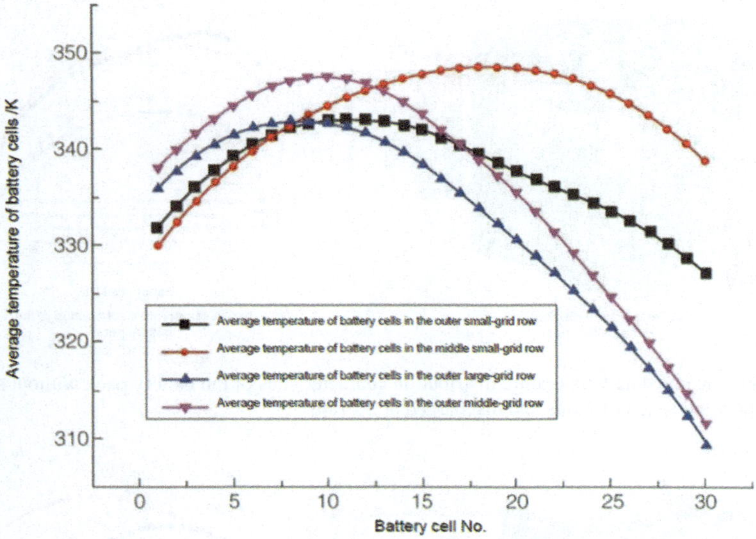

Fig. 4.41 Comparison of average temperature of steady-state heat dissipation of battery packs with two grid specifications

Summary

In this chapter, the battery packs are taken as the research objects, and a quarter thermal field-flow field coupling heat dissipation model is established adopting the method of slotted aluminum cooling plates + fans, the steady-state and transient simulated analysis of battery pack heat dissipation is carried out through the head heat dissipation models, and the aluminum cooling plates and battery box structure are optimized to improve the heat dissipation effect. The main conclusions are as follows:

(1) Increasing the aluminum cooling plates between battery cells can significantly reduce the temperature of battery pack, and increasing the slot widths of the aluminum cooling plates can reduce the temperature of some batteries, but only increasing the slot widths of aluminum cooling plates cannot effectively reduce the temperature difference of the battery pack.

(2) Merely adjusting the air inlet position can change the temperature distribution of the battery pack, but can not reduce the temperature difference of the battery pack; In case that other conditions are unchanged, adjusting the air inlet position and increasing the air inlet area can reduce the temperature difference of the battery pack, but the air outlet area has not changed correspondingly, so the effect is limited. At the same time, the max. temperature of the battery pack and the temperature difference of the battery pack can be effectively reduced by increasing the areas of the air inlet and the air outlet and increasing the wind speed of the cooling fans.

(3) Reducing the height of the battery box and the interval at the top of the battery pack can improve the heat dissipation effect. Therefore, the interval between the battery pack and the top of the battery box should be minimized as far as possible if the internal wiring and structure of the battery box allow.
(4) Through the transient simulated analysis of battery pack heat dissipation, it can be seen that with the increase of temperature difference between cooling air and battery pack, the uniformity of battery pack temperature distribution becomes worse. Therefore, when designing the cooling system of the battery pack, the temperature at which the battery pack starts to dissipate heat should be set to avoid excessive temperature difference between the cooling air and the battery pack.
(5) The air-cooled heat dissipation of a battery box composed of three battery modules is analyzed as an example, a three-dimensional finite element model is established based on geometric dimensions and boundary conditions, and simulated analysis on air cooling heat dissipation is carried out. According to the steady-state analysis results, the structural design of the battery box is unreasonable whether it has separators or not, and the long-time operation of the battery may easily cause uneven temperature distribution. Unsteady simulation further verifies this conclusion. Furthermore, the influence of wind speed and grid area on heat dissipation is analyzed by simulation, and it is found that neither of them can effectively solve the problem of uneven temperature distribution.

References

Ahmad AP, Burch S, Keyser M (1999) An approach for designing thermal management systems for electric and hybrid vehicle battery packs. In: Proceeding of the 4th vehicle thermal management systems conference and exhibition. London, UK, pp 1–16

Kays W (2007) Convective heat and mass transfer. Higher Education Press, Translated by Zhao Z N. Beijing

Kenneth JK, Mark M, Matthew Z (2002) Battery usage and thermal performance of the Toyota Prius and Honda Insight during chassis dynamometer testing. In: The seventeenth annual battery conference on applications and advances. Long Beach, California, pp 247–252

Li JL, Li CX, Hu RX (2009) Proficient in Fluent6.3 flow field analysis. Chemical Industry Press, Beijing

Liu J (2005) Principles of heat and mass transfer and their applications in electric power science and technology. China Electric Power Press, Beijing, pp 46–51

Pesaran AA (2001) Battery thermal management in EVs and HEVs: issues and solutions [EB/OL]. http://www.nrel.gov/vehiclesandfuels/energystorage

Rami S, Kizilel R, Selman JR et al (2008) Active (air-cooled) versus passive (phase change material) thermal management of high power lithium-ion packs: limitation of temperature rise and of temperature distribution. J Power Sources 182:630–638

Tao W (2006) Heat transfer. Northwestern Polytechnical University Press, Xi'an

Chapter 5
Modeling and Optimization of Liquid Cooling Heat Dissipation of Lithium-ion Battery Packs

Compared with air cooling, liquid cooling can achieve better cooling effect because of the use of liquid medium with higher convective heat transfer coefficient. Based on the flow field theory in Chap. 4, a liquid cooling heat dissipation model of battery packs is established, and the simulation research of liquid cooling heat dissipation of battery pack is carried out according to the environmental temperature, battery charge and discharge rate and other factors.

5.1 Liquid Cooling Scheme for Lithium-ion Battery Packs

According to whether the liquid medium is in direct contact with the battery, liquid cooling can be divided into contact type and non-contact type, where the contact cooling liquid directly contacts the battery cells and takes away the heat, while the non-contact cooling liquid flows through the channel to take away the heat conducted from the battery to the channel. The structural schematic diagram is shown in Fig. 5.1.

Figure 5.2 shows four heat dissipation methods: air cooling, fin cooling, non-contact liquid cooling and contact liquid cooling (Chen 2017). It can be seen that these four methods all radiate heat from the largest surface of the battery. Figure 5.2a shows the structure of direct air cooling, in which air flows through the gap between two batteries and directly contacts the side surfaces of the batteries. This method will not bring extra weight to the lithium-ion battery system, and the design of the air circulation system is relatively simple. Figure 5.2b shows the cooling structure of radiating fins. A heat-conducting sheet is sandwiched between two battery cells, and a cooling plate is arranged on the side of the battery. The heat-conducting sheet can conduct the heat on the battery surface to the cooling plate, and finally the heat is taken away by the cooling plate. This method is simple in design, but the weight of the battery system will be greatly increased due to the addition of heat conducting sheets and cooling plates. Figure 5.2c shows the structure of non-contact cooling (fin

J. Li, *Modeling and Simulation of Lithium-ion Power Battery Thermal Management*,
Key Technologies on New Energy Vehicles,
https://doi.org/10.1007/978-981-19-0844-6_5

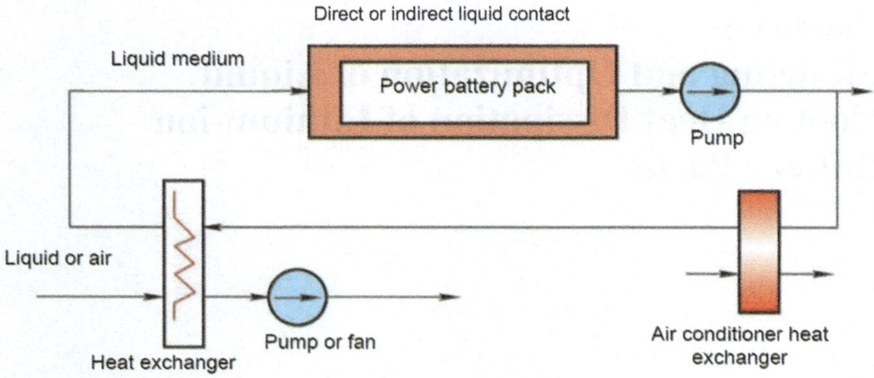

Fig. 5.1 Schematic diagram of a liquid cooling mechanism (He 2020)

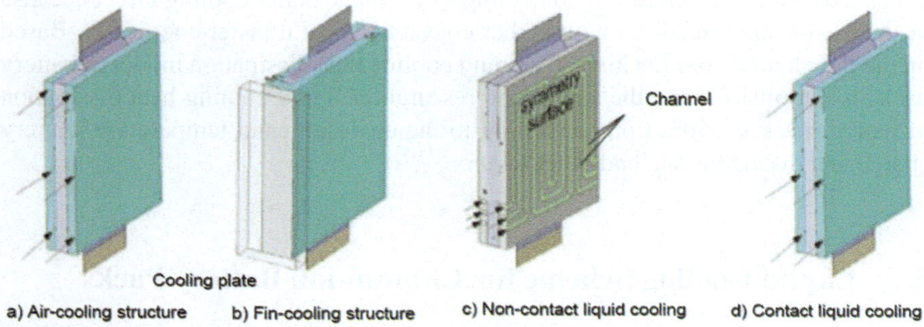

Fig. 5.2 Heat dissipation modes of lithium-ion batteries (Chen 2017)

cooling). During charge and discharge, the heat generated is conducted to the fins through the largest side surface, and then the heat is conducted from the fins to the liquid cooling channel, and the cooling liquid takes the heat out of the battery system. However, the production of such fins is complicated and costly. Figure 5.2d shows the contact liquid cooling structure, which is similar to the way shown in Fig. 5.2a, in which the fluid directly flows between the battery cells to take away the heat. In general, the coolant is electrolyte mineral oil, and because the liquid medium usually has high viscosity, the liquid flow rate is low and the heat exchange effect is limited. This kind of design is too difficult, which needs to consider the leakage of coolant, and will also increase the weight of the battery system.

5.2 Finite Element Simulated Modeling of Liquid Cooling Heat Dissipation of Lithium-ion Batteries

In order to better analyze the heat dissipation of battery packs, this section establishes the thermal model of battery modules with liquid cooling by using the flow field theory. Where, the flow field theory is based on the mass conservation equation, momentum conservation equation and energy conservation equation introduced in Chap. 4, and the turbulence standard two-equation k-ε model is adopted for convection type.

5.2.1 Geometric Model

For a battery pack with the structure of 4P33S, the liquid-cooled flow channel is arranged on the side of the battery pack, and nearly 1/8 of the modules are selected in the model, that is, 17 cells are used to build a battery module with liquid cooling. The geometric model is shown in Fig. 5.3. The blue part in Fig. 5.3 is the flow channel established by the model. The flow channel is located in the cooling plate, and its two ends are connected with the water inlet and water outlet respectively. Heat conducting fins are designed between the battery cells, and the excess heat of the battery is conducted by the heat conducting fins first, and finally brought out of the battery system through the convection heat exchange between the cooling plates and the cooling liquid. For convenience of description, the battery cells are named as Cells 1–17 from left to right.

Fig. 5.3 Geometric model with liquid cooling modules (unit: m)

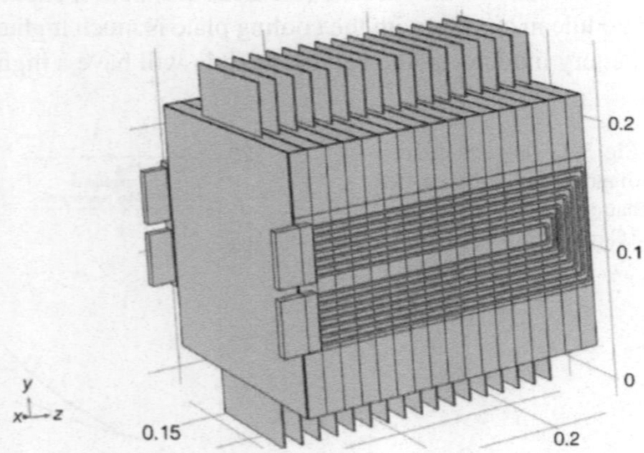

5.2.2 Model Settings

When establishing a finite element model of liquid cooling, two physical fields, heat transfer and turbulence, should be considered and coupled in the liquid cooling module. And the physical field of solid heat transfer is used to simulate conduction heat transfer, convection heat transfer and radiation heat transfer. In addition to the solid model, a fluid model is added to the interface, and the dependent variable is temperature T. Turbulence field is used to calculate the velocity field and pressure field of single-phase fluid flow in laminar flow, and the dependent variables are velocity field u and pressure p. When the fluid temperature changes, its material properties (such as density and viscosity) will change accordingly.

The model cell is wrapped by a 1 mm thick heat conducting plate, and the channel is located in the middle of a 10 mm thick aluminum plate. The flow rate of a single battery box is set at 12L/min, because 1/8 box is selected in the simulation, the calculated flow rate at the entrance and exit is 0.05 m/s.

5.2.3 Simulated Analysis

The simulation results of liquid cooling modules under two working conditions are shown in Figs. 5.4, 5.5, 5.6, 5.7, 5.8 and 5.9. As shown in Fig. 5.4, during 1C charging, the temperature of the lithium-ion battery pack increases from 20 to 24.5 °C. As shown in Fig. 5.6, the surface temperature difference of the lithium-ion battery pack is high, and the temperature difference is close to 5 °C. This is because there is basically no temperature rise when the cooling plate flows through its body, which leads to the fact that the heat dissipation capacity of the part of the battery module in contact with the cooling plate is much higher than that of the center of the battery module, so the battery module will have a high temperature difference. The

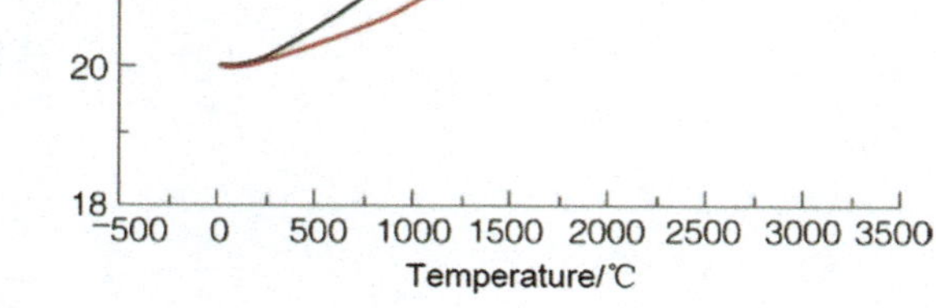

Fig. 5.4 Temperature rise diagram of a liquid-cooled battery module charged at 1C rate

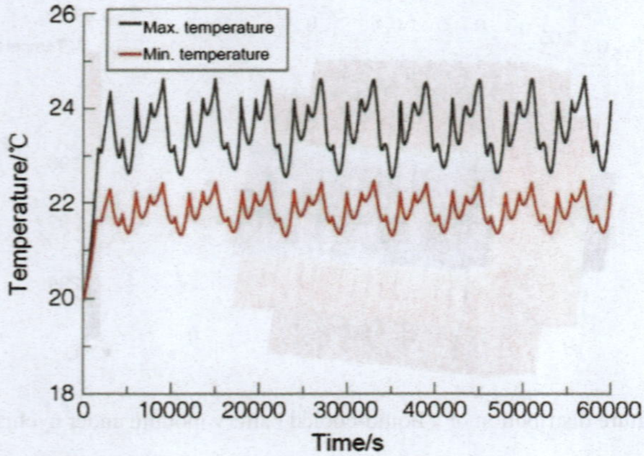

Fig. 5.5 Temperature rise diagram of a liquid-cooled battery module under cyclic condition

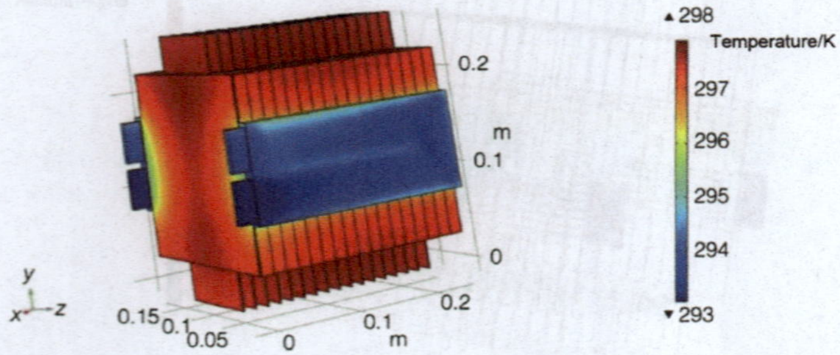

Fig. 5.6 Temperature distribution of liquid-cooled battery module during 1C rate charging

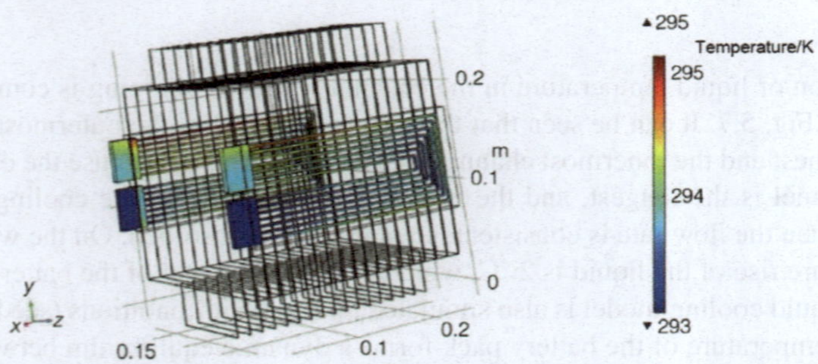

Fig. 5.7 Flow channel temperature distribution of liquid-cooled battery module during 1C rate charging

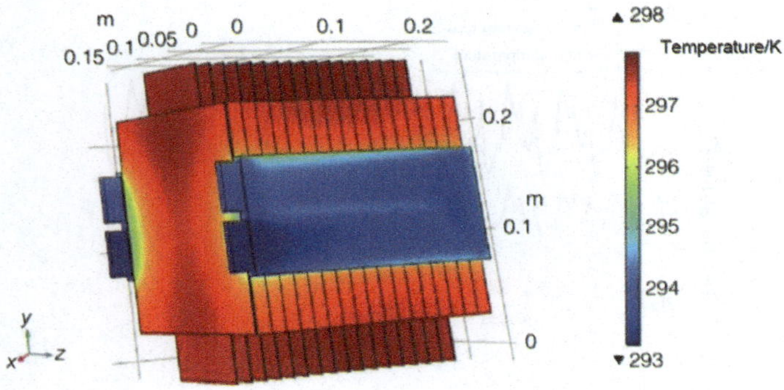

Fig. 5.8 Temperature distribution of a liquid-cooled battery module under cyclic condition

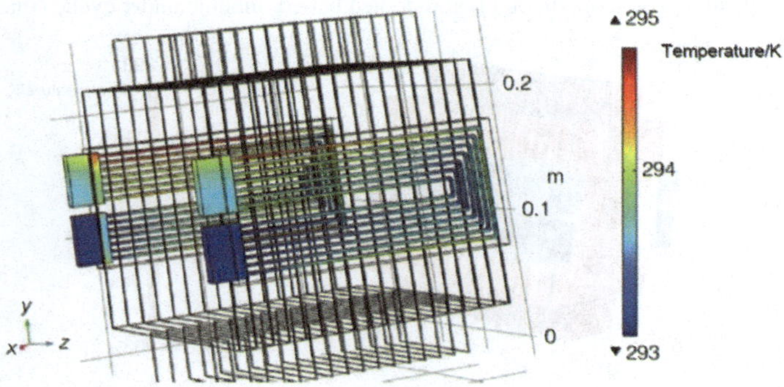

Fig. 5.9 Flow channel temperature distribution of a liquid-cooled battery module under cyclic condition

distribution of liquid temperature in the channel when the charging is completed is shown in Fig. 5.7. It can be seen that the temperature rise of the outermost channel is the highest and the innermost channel is the lowest. This is because the outermost flow channel is the longest, and the heat exchange time with the cooling plate is longer when the flow rate is consistent, so more heat is absorbed. On the whole, the temperature rise of the liquid is 2 °C, which is lower than that of the battery.

The liquid cooling model is also simulated under cyclic conditions (see Fig. 5.5), and the temperature of the battery pack forms a dynamic equilibrium between 21.5 and 24.5 °C under cyclic conditions. Therefore, the distribution of the battery temperature and the liquid cooling channel at the highest temperature is almost the same as that at 1C rate charging, and the highest temperature of liquid in channel chamber is 3 °C lower than that of the battery.

5.3 Simulated Analysis of a Liquid Cooling Scheme for Lithium-ion Battery Packs

Based on the liquid cooling heat dissipation model of battery packs established in Sect. 5.2, this section conducts simulated analysis from the aspects of ambient temperature, battery charge and discharge rate, coolant flow rate and coolant type. It can be seen from the simulation cycle conditions in Sect. 5.2.3 that the temperature of the battery pack will eventually form a dynamic equilibrium under the cycle charge and discharge conditions, so considering the calculation cost of simulation, the simulation conditions are 4 charge and discharge cycles for all cases.

5.3.1 Influence of Temperature on Liquid Cooling Heat Dissipation of Battery Packs

In order to explore the influence of ambient temperature on the liquid cooling effect of the battery pack, from the perspective of ambient temperature change, this section simulates and analyzes the ambient temperature of the battery pack under the conditions that the battery pack is charged and discharged at 1C rate, the inlet velocity of coolant is 0.05 m/s, and the coolant medium is water. In the simulation, the initial temperature of the coolant is set at 20 °C, and the ambient temperature is set at 20 °C, 30 °C, 40 °C and 50 °C, so the initial temperature of the battery pack is consistent with the ambient temperature. In this section, the simulation results at various ambient temperatures are analyzed first, and finally, the temperature rise and max. temperature difference of the battery pack at different ambient temperatures are comprehensively analyzed.

1. Analysis on simulation results of the battery pack at 20 °C

The simulation results of the battery pack at 20 °C are shown in Fig. 5.10. With the charge and discharge cycle of the battery pack going, the temperature of the battery pack gradually rises and reaches dynamic equilibrium. At this time, the max. temperature of the battery pack reaches 24.97 °C, the minimum temperature is 20.88 °C, and the average temperature of the battery pack is between 22 and 24.35 °C. The max. temperature rise of the battery pack is 4.35 °C, and the max. internal temperature difference is 3.07 °C. In combination with Fig. 5.10, moments when internal temperature difference of the battery pack is high in dynamic equilibrium are selected to draw the temperature distribution cloud atlas of battery pack and the channel, as shown in Fig. 5.11. It can be seen from the figure that the temperature distribution of the battery pack is relatively uniform except the parts in contact with the channel. The outlet temperature of the upper part of the cooling channel is obviously higher than the inlet temperature of the lower part. It can be seen that the cooling liquid takes away the heat of the battery pack, but it also causes the uneven temperature distribution of the battery pack.

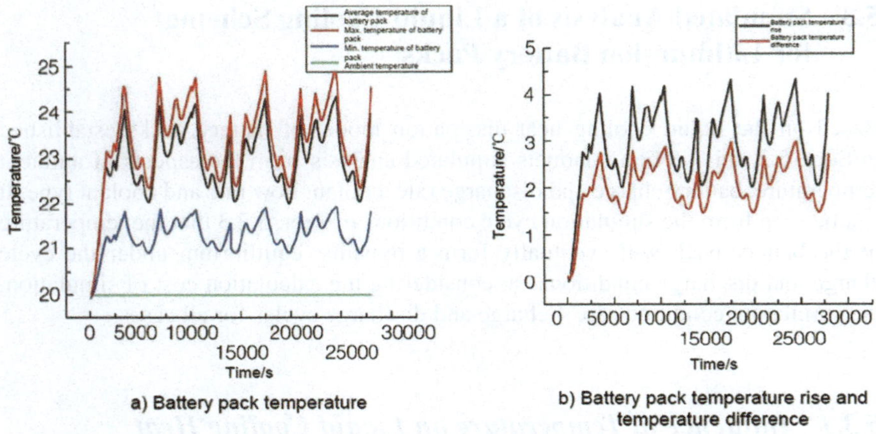

Fig. 5.10 Temperature curve of the battery pack at 20 °C

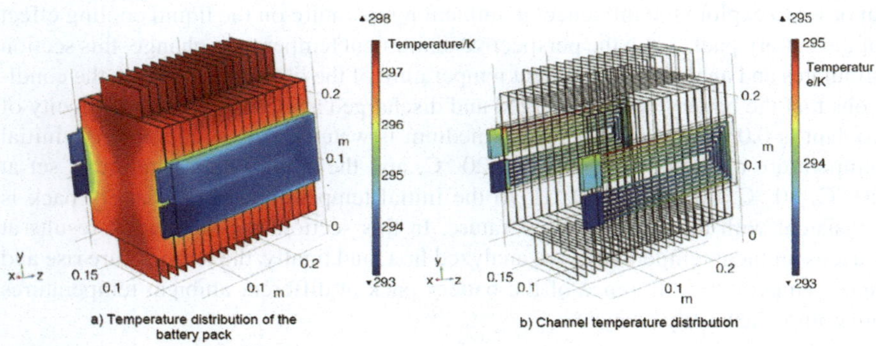

Fig. 5.11 Temperature distribution cloud atlas at 20 °C for 17500 s under simulated working condition

2. Analysis on simulation results of the battery pack at 30 °C

The simulation results of the battery pack at 30 °C are shown in Fig. 5.12, and the temperature distribution at 17500 s is shown in Fig. 5.13. With the charge and discharge cycle of the battery pack going, the temperature of the battery pack gradually rises and reaches dynamic equilibrium. At this time, the max. temperature of the battery pack reaches 26.75 °C, the minimum temperature is 21.64 °C, and the average temperature of the battery pack is between 23.38 and 25.86 °C. The max. temperature rise of the battery pack is −4.14 °C (temperature decline), and the max. internal temperature difference is 3.98 °C. At this time, due to the existence of the liquid cooling system, the overall temperature of the battery pack is below the

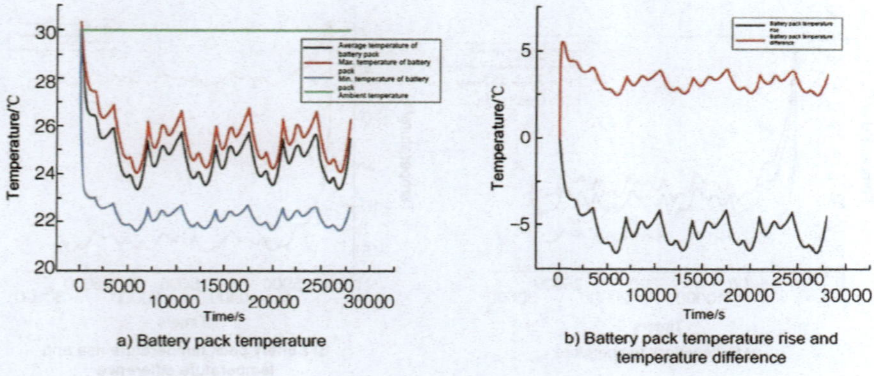

Fig. 5.12 Temperature curve of the battery pack at 30 °C

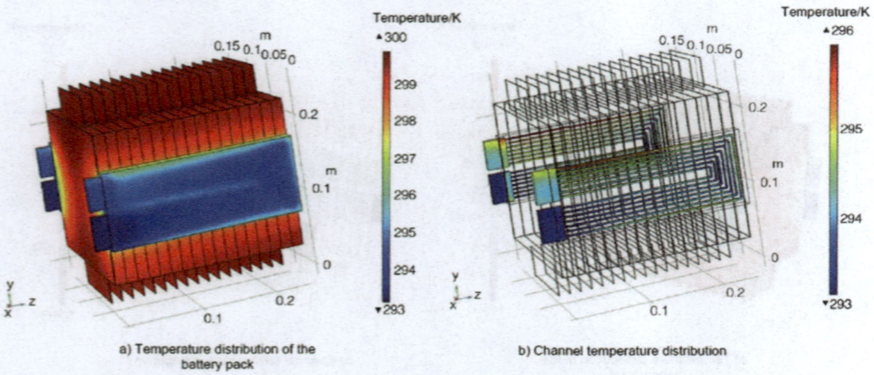

Fig. 5.13 Temperature distribution at 30 °C for 17500 s under simulated working condition

ambient temperature. Compared with the ambient temperature of 20 °C, the temperature difference inside the battery pack increases and the uniformity of temperature distribution inside the battery pack becomes worse.

3. Analysis on simulation results of the battery pack at 40 °C

The simulation results of the battery pack at 40 °C are shown in Fig. 5.14, and the temperature distribution at 17500 s is shown in Fig. 5.15. With the charging cycle of the battery pack going, the temperature of the battery pack gradually rises and reaches dynamic equilibrium. At this time, the max. temperature of the battery pack reaches 28.65 °C, the minimum temperature is 22.37 °C, and the average temperature of the battery pack is between 24.8 and 27.25 °C. The max. temperature rise of the battery pack is −12.75 °C (temperature decline), and the max. internal temperature difference is 5.12 °C. At this time, due to the existence of the liquid cooling system, the overall temperature of the battery pack is below the ambient temperature. Compared with the ambient temperature of 30 °C, the temperature difference inside the battery

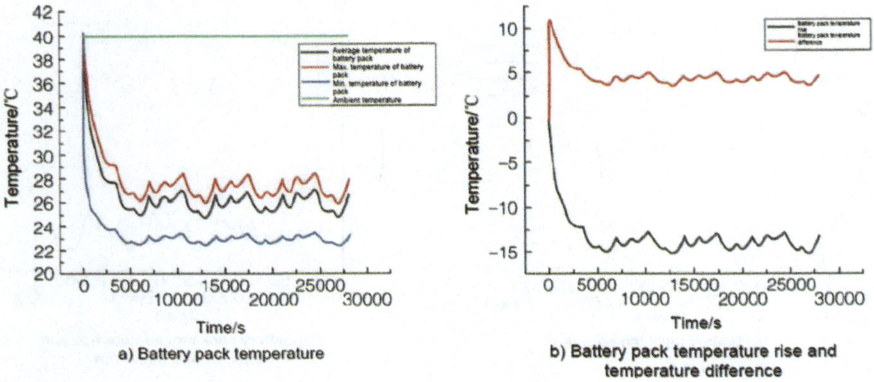

Fig. 5.14 Temperature curve of battery pack at 40 °C

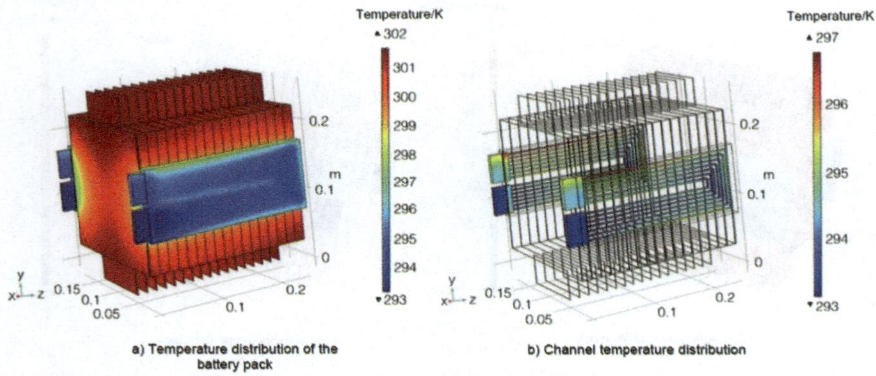

Fig. 5.15 Temperature distribution at 40 °C for 17500 s under simulated working condition

pack increases and the uniformity of temperature distribution inside the battery pack becomes further worse.

4. Analysis on simulation results of the battery pack at 50 °C

The simulation results of the battery pack at 50 °C are shown in Fig. 5.16, and the temperature distribution at 17500 s is shown in Fig. 5.17. With the charge and discharge cycle of the battery pack going, the temperature of the battery pack gradually rises and reaches dynamic equilibrium. At this time, the max. temperature of the battery pack reaches 30.58 °C, the minimum temperature is 23.04 °C, and the average temperature of the battery pack is between 26.22 and 28.69 °C. The max. temperature rise of the battery pack is −21.3 °C (temperature decline), and the max. internal temperature difference is 6.3 °C. At this time, due to the existence of the liquid cooling system, the overall temperature of the battery pack is below the

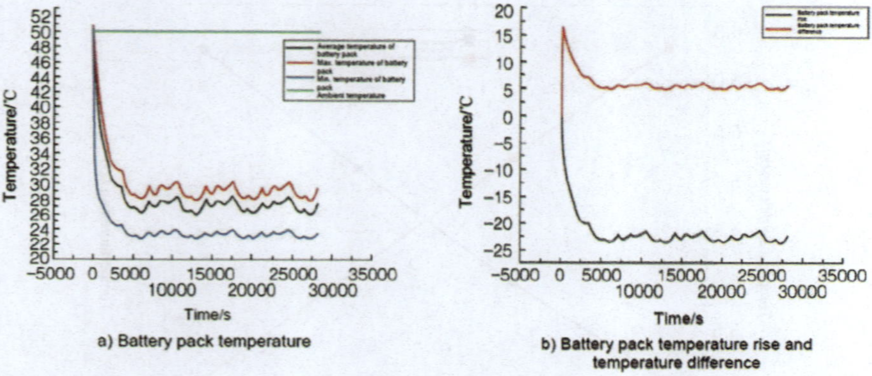

Fig. 5.16 Temperature curve of the battery pack at 50 °C

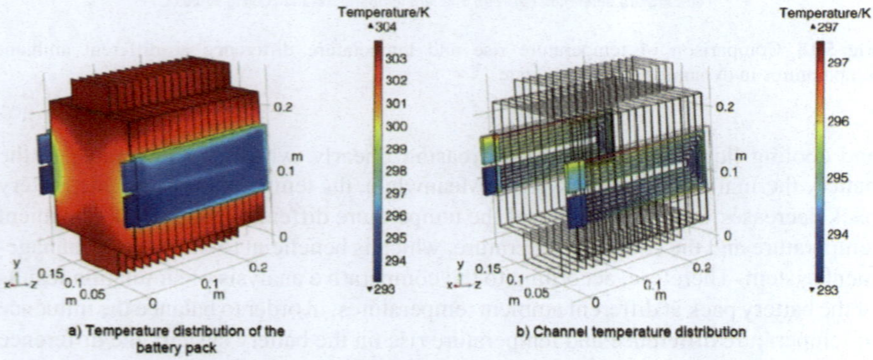

Fig. 5.17 Temperature distribution at 50 °C for 17500 s under simulated working condition

ambient temperature. Compared with the ambient temperature of 40 °C, the temperature difference inside the battery pack increases and the uniformity of temperature distribution inside the battery pack becomes worse.

5. Comparative analysis on simulation results of the battery pack at different ambient temperatures

In dynamic equilibrium state, the comparison of temperature difference and temperature rise of the battery pack at different ambient temperatures is shown in Fig. 5.18. The greater the temperature difference in the battery pack, the greater the impact on the battery consistency, which will lead to the performance degradation of the battery pack. In a reasonable range of the operating temperature of the battery pack, the thermal management system hopes to dissipate the excess heat as much as possible. Figure 5.18 shows that with the increase of the ambient temperature, under the condition of liquid cooling and heat dissipation, the temperature difference in the battery pack increases with the temperature difference between ambient temperature

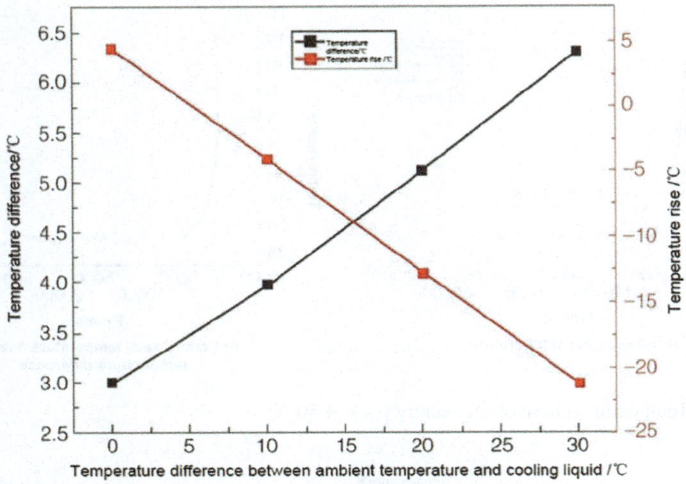

Fig. 5.18 Comparison of temperature rise and temperature difference at different ambient temperatures in dynamic equilibrium state

and cooling liquid, and basically increasing linearly, which is unfavorable to the battery thermal management system. Meanwhile, the temperature rise of the battery pack decreases with the increase of the temperature difference between the ambient temperature and the coolant temperature, which is beneficial to the thermal management system. Therefore, according to the comparative analysis of simulation results of the battery pack at different ambient temperatures, in order to balance the influence of temperature difference and temperature rise on the battery system, the difference between the ambient temperature and the coolant temperature should not be too low or too high, and it is necessary to balance the influence of temperature difference and temperature rise inside the battery pack according to the actual situation.

5.3.2 *Influence of Charge–Discharge Ratio on Liquid Cooling Heat Dissipation of the Battery Pack*

In order to explore the influence of battery charge/discharge ratio on the liquid cooling effect of a battery pack, this section simulates and analyzes the ambient temperature at 20 °C and 50 °C respectively from the angle of changing the battery charge/discharge ratio. In the simulation, the inlet flow rate of the coolant is 0.05 m/s, the coolant medium is water, the initial temperature of the coolant is set at 20 °C, the initial temperature of the battery pack is consistent with the ambient temperature, and the battery charge and discharge rate is considered to be 0.5C, 1C, 1.5C and 2C, among which the working condition of 1C has been analyzed in 5.3.1. In this section, the simulation results at various rates are analyzed first, and finally, the temperature rise and max. temperature difference of the battery pack at different rates are comprehensively analyzed.

1. Simulation Result Analysis on the Battery Pack at Different Charge and Discharge Rates at the Ambient Temperature of 20 °C

(1) Simulated condition of the battery pack at 0.5C charging/discharging rate

The simulation results of the battery pack at 0.5C rate are shown in Fig. 5.19, and the temperature distribution at 42500 s is shown in Fig. 5.20. With the charging cycle of the battery pack going, the temperature of the battery pack gradually rises and reaches dynamic equilibrium. At this time, the max. temperature of the battery pack reaches 23.5 °C, the minimum temperature is 20.61 °C, and the average temperature

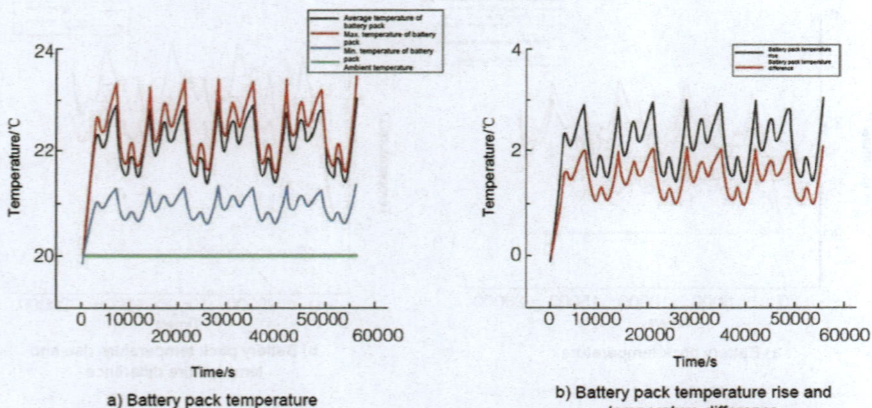

Fig. 5.19 Temperature curve of the battery pack at 0.5c rate

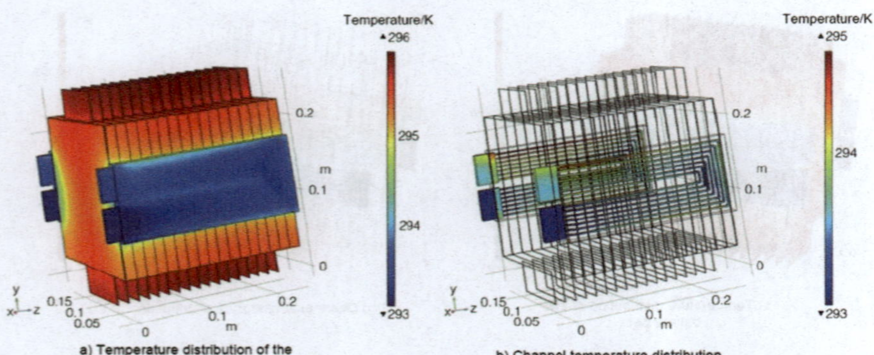

Fig. 5.20 Temperature distribution at 42500 s of the simulated working condition at 0.5c rate

of the battery pack is between 21.38 and 23.7 °C. The max. temperature rise of the battery pack is 3.07 °C, and the max. internal temperature difference is 2.12 °C.

(2) Simulated condition of the battery pack at 1.5C charging/discharging rate

The simulation results of the battery pack at 1.5C rate are shown in Fig. 5.21, and the temperature distribution at 12500 s is shown in Fig. 5.22. With the charging cycle of the battery pack going, the temperature of the battery pack gradually rises and reaches dynamic equilibrium. At this time, the max. temperature of the battery pack reaches 27.17 °C, the minimum temperature is 21.56 °C, and the average temperature of the battery pack is between 23.51 and 26.29 °C. The max. temperature rise of the battery pack is 6.29 °C, and the max. internal temperature difference is 4.33 °C.

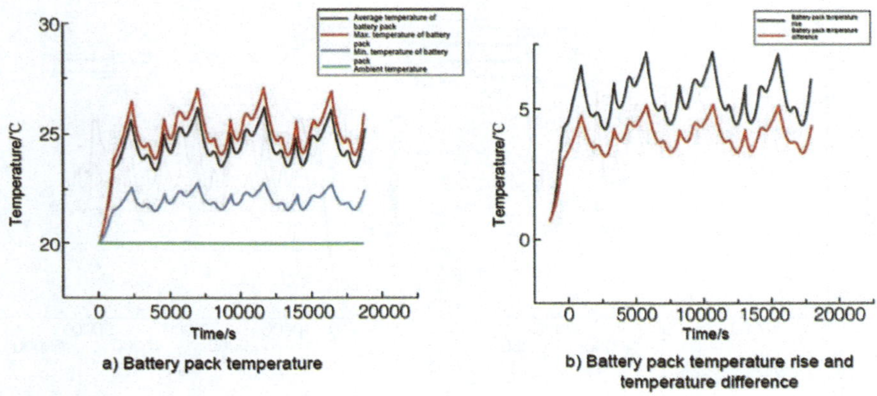

Fig. 5.21 Temperature curve of the battery pack at 1.5c rate working condition

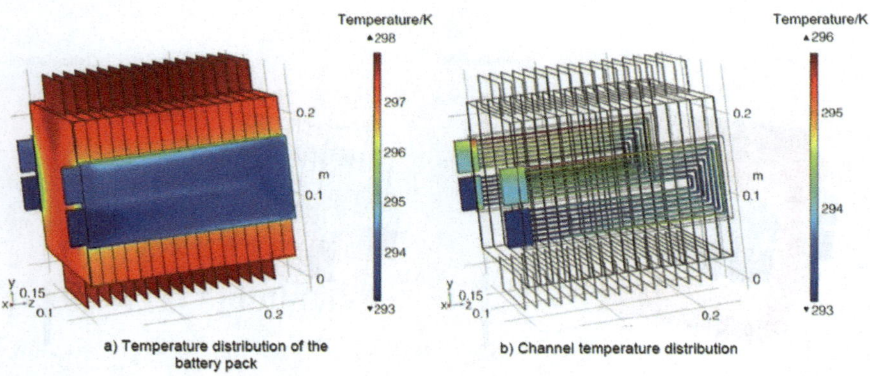

Fig. 5.22 Temperature distribution 12500 s of the simulated working condition at 1.5c rate

(3) Simulated condition of the battery pack at 2C charging/discharging rate

The simulation results of the battery pack at 2C rate are shown in Fig. 5.23, and the temperature distribution at 8750 s of the simulated working condition is shown in Fig. 5.24. With the charging cycle of the battery pack going, the temperature of the battery pack gradually rises and reaches dynamic equilibrium. At this time, the max. temperature of the battery pack reaches 29.9 °C, the minimum temperature is 22.4 °C, and the average temperature of the battery pack is between 25.42 and 28.68 °C. The max. temperature rise of the battery pack is 8.68 °C, and the max. internal temperature difference is 5.98 °C.

(4) Simulation and comparative analysis of battery packs at different rates

See Fig. 5.25 for the internal temperature difference and temperature rise of the battery pack after reaching the dynamic equilibrium at the charge and discharge

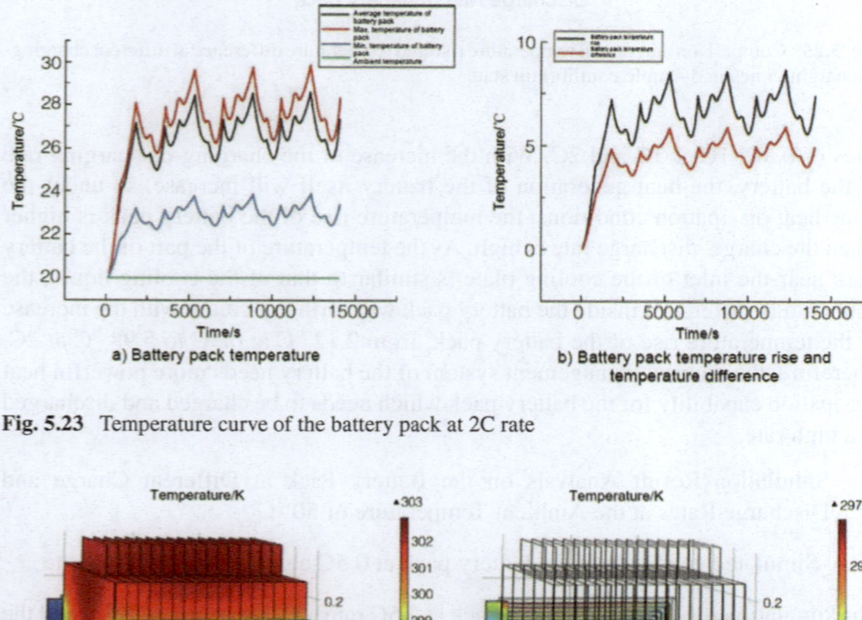

Fig. 5.23 Temperature curve of the battery pack at 2C rate

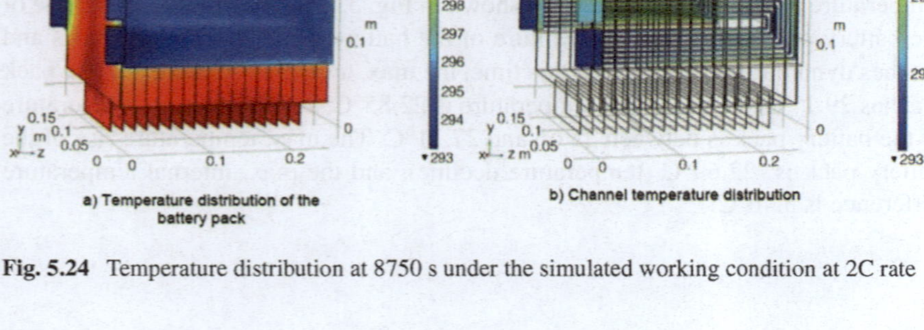

Fig. 5.24 Temperature distribution at 8750 s under the simulated working condition at 2C rate

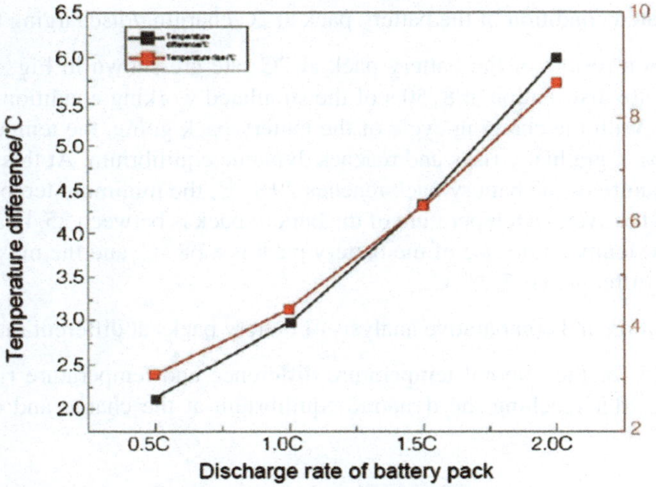

Fig. 5.25 Comparison diagram of temperature rise and temperature difference at different charging-discharging rates in dynamic equilibrium state

rates of 0.5C, 1C, 1.5C and 2C. With the increase of the charging-discharging rate of the battery, the heat generation of the battery itself will increase, so under the same heat dissipation conditions, the temperature rise of the battery pack is higher when the charge–discharge rate is high. As the temperature of the part of the battery pack near the inlet of the cooling plate is similar to that of the cooling liquid, the temperature difference inside the battery pack will further increase with the increase of the temperature rise of the battery pack, from 2.12 °C at 0.5C to 5.98 °C at 2C. Therefore, the thermal management system of the battery needs more powerful heat dissipation capability for the battery pack which needs to be charged and discharged at a high rate.

2. Simulation Result Analysis on the Battery Pack at Different Charge and Discharge Rates at the Ambient Temperature of 50 °C

(1) Simulated condition of the battery pack at 0.5C charging/discharging rate

The simulation results of the battery pack at 0.5C rate are shown in Fig. 5.26, and the temperature distribution at 42500 s is shown in Fig. 5.27. With the charging cycle of the battery pack going, the temperature of the battery pack gradually declines and reaches dynamic equilibrium. At this time, the max. temperature of the battery pack reaches 29.1 °C, the minimum temperature is 22.85°C, and the average temperature of the battery pack is between 25.66 and 27.31°C. The max. temperature rise of the battery pack is -22.69°C (temperature decline), and the max. internal temperature difference is 5.46°C.

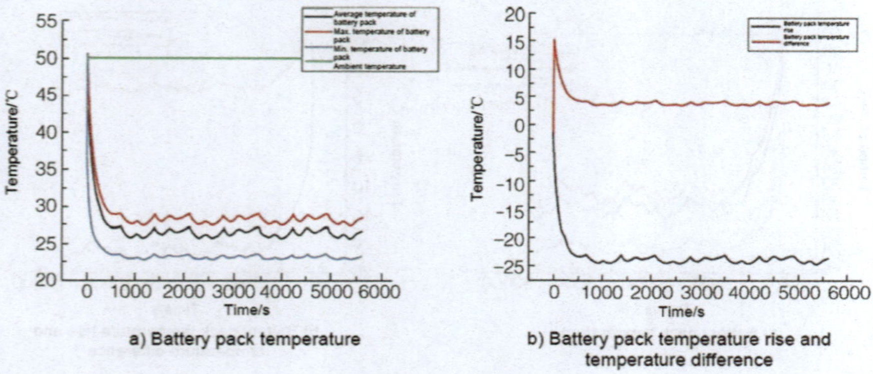

Fig. 5.26 Temperature curve of the battery pack at 0.5C rate

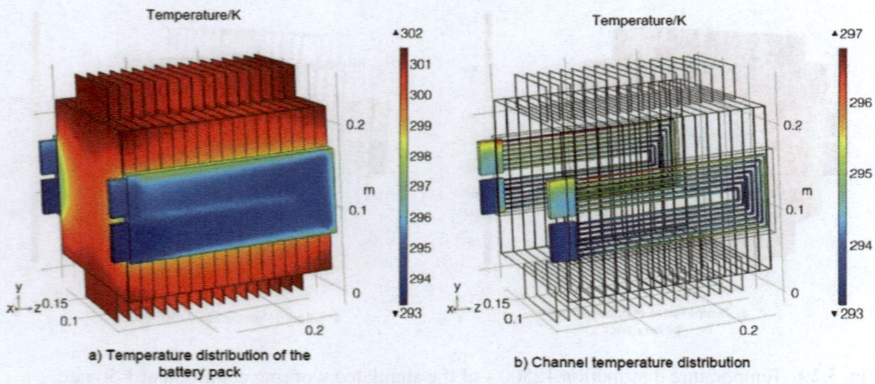

Fig. 5.27 Temperature distribution at 42500 s of the simulated working condition at 0.5c rate

(2) Simulated condition of the battery pack at 1.5C charging/discharging rate

The simulation results of the battery pack at 1.5C rate are shown in Fig. 5.28, and the temperature distribution at 12500 s is shown in Fig. 5.29. With the charging cycle of the battery pack going, the temperature of the battery pack gradually declines and reaches dynamic equilibrium. At this time, the max. temperature of the battery pack reaches 32.61 °C, the minimum temperature is 23.75 °C, and the average temperature of the battery pack is between 27.62 and 30.55 °C. The max. temperature rise of the battery pack is −19.46 °C (temperature decline), and the max. internal temperature difference is 7.46 °C.

(3) Simulated condition of the battery pack at 2C charging/discharging rate

The simulation results of the battery pack at 2C rate are shown in Fig. 5.30, and the temperature distribution at 8750 s is shown in Fig. 5.31. With the charging cycle of the battery pack going, the temperature of the battery pack gradually declines and

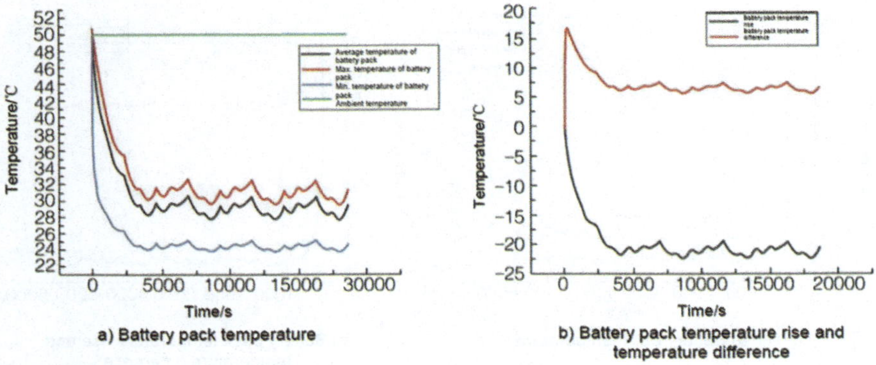

Fig. 5.28 Temperature curve of the battery pack at 0.5c rate

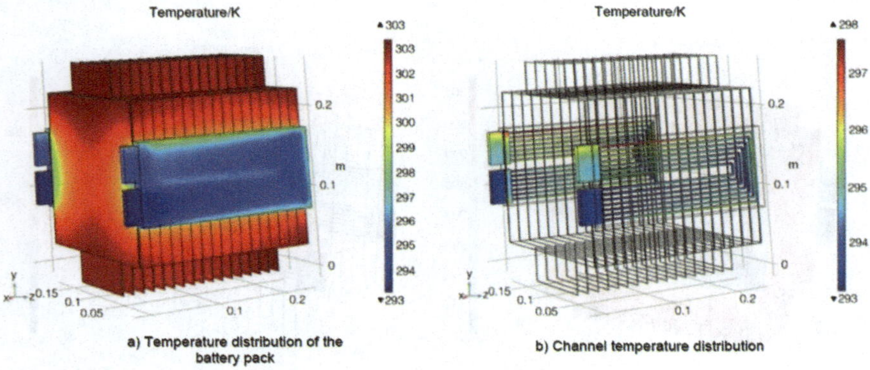

Fig. 5.29 Temperature distribution 12500 s of the simulated working condition at 1.5c rate

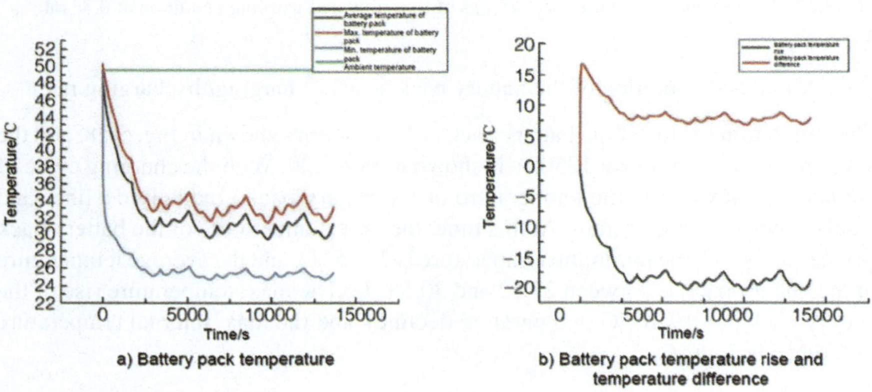

Fig. 5.30 Temperature curve of the battery pack at 2C rate

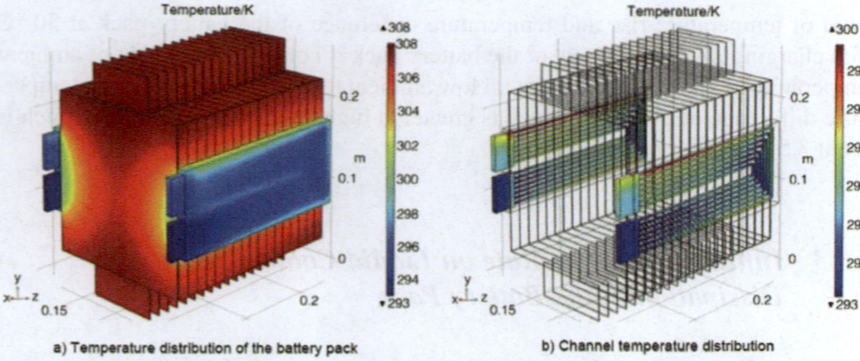

Fig. 5.31 Temperature distribution at 8750 s under the simulated working condition at 2C rate

reaches dynamic equilibrium. At this time, the max. temperature of the battery pack reaches 35.05 °C, the minimum temperature is 24.65 °C, and the average temperature of the battery pack is between 29.56 and 32.81 °C. The max. temperature rise of the battery pack is −17.19 °C (temperature decline), and the max. internal temperature difference is 8.87 °C.

(4) Simulation and comparative analysis of battery packs at different rates.

At an ambient temperature of 50 °C, the internal temperature difference and temperature rise of the battery pack after reaching the dynamic equilibrium at the charge and discharge rates of 0.5C, 1C, 1.5C and 2C is shown in Fig. 5.32. The variation

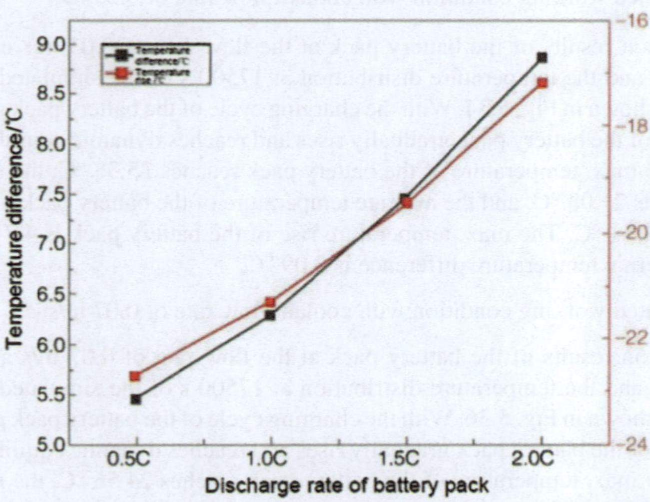

Fig. 5.32 Comparison diagram of temperature rise and temperature difference at different charging-discharging rates in dynamic equilibrium state

trend of temperature rise and temperature difference of the battery pack at 50 °C with charging-discharging rate of the battery pack is consistent with that of ambient temperature at 20 °C. Compared with low ambient temperature, the internal temperature difference of the battery pack is greater at high ambient temperature, which is about 3.5 °C higher at different rates.

5.3.3 *Influence of Flow Rate on Liquid Cooling Heat Dissipation of the Battery Pack*

In order to explore the influence of coolant flow rate on the heat dissipation effect of liquid cooling battery pack, simulations are carried out for ambient temperatures of 20 and 50 °C, respectively, from the perspective of varying the coolant flow rate herein. In the simulation, the charge and discharge rate of the battery is 1C, the coolant medium is water, the initial temperature of the coolant is all set to 20 °C, the initial temperature of the battery pack is consistent with the ambient temperature, and the flow rate of the battery is considered to be 0.03, 0.05 and 0.07 m/s, of which the working condition of 0.05 m/s has been analyzed in Sect. 5.3.1. In this section, the simulation results at various flow rates are analyzed first, and finally, the temperature rise and max. temperature difference of the battery pack at different flow rates are comprehensively analyzed.

1. Analysis on the simulation results of the battery pack at different flow rates at 20 °C

(1) Simulated working condition with coolant flow rate of 0.03 m/s

The simulation results of the battery pack at the flow rate of 0.03 m/s are shown in Fig. 5.33, and the temperature distribution at 17500 s of the simulated working condition is shown in Fig. 5.34. With the charging cycle of the battery pack going, the temperature of the battery pack gradually rises and reaches dynamic equilibrium. At this time, the max. temperature of the battery pack reaches 25.38 °C, the minimum temperature is 21.08 °C, and the average temperature of the battery pack is between 22.27 and 24.74 °C. The max. temperature rise of the battery pack is 4.74 °C, and the max. internal temperature difference is 3.09 °C.

(2) Simulated working condition with coolant flow rate of 0.07 m/s

The simulation results of the battery pack at the flow rate of 0.07 m/s are shown in Fig. 5.35, and the temperature distribution at 17500 s of the simulated working condition is shown in Fig. 5.36. With the charging cycle of the battery pack going, the temperature of the battery pack gradually rises and reaches dynamic equilibrium. At this time, the max. temperature of the battery pack reaches 24.56 °C, the minimum temperature is 20.80 °C, and the average temperature of the battery pack is between 21.89 and 24.24 °C. The max. temperature rise of the battery pack is 4.24 °C, and the max. internal temperature difference is 3.04 °C.

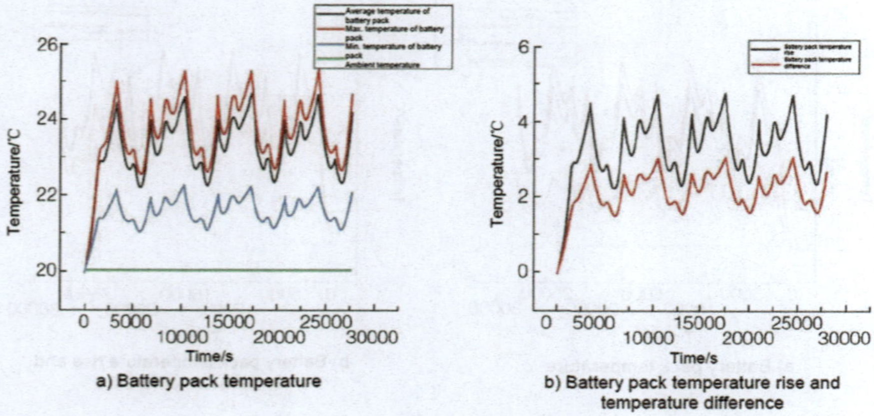

Fig. 5.33 Temperature curve of the battery pack at the coolant flow rate of 0.03 m/s

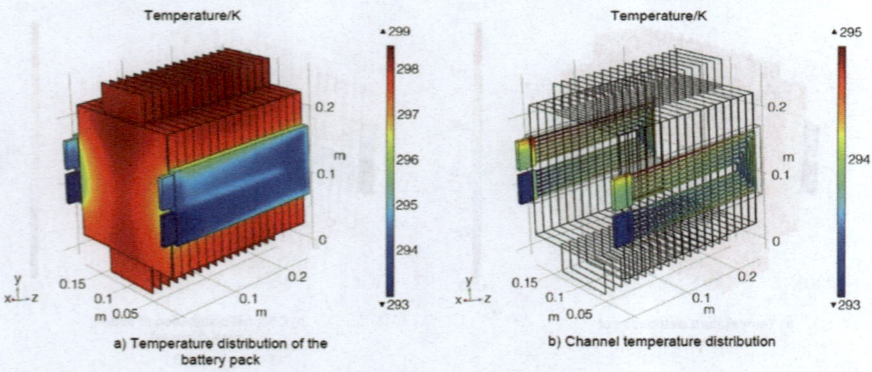

Fig. 5.34 Temperature distribution at 17500 s of the simulated working condition at the coolant flow rate of 0.03 m/s

(3) Simulation and comparative analysis of the battery pack at different flow rates

When the cooling fluid flow rate of the battery pack is 0.03 m/s, 0.05 m/s and 0.07 m/s, the internal temperature difference and temperature rise of the battery pack after reaching the dynamic equilibrium are shown in Fig. 5.37, and the corresponding flow rates of the battery box are 7.2L/min, 12L/min and 16.8L/min respectively. According to the simulation results in the figure, in the current state, the flow rate has little influence on the liquid cooling system. When the flow rate increases from 0.03 to 0.07 m/s, the temperature rise of the battery drops by about 0.6 °C, and the temperature difference is basically close.

2. Analysis on the simulation results of the battery pack at different flow rates at 50 °C

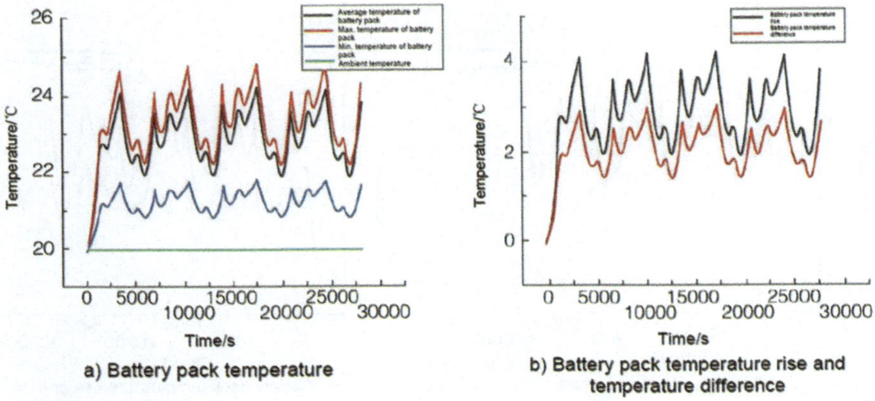

Fig. 5.35 Temperature curve of the battery pack at the coolant flow rate of 0.07 m/s

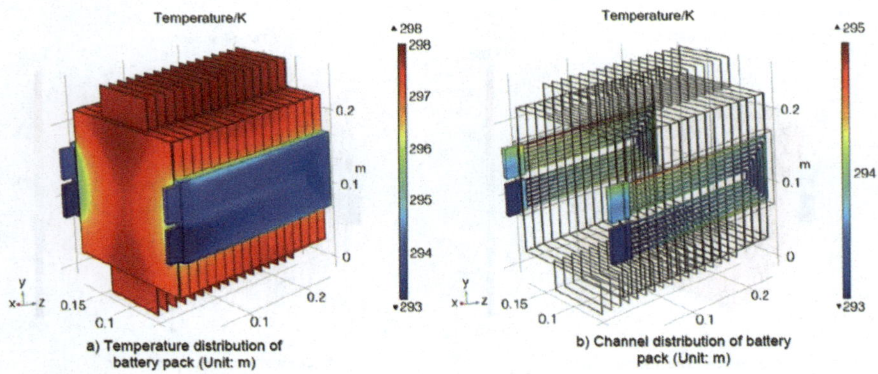

Fig. 5.36 Temperature distribution at 17500 s of the simulated working condition at the coolant flow rate of 0.07 m/s

(1) Simulated working condition with coolant flow rate of 0.03 m/s

The simulation results of the battery pack at the flow rate of 0.03 m/s are shown in Fig. 5.38, and the temperature distribution at 17500 s of the simulated working condition is shown in Fig. 5.39. With the charging cycle of the battery pack going, the temperature of the battery pack gradually declines and reaches dynamic equilibrium. At this time, the max. temperature of the battery pack reaches 31.35 °C, the minimum temperature is 23.67 °C, and the average temperature of the battery pack is between 26.96 and 29.44 °C. The max. temperature rise of the battery pack is −20.56 °C (temperature decline), and the max. internal temperature difference is 6.43 °C.

(2) Simulated working condition with coolant flow rate of 0.07 m/s

The simulation results of the battery pack at the flow rate of 0.07 m/s are shown in Fig. 5.40, and the temperature distribution at 17500 s of the simulated working

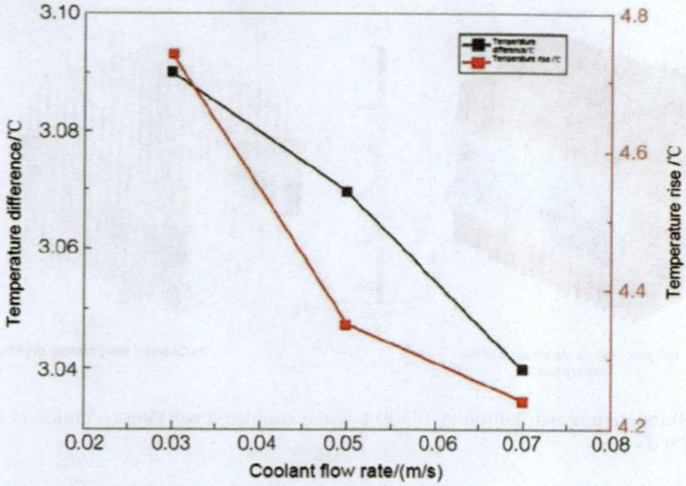

Fig. 5.37 Comparison of temperature rise and temperature difference in dynamic equilibrium state and at different coolant flow rates

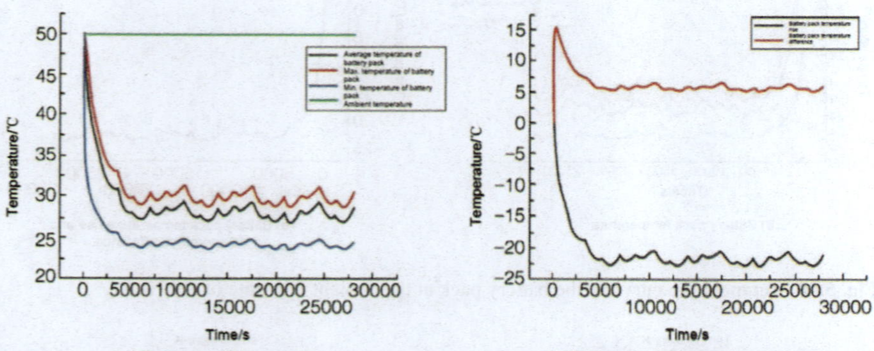

a) Battery pack temperature

b) Battery pack temperature rise and temperature difference

Fig. 5.38 Temperature curve of the battery pack at the coolant flow rate of 0.03 m/s

condition is shown in Fig. 5.41. With the charging cycle of the battery pack going, the temperature of the battery pack gradually declines and reaches dynamic equilibrium. At this time, the max. temperature of the battery pack reaches 30.19 °C, the minimum temperature is 22.88 °C, and the average temperature of the battery pack is between 25.98 and 28.29 °C. The max. temperature rise of the battery pack is −21.71 °C (temperature decline), and the max. internal temperature difference is 6.26 °C.

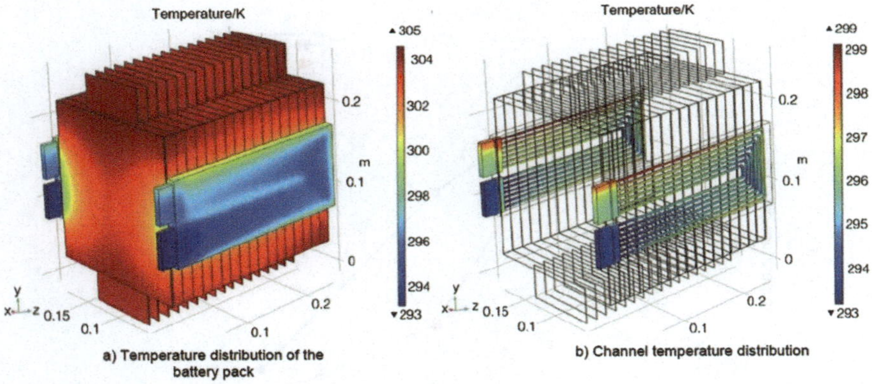

Fig. 5.39 Temperature distribution at 17500 s of the simulated working condition at the coolant flow rate of 0.03 m/s

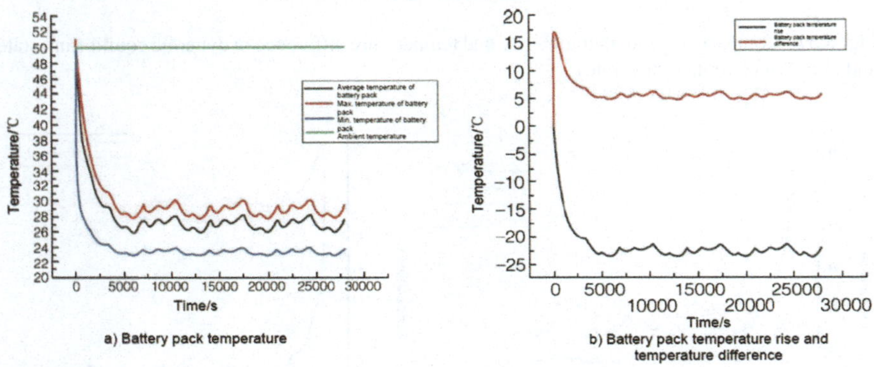

Fig. 5.40 Temperature curve of the battery pack at the coolant flow rate of 0.07 m/s

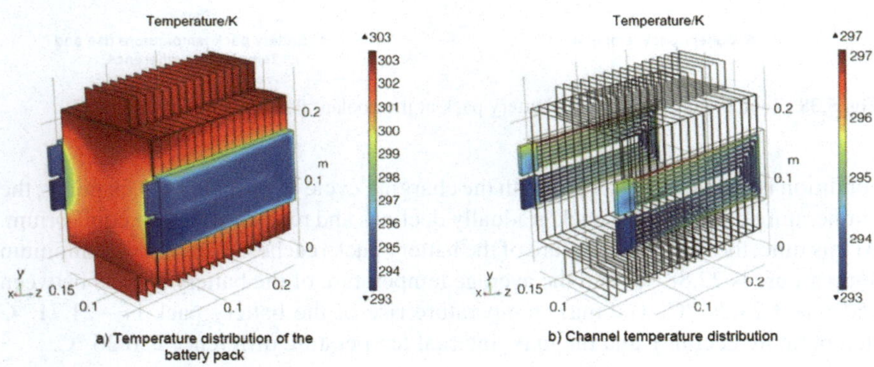

Fig. 5.41 Temperature distribution at 17500 s of the simulated working condition at the coolant flow rate of 0.07 m/s

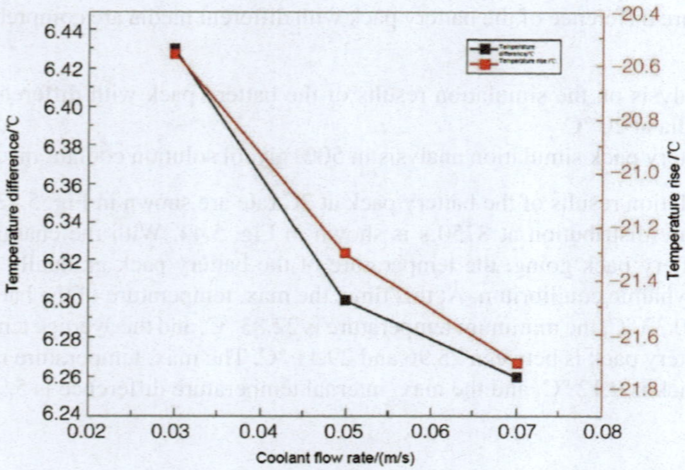

Fig. 5.42 Comparison of temperature rise and temperature difference in dynamic equilibrium state and at different coolant flow rates

(3) Simulation and comparative analysis of the battery pack at different flow rates

The internal temperature difference and temperature rise of the battery pack after reaching dynamic equilibrium at the cooling fluid flow rates of 0.03 m/s, 0.05 m/s and 0.07 m/s and the ambient temperature of 50 °C are shown in Fig. 5.42. Compared with the simulation results with the ambient temperature of 20 °C, the internal temperature difference of the battery has increased by about 3 °C, and the change of temperature rise is more obvious. Under the simulated condition of the ambient temperature of 50 °C, the influence of coolant flow rate on battery temperature rise and internal temperature difference is also little.

5.3.4 Influence of the Medium on the Liquid Cooling Heat Dissipation of the Battery Pack

In order to explore the influence of coolant medium on the heat dissipation effect of liquid cooling battery pack, this section carries out simulated analysis on the ambient temperature of 20 °C and 50 °C respectively from the angle of changing the coolant medium. In the simulation, the battery charge and discharge rate is 2C, the coolant flow rate is 0.03 m/s, the initial temperature of the coolant is all set at 20 °C, the initial temperature of the battery pack is consistent with the ambient temperature, and the coolant media are water and 50% glycol solution, where, the case with the medium being water has been analyzed in Sect. 5.3.2. In this section, the simulation results of 50% glycol solution are analyzed, and finally, the temperature rise and max.

temperature difference of the battery pack with different media are comprehensively analyzed.

1. Analysis on the simulation results of the battery pack with different coolant media at 20 °C
(1) Battery pack simulation analysis in 50% glycol solution coolant medium

The simulation results of the battery pack at 2C rate are shown in Fig. 5.43, and the temperature distribution at 8750 s is shown in Fig. 5.44. With the charging cycle of the battery pack going, the temperature of the battery pack gradually rises and reaches dynamic equilibrium. At this time, the max. temperature of the battery pack reaches 30.33 °C, the minimum temperature is 22.83 °C, and the average temperature of the battery pack is between 25.96 and 29.11 °C. The max. temperature rise of the battery pack is 9.12 °C, and the max. internal temperature difference is 5.93 °C.

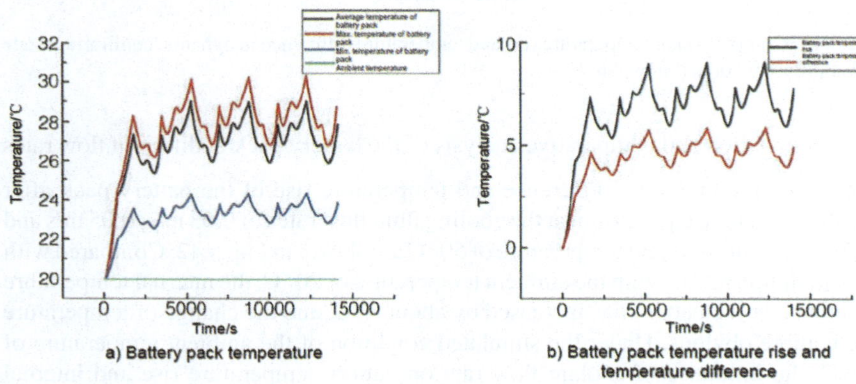

Fig. 5.43 Temperature curve of the battery pack in the coolant medium of 50% glycol solution

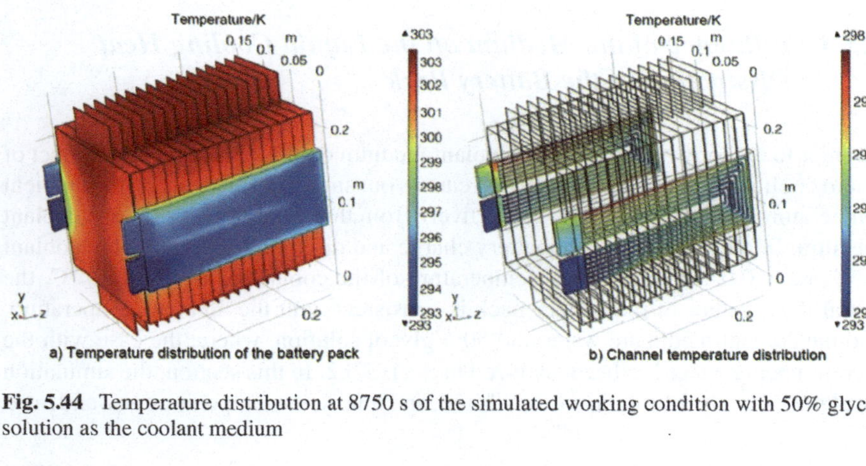

Fig. 5.44 Temperature distribution at 8750 s of the simulated working condition with 50% glycol solution as the coolant medium

(2) Comparative analysis of the battery pack simulation with different coolant media

After the battery pack achieved dynamic equilibrium in the cooling medium of water and 50% glycol solution respectively, the internal temperature difference and temperature rise of the battery pack are shown in Fig. 5.45. See Table 5.1 for the related physical parameters of the two media. Under the two simulated conditions, the difference in the internal temperature difference of the battery is within 0.1 °C, and the difference in the temperature rise of the battery is within 0.5 °C, so the cooling effects of water and 50% ethylene glycol solution are similar. However, considering the high boiling point, high flash point and low freezing point of 50% glycol solution, 50% glycol solution is more suitable for liquid cooling systems of battery packs.

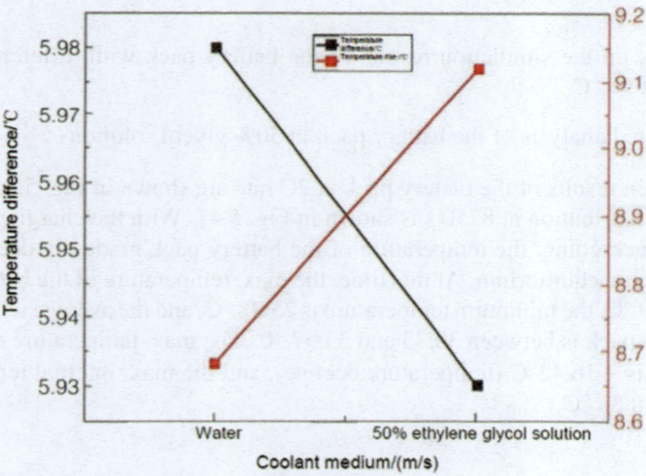

Fig. 5.45 Comparison of temperature rise and temperature difference with different coolant media in dynamic equilibrium state

Table 5.1 Physical parameters of coolants

Coolant	Density/(kg/m3)	Specific heat capacity/[J/(kg·K)]	Thermal conductivity coefficient/[W/(m·K)]	Dynamic viscosity/(mPa·s)
Water	998.2054	4186.9	0.5942	1.0093
50% ethylene glycol solution	1073.4	3281	0.38	3.94

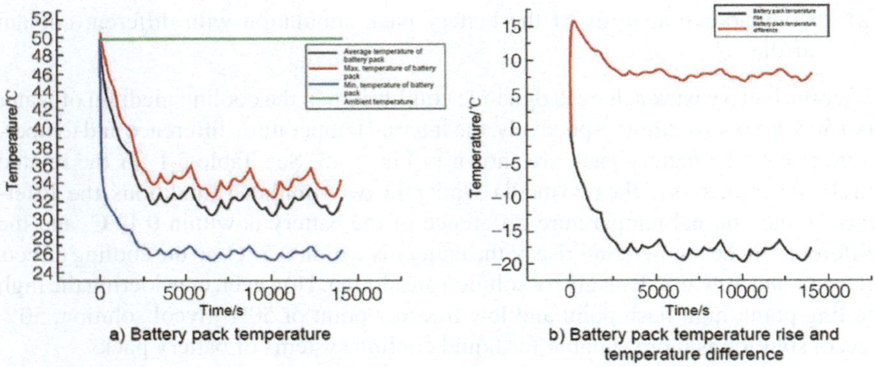

Fig. 5.46 Temperature curve of the battery pack in the coolant medium of 50% glycol solution

2. Analysis on the simulation results of the battery pack with different coolant media at 50 °C

(1) Simulated analysis of the battery pack in 50% glycol solution

The simulation results of the battery pack at 2C rate are shown in Fig. 5.46, and the temperature distribution at 8750 s is shown in Fig. 5.47. With the charging cycle of the battery pack going, the temperature of the battery pack gradually declines and reaches dynamic equilibrium. At this time, the max. temperature of the battery pack reaches 35.79 °C, the minimum temperature is 25.28 °C, and the average temperature of the battery pack is between 30.33 and 33.57 °C. The max. temperature rise of the battery pack is −16.42°C (temperature decline), and the max. internal temperature difference is 8.85 °C.

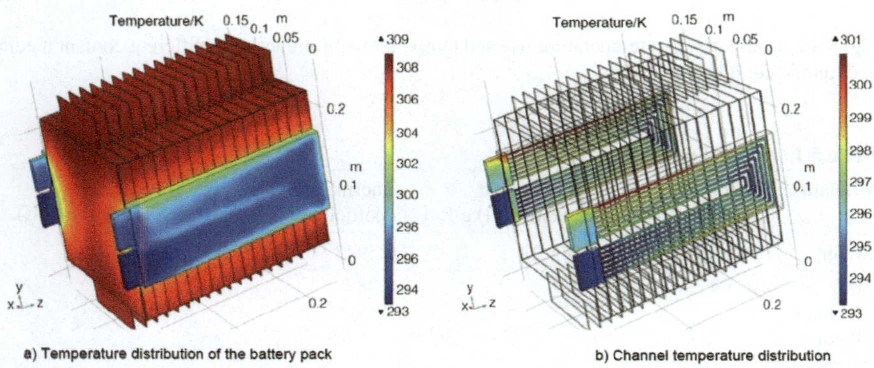

Fig. 5.47 Temperature distribution at 8750 s of the simulated working condition with 50% glycol solution as the coolant medium

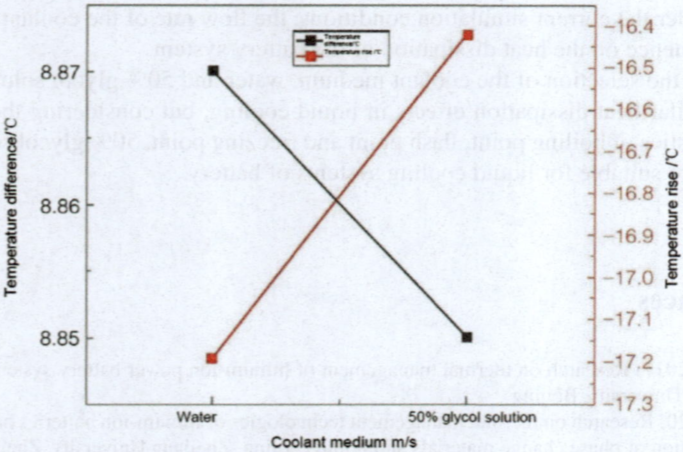

Fig. 5.48 Comparison of temperature rise and temperature difference with different coolant media in dynamic equilibrium state

(2) Comparative analysis of the battery pack simulation with different coolant media

After the battery pack achieved dynamic equilibrium at the ambient of 50 °C in the cooling medium of water and 50% glycol solution respectively, the internal temperature difference and temperature rise of the battery pack are shown in Fig. 5.48. Similar to the results at 20 °C, water and 50% glycol solution have similar heat dissipation effects in liquid cooling.

Summary

In this chapter, battery packs are taken as the research objects, aiming at the heat dissipation of the battery pack by liquid cooling, a coupled heat dissipation model of one eighth thermal field and flow field of the battery pack is established. The heat dissipation of battery pack is simulated and analyzed through the heat dissipation model, and the main conclusions are as follows:

(1) According to the comparative analysis of simulation results of the battery pack at different ambient temperatures, the temperature difference and temperature rise of batteries are contradictory to the battery thermal management system, so it is necessary to balance the influence of temperature difference and temperature rise inside the battery packs according to the actual situation.

(2) With the increase of the charging-discharging rate of the battery, the temperature rise and internal temperature difference of the battery pack will increase regardless of the ambient temperature. Therefore, a thermal management system with higher heat dissipation capacity is needed for battery packs that need to be charged and discharged at a high rate.

(3) Under the current simulation conditions, the flow rate of the coolant has little influence on the heat dissipation of the battery system.
(4) For the selection of the coolant medium, water and 50% glycol solution have similar heat dissipation effects in liquid cooling, but considering the characteristics of boiling point, flash point and freezing point, 50% glycol solution is more suitable for liquid cooling systems of battery.

References

Chen DF (2017) Research on thermal management of lithium-ion power battery systems. Beijing Jiaotong University, Beijing

He XF (2020) Research on thermal management technologies of lithium-ion batteries based on the combination of phase change materials and liquid cooling. Zhejiang University, Zhejiang

Chapter 6
External Heating Technology for Lithium-ion Batteries

With the gradual promotion of electric vehicles, the low-temperature performance of lithium-ion batteries is attracting more and more attention. The problems of difficult charging, discharge capacity degradation and reduced driving range of battery packs at low temperatures are gradually exposed, so it is necessary to study the low-temperature heating of lithium-ion battery packs.

To heat a lithium-ion battery pack, two issues must be considered: firstly, it needs to be determined whether the battery is to be heated externally or internally. The advantages of heating the battery externally include safety and no modifications to the battery itself. The disadvantages include long heating time, high heating energy loss and uneven heating. On the contrary, heating the battery internally reduces the heating time, makes full use of the heating energy and heats the battery evenly, but there are such problems as difficult installation of the heating unit and the poor battery safety. Second, the energy required for heating needs to be considered. If no external heating source is available, then the lithium-ion battery itself must be considered as a heating source to power the heater for self-heating and restoring charge and discharge performance, which is essential for electric drive vehicles. This chapter describes two external heating methods, namely, PTC thermistors (PTC for short) and wide wire metal films.

6.1 Study on the Characteristics of Heating Batteries with PTC

6.1.1 PTC Heating Principle

The heating material of the PTC (positive temperature coefficient thermistor) features constant temperature heating. The principle is that the PTC heats itself up after being

J. Li, *Modeling and Simulation of Lithium-ion Power Battery Thermal Management*,
Key Technologies on New Energy Vehicles,
https://doi.org/10.1007/978-981-19-0844-6_6

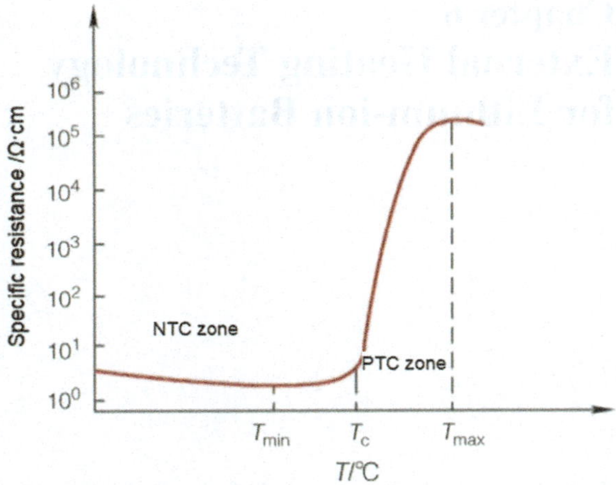

Fig. 6.1 PTC resistance-temperature characteristic curve

charged so that the resistance value enters the jump zone, where the resistance value varies greatly, and after entering the jump zone, the PTC has constant surface temperature, i.e. the resistance remains the same. The high resistance of the PTC material and its slow rise in temperature above the Curie temperature are used to generate heat in the cell. The resistance–temperature characteristics of PTC materials are their most important feature. The resistance of this material increases exponentially when the temperature rises to its Curie temperature point (the starting point of the positive temperature characteristic is called the Curie point), thus limiting and protecting the circuit from short-circuit currents. A typical resistance–temperature characteristic curve is shown in Fig. 6.1.

6.1.2 PTC Heating Experimental Programme

In order to fully analyze the results achieved by this heating method, the following experiments were carried out.

(1) Battery module pre-heating experiments at −40 °C and −30 °C.
(2) Temperature rise and discharge performance experiment of batteries with different discharge rates after pre-heating at −40 °C.
(3) Temperature rise and discharge performance experiment of batteries with different discharge rates after pre-heating at −30 °C.

In this case, the heating aluminum plates between the battery modules are used to heat these modules with a constant power (35 W), during which a 220 V AC voltage is applied directly to both ends of the PTC material.

The battery test system Digatron EVT500-500 enables real-time measurement of parameters such as battery voltage, current and temperature. The main control

Table 6.1 Technical data of Digatron EVT500-500

Technical parameters	Value
Max. discharge current	500 A
Max. charging current	500 A
Max. voltage	500 V
Voltage measurement error	5‰
Current measurement error	5‰
Temperature measurement error	±0.5 °C

computer is connected to the EVT500-500 via the CAN interface. The test software "BTS-600", installed in the main control computer, can be used to write the battery test procedure and to collect data such as test time, battery voltage, current, power, temperature, charge/discharge energy (W–h) and charge/discharge capacity (A-h). Meanwhile, the system sampling time is also set and the maximum sampling frequency is set to 10 kHz. See Table 6.1 for parameter indexes.

The voltage and surface temperature of the battery module are collected using an NI data acquisition system. The NI data acquisition system is connected to the computer, and the data is recorded, processed and saved in real time using the virtual instrument software developed in LabVIEW, thus completing the information processing tasks of the experiment. The LabVIEW software has a good human–machine interface, so the system also has some monitoring functions. The battery module under test is placed in a thermostat and tested at an ambient temperature of −40 to 55 °C.

In the experiment, in order to heat the battery module, aluminum plates with slots are added to the ends of the battery cells and the slots are embedded with heating elements (PTC heating material). These aluminum plates are heated by the heat generated by the PTC heating material, thus heating the battery module.

The battery module for the experiment is shown in Fig. 6.2 and consists of 3 battery cells with 4 aluminum plates and 6 PTC heating materials etc. The 3 battery cells are connected in series and the battery module has a total rated voltage of 11.1 V and a capacity of 35A-h. The aluminum plates are distributed on both sides of the battery

Fig. 6.2 Battery module for experiment

Fig. 6.3 Battery box for experiment in a thermostat

cell and the PTC heating material is embedded in the slotted aluminum plates. The battery module is placed in a battery box for testing in a low temperature environment. The external temperature is controlled by a thermostat. The battery box for the experiment in the thermostat is shown in Fig. 6.3. The cables for temperature testing, charge and discharge and heating elements (PTC heating material) are connected to the charge and discharge machine, temperature acquisition system, etc. via this thermostat so that the heating experiment of aluminum plates can be done at low temperatures (−40 °C and −30 °C respectively).

The aluminum plates used for heat transfer in the experiments are shown in Fig. 6.4. The aluminum plate has the same dimensions and length as the battery cells and is in close contact with the individual cells, ensuring that most of the heat is transferred to the cells. The aluminum plate has several slots on one side for inserting part of the PTC heating material, which ensures easy assembly. These slots also facilitate heat dissipation of the battery module when it is operating at higher temperatures.

In the experiment, the PTC heating material is embedded in slots of the aluminum plate and the resistance of the PTC heating material increases sharply when a 220 V voltage is applied to both ends of the PTC heating material. If this resistance remains constant after a short period of time, the PTC heating material is maintained at a constant power to complete the heating of the battery module. The PTC heating

Fig. 6.4 Aluminum sheets for heat transfer

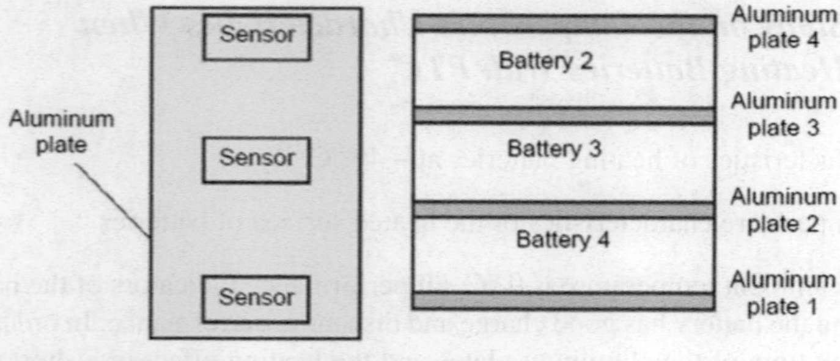

Fig. 6.5 Layout of the battery module temperature sensors

material is embedded in the slots of the aluminum plates, which allows the battery module to be heated with an even temperature distribution and a better heating effect.

Temperature sensors are attached to the side of each of the three battery cells experimented to monitor changes in cell temperature in real time. The layout of each temperature sensor is illustrated in Fig. 6.5.

In the experiment, 14 temperature sensors are placed at various locations of the aluminum plate and the battery. Sensor 2 is placed at the connection between battery 4 and the cathode and anode lugs of battery 3. Sensor 3 is positioned at the connection between the anode of battery 3 and the cathode lug of battery 2. Sensor 4 is positioned at the anode of battery 2 (i.e. the anode of the 3 batteries in series). Sensor 5 is positioned at the cathode of battery 4 (i.e. the cathode of the 3 batteries in series). Sensors 1 and 6 are arranged in the centre of the outermost aluminum plate. The remaining 8 sensors are arranged on the two middle aluminum plates, including 3 on each side (near the cathode and anode of battery, and in the centre of the battery) and one in the centre.

During the pre-heating of the battery module, the surface temperature of each cell is monitored and recorded by a temperature sensor in order to analyze the temperature and heating uniformity of the battery module in the heating process. At the end of the heating process, when the battery module is about to start discharging, the total voltage change and the temperature at each position of the battery module are monitored in order to analyze and compare the discharging and the temperature rise characteristics of the battery at different rates and thus evaluate the heating effect of this heating method.

6.1.3 Study on the Temperature Characteristics When Heating Batteries With PTC

1. Characteristics of heating batteries at −40 °C

(1) Temperature characteristics of the heated surface of batteries

When the ambient temperature is 0 °C, all performance indicators of the battery are normal and the battery has good charge and discharge performance. In order to study the heating time of the aluminum plates and the heating effect in a short period of time, 0 °C is used as the heating termination temperature of the battery module in the experiment. In the experiment, an AC 220 V power supply is used to power the PTC heating resistor embedded in the aluminum plate. The current variation curve after 25 min of heating is shown in Fig. 6.6.

Initially the PTC heating resistor has a low resistance value and produces a large heating power for a short period of time. As the heating time increases, the resistance value increases and the heating power decreases. When the temperature rises to its constant value, the heating power remains constant. In the experiment, the battery module is heated using the PTC heating resistor which maintains a constant power over a long period of time. After 25 min of heating, the temperature changes in each part of the battery are shown in Figs. 6.7, 6.8 and 6.9.

As shown in Fig. 6.7, the temperature at the connection of the cathode and anode of battery cell (sensor 2 and sensor 3) increases from −38 to −7 °C, and the temperature at the cathode and anode of battery module (sensor 4 sensor 5) increases from −38 to −15 °C. The temperature rise at the cathode and anode of battery module is

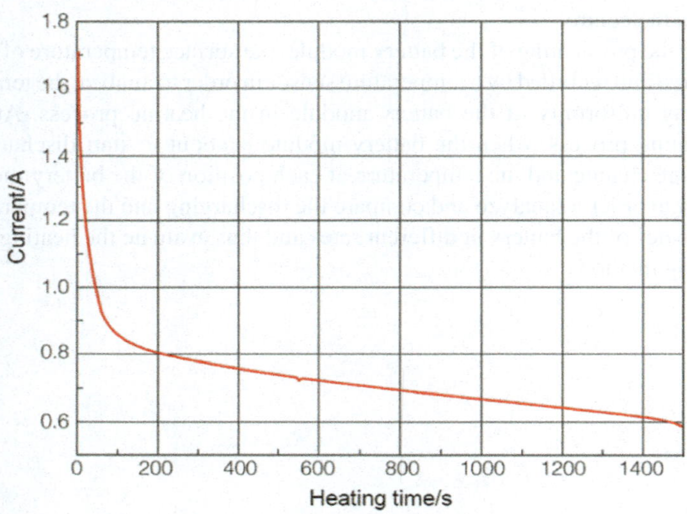

Fig. 6.6 Current change curve of PTC heating resistance after 25 min heating

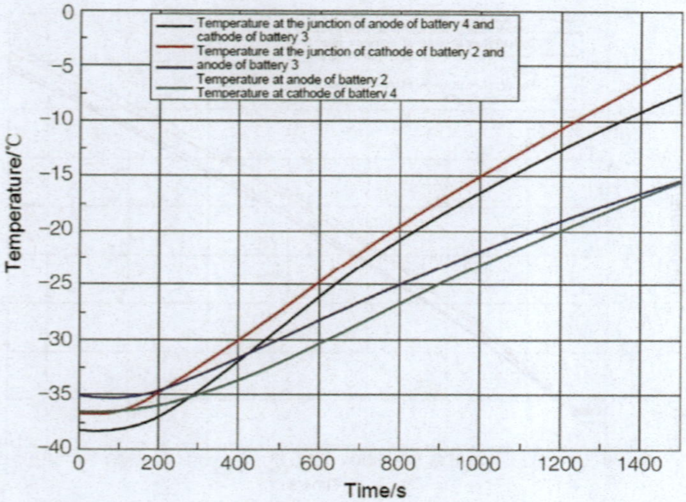

Fig. 6.7 Temperature change curve at the pole lug during heating

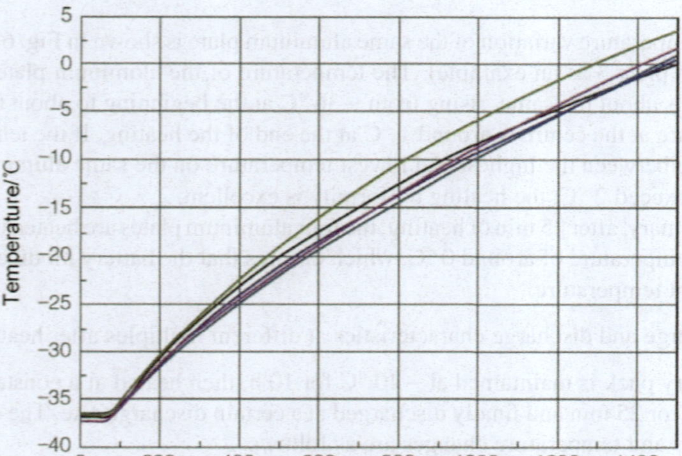

Fig. 6.8 Temperature variation at the centre of the four heating aluminum plates

lower than at the battery cell connection, mainly because the heat at the battery cell connection comes from both batteries.

As shown in Fig. 6.8, the temperature rise at the centre of the four heating aluminum plates is basically the same. The temperature here rises from −38 °C to about 1 °C, indicating that the temperature distribution of the different aluminum plates is relatively uniform after 25 min of heating.

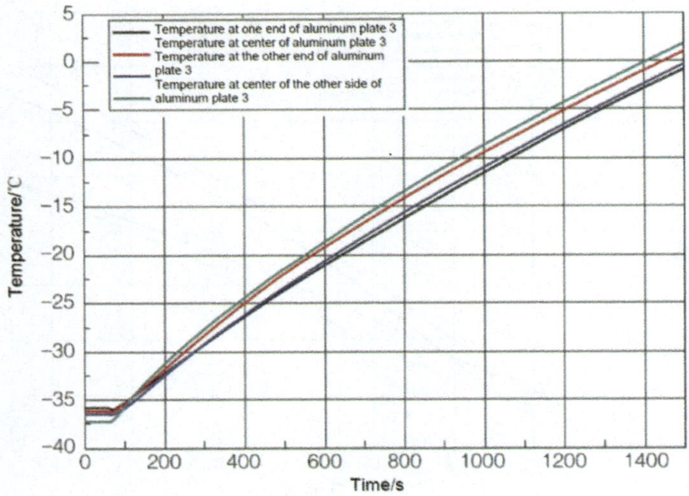

Fig. 6.9 Temperature change curve at each typical position after heating of the same aluminum plate

The temperature variation of the same aluminum plate is shown in Fig. 6.9 (taking aluminum plate 3 as an example). The temperature of the aluminum plate near the two ends is about the same, rising from −36 °C at the beginning to about 0 °C. The temperature at the centre is around 1 °C at the end of the heating. If the temperature difference between the highest and lowest temperature on the same aluminum plate does not exceed 2 °C, the heating uniformity is excellent.

In summary, after 25 min of heating, the four aluminum plates are heated uniformly and at a temperature of around 0 °C, which ensures that the battery module operates at the right temperature.

(2) Charge and discharge characteristics at different multiples after heating

The battery pack is maintained at −40 °C for 10 h, then heated at a constant power of 140 W for 25 min and finally discharged at a certain discharge rate. The discharge conditions and temperature changes are as follows:

The heated batteries are discharged at 0.3C, 0.5C, 1C and 2C rates. Then, the discharges are shown in Figs. 6.10 and 6.11.

As can be seen from Figs. 6.10 and 6.11, the voltage drops quickly at the beginning of the discharge. As the discharge process proceeds, the voltage plateaus. This is mainly because at the beginning of discharge, the internal resistance of the battery is large, which makes the discharge voltage drop quickly. As the discharge process proceeds, the temperature of the battery rises due to self-heating and the internal resistance decreases, making the discharge voltage stabilize.

The battery has a low voltage platform when discharged at large multiples. For example, the voltage platform is approximately 10.6 V at 2C rate discharge and 11 V at 1C rate discharge. Increasing the voltage platform helps to reduce the discharge

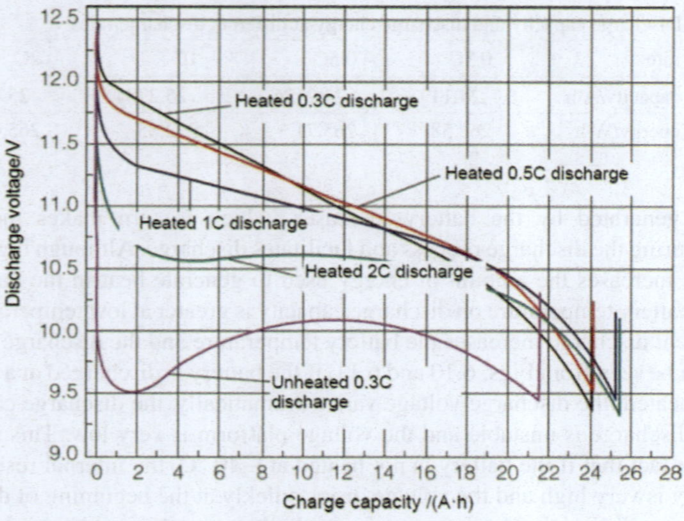

Fig. 6.10 Discharge curves at different discharge rates

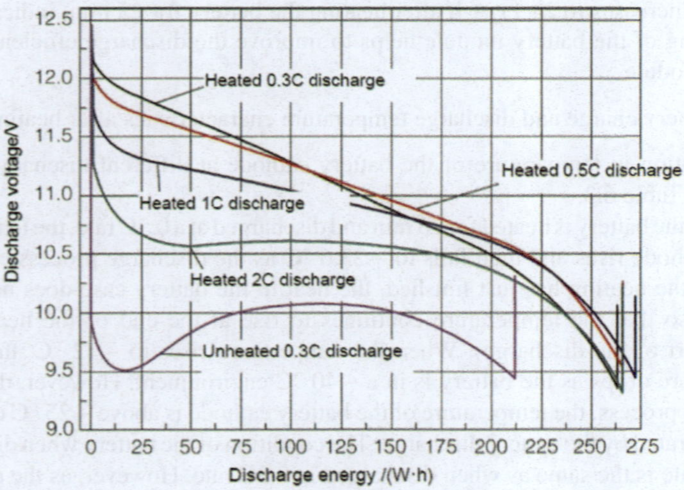

Fig. 6.11 Variation of discharge voltage with discharge energy

current at constant power output, lower the energy consumption caused by the internal resistance of the battery, and increase the discharge efficiency of the battery.

As the discharge rate increases, the discharge process becomes smoother and the discharge capacity increases. The results are shown in Table 6.2.

In low temperature environments, the increase in battery temperature is conducive to higher battery discharge efficiency. Under the condition of high current discharge,

Table 6.2 Discharge capacity and discharge energy at different discharge rates

Discharge rate	0.3C	0.5C	1C	2C
Discharge capacity/A·h	24.11	24.2126	25.2784	25.449
Discharge energy/W·h	263.58	265.71	272.65	265.47

the heat generated by the battery increases, which in turn makes the battery warmer during the discharge process and facilitates discharge. Although high current discharge increases the amount of energy used to generate heat in the battery, the effect of battery temperature on discharge capacity is greater at low temperatures, i.e. high current discharge increases the battery temperature and the discharge capacity.

As can be seen from Figs. 6.10 and 6.11, if the battery is discharged at a 0.3C rate when unheated, the discharge voltage varies dramatically, the discharge capacity is low, the discharge is unstable and the voltage platform is very low. This is mainly due to the fact that if the battery is not heated at −40 °C, the internal resistance of the battery is very high and the voltage drops quickly at the beginning of discharge. Furthermore, during the discharge process, the battery temperature is always low and the battery discharge performance deteriorates. In comparison, the discharge capacity is 21.9 A-h at 0.3C rate after heating the battery for 25 min. The discharge capacity increases to 24.11 A-h after heating the battery for 25 min, indicating that the heating of the battery module helps to improve the discharge efficiency of the battery module.

(3) Battery charge and discharge temperature characteristics after heating

The variation in temperature of the battery cathode at different discharge rates is shown in Table 6.3.

When the battery is heated for 25 min and discharged at a 0.3C rate, the temperature of the cathode rises and then falls to −23.6 °C as the discharge process continues. Because the heating has just finished, the heat in the battery case does not escape in time, so that the temperature continues to rise at the end of the heating (i.e. at the start of the discharge). When the temperature rises to −12 °C, the battery temperature drops as the battery is in a −40 °C environment. However, during the discharge process, the temperature of the battery cathode is above −25 °C due to the heat generated by the battery discharge. The condition of the battery when discharged at 0.5C rate is the same as when discharged at 0.3C rate. However, as the discharge current is higher than 0.3C, the temperature at the end of the discharge is higher. When the battery is discharged at 1C and 2C rates, the battery temperature is high

Table 6.3 Temperature variation of the battery cathode at different discharge rates

Discharge rate	0.3C	0.5C	1C	2C
Temperature at initial discharge /°C	−15.6	−15.5	−15.4	−15.2
Maximum temperature/°C	−12	−11	−8	6.4
Temperature at the end of discharge/°C	−23.6	−16.2	−8	6.4

Table 6.4 Temperature variation of the battery anode at different discharge rates

Discharge rate	0.3C	0.5C	1C	2C
Temperature at initial discharge/°C	−16.8	−16.5	−16.3	−15.2
Maximum temperature/°C	−15	−13	−15.3	−10
Temperature at the end of discharge/°C	−25.8	−16.2	−15.3	−10

at the end of the discharge, namely, −8 °C and 6.4 °C respectively, due to the high discharge current and the high heat generated by the battery.

The temperature variation of the battery anode at different discharge rates is shown in Table 6.4. The trend in temperature change at each discharge rate and the reasons for it are the same as the temperature change at the cathode.

The temperature variation at the centre of the battery (taking battery 3 as an example) is shown in Fig. 6.12.

At the beginning of the discharge, the temperature at each discharge rate is approximately the same, about 1 °C. As the discharge progresses, the temperature tends to decrease at 0.3C rate discharge, and drops to approximately −16 °C when the discharge ends. This is the same trend as for 0.5C rate discharge, except that for the latter the temperature drops to approximately −6.7 °C when the discharge ends. However, the temperature at 1C and 2C rate discharge shows an increasing trend, and rises to 4 °C and 15 °C respectively when the discharge ends. This is mainly because, when being discharged at low currents, the battery generates less heat, which makes the temperature of the battery in a low temperature environment tend to decrease. The opposite is true when the battery is discharged at high currents.

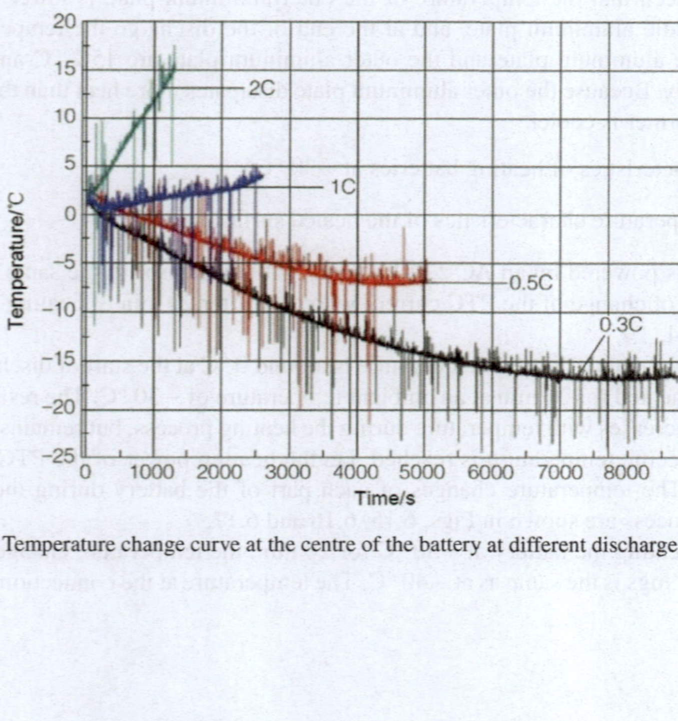

Fig. 6.12 Temperature change curve at the centre of the battery at different discharge rates

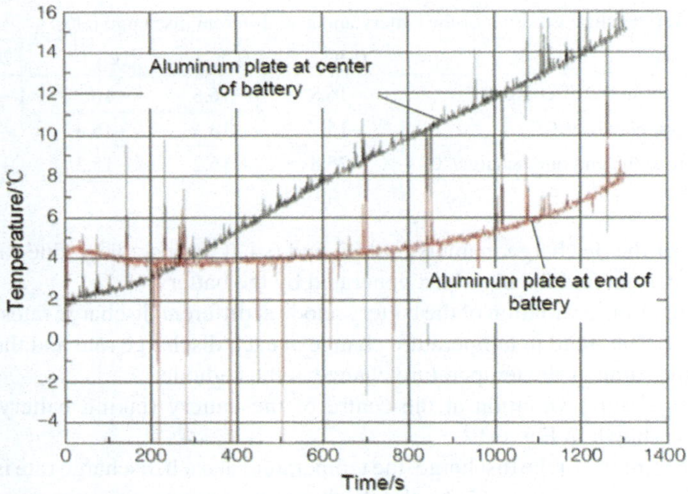

Fig. 6.13 Temperature change curve of aluminum plates at different positions during discharge

The temperature change curves of the aluminum plates at different positions (central position) during discharge are shown in Fig. 6.13 (taking aluminum plates 3 and 4 at 2C discharge rates as examples).

At the beginning of the discharge, the temperatures of the two aluminum plates are 2 °C and 4 °C respectively, which is not a big difference. And because of the high current discharge, the temperature is on the rise. As the discharge proceeds, it can be seen that the temperature of the outer aluminum plate is lower than that of the middle aluminum plate, and at the end of the discharge the temperature of the middle aluminum plate and the outer aluminum plate are 15.2 °C and 7.9 °C respectively. Because the outer aluminum plate dissipates more heat than the middle one, the former is cooler.

2. Characteristics of heating batteries at −30 °C

(1) Temperature characteristics of the heated surface of batteries

The PTC is powered by an AC 220 V supply with the sensor in the same position. The curve of change of the PTC current with time after 20 min of heating is shown in Fig. 6.14.

To ensure that the battery temperature is around 0 °C at the start of discharge, the battery is heated for 20 min at an ambient temperature of −30 °C. The resistance of the PTC increases with temperature during the heating process, but remains constant when a specific temperature is reached, i.e. the heating power of the PTC remains constant. The temperature changes of each part of the battery during the 20 min heating process are shown in Figs. 6.15, 6.16 and 6.17.

After heating the battery at −30 °C for 20 min, the temperature change trend at the battery lugs is the same as at −40 °C. The temperature at the connection between

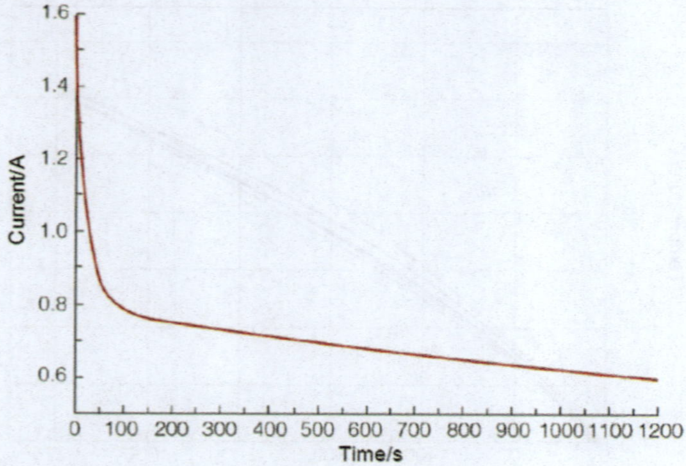

Fig. 6.14 Change of the PTC current with time

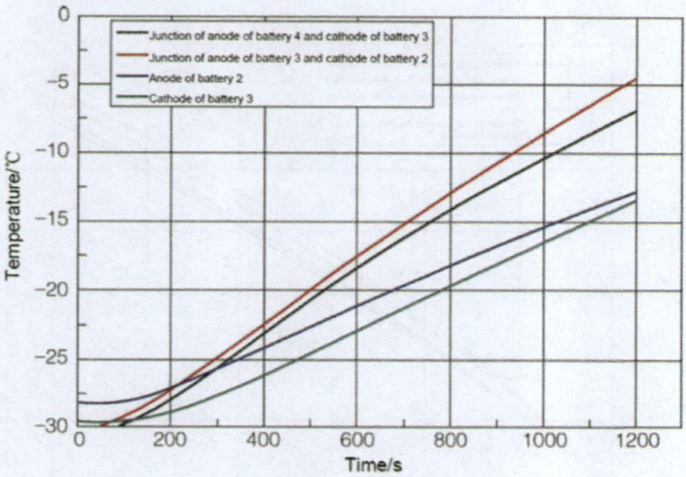

Fig. 6.15 Change of the temperature with heating time at the lug

the cathode and anode of battery cell rises from −30 °C to around −7 °C, while the temperature at the cathode and anode of battery module rises from −30 to −12 °C. In other words, the temperature rise of the latter is lower than that of the former. Because the heat at the connection of the cathode and anode of battery cell comes from both batteries, the temperature rise here is greater.

The temperature change at the centre of the aluminum plate after 20 min of heating at different positions is shown in Fig. 6.16. The aluminum plates 1 and 4 have sensors on one side only and the aluminum plates 2 and 3 have sensors on both sides.

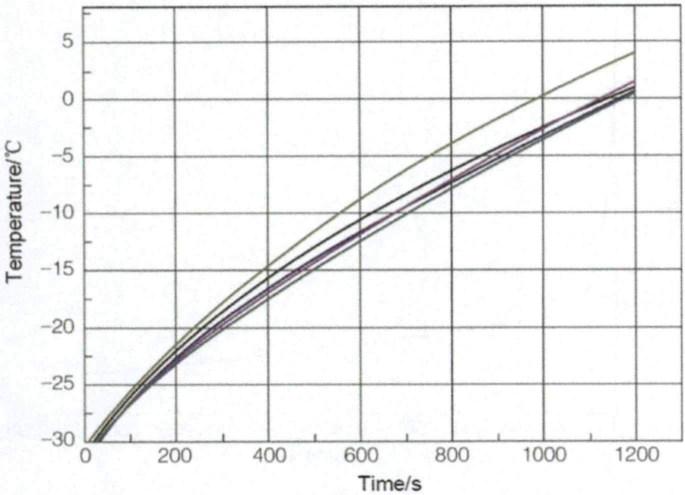

Fig. 6.16 Temperature variation at the centre of the four heating aluminum plates

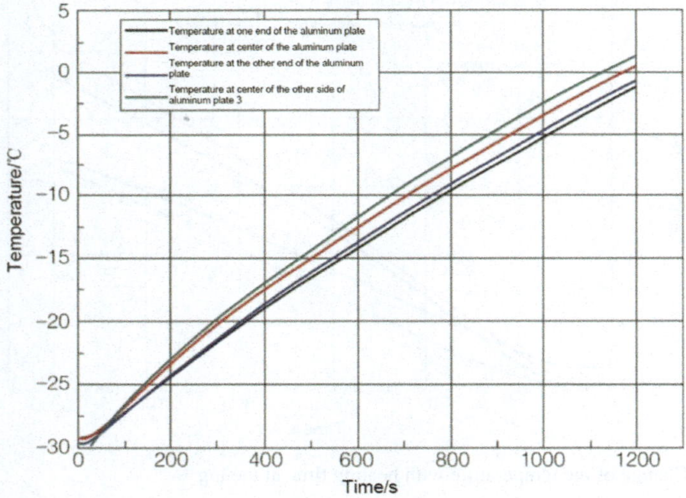

Fig. 6.17 Temperature change curve after heating of the same aluminum plate

As can be seen from the figure, the temperature rise of the four aluminum plates is basically the same after 20 min of heating, with the temperature rising from −30 °C to about 2 °C. This indicates that the temperature of the battery module is evenly distributed under this heating method.

The temperature variation of the same aluminum plate is shown in Fig. 6.17 (taking aluminum plate 3 as an example).

The temperature of the aluminum plate near the two ends is about the same, rising from −30 °C at the beginning to about 0 °C. The temperature at the centre is around 1 °C at the end of the 20 min heating. If the temperature difference between the highest (1 °C) and lowest (0°C) temperature on the same aluminum plate does not exceed 2 °C, the heating uniformity is excellent.

As a result, the four aluminum plates are heated uniformly and at around 0 °C at the end of the heating process, which are qualified for the operation of the battery.

(2) Charge and discharge characteristics of the battery at different rates after heating

The battery pack is maintained at −30 °C for 10 h, then heated at a constant power of 140 W for 20 min and finally discharged at a certain discharge rate. The discharge conditions and temperature changes are as follows:

The heated batteries are discharged at 0.5C, 1C and 2C rates. Then, the discharges are shown in Figs. 6.18 and 6.19.

As can be seen from Figs. 6.18 and 6.19, the voltage drops quickly at the beginning of the discharge. However, the voltage tends to be stable as the discharge proceeds. This is mainly because at the beginning of discharge, the internal resistance of the battery is large, which makes the discharge voltage drop significantly. As the discharge process proceeds, the temperature of the battery rises due to self-heating and the internal resistance decreases, making the discharge voltage stabilize.

The battery has a low voltage platform when discharged at large multiples. The voltage platform is approximately 10.6 V at 2C rate discharge and 11 V at 1C rate discharge. Increasing the voltage platform helps to reduce the discharge current at constant power output, lower the energy consumption caused by the internal resistance of the battery, and increase the discharge efficiency of the battery.

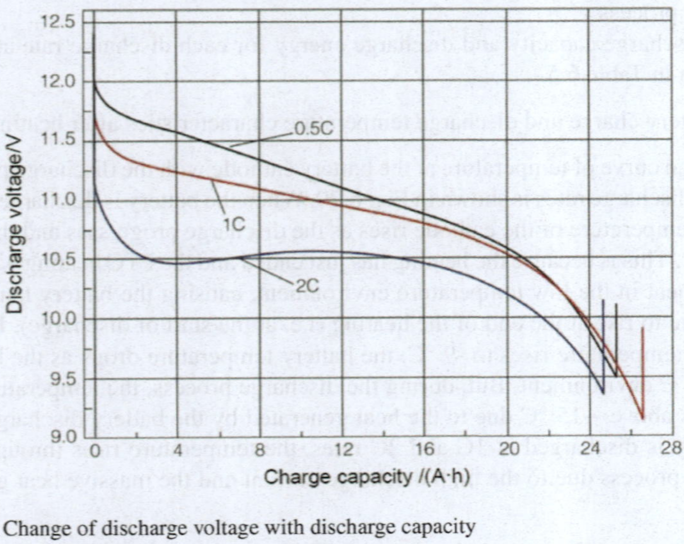

Fig. 6.18 Change of discharge voltage with discharge capacity

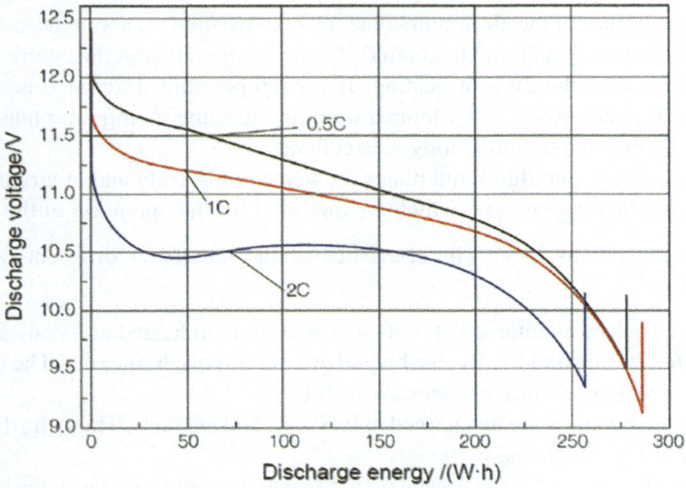

Fig. 6.19 Variation of discharge voltage with discharge energy

Table 6.5 Discharge capacity and discharge energy at different discharge rates at −30 °C

Discharge rate	0.5C	1C	2C
Discharge capacity/A·h	25.29	26.55	24.66
Discharge energy/W·h	278.02	285.79	256.61

The discharge process is smoother as the discharge rate increases, mainly because the high current discharge generates more heat, and the battery can be discharged in a more suitable environment, resulting in smoother voltage changes during the discharge process.

The discharge capacity and discharge energy for each discharge rate at −30 °C are shown in Table 6.5.

(3) Battery charge and discharge temperature characteristics after heating

The change curve of temperature at the battery cathode with the discharge process at different discharge rates is shown in Fig. 6.20. When the battery is discharged at 0.5C rate, the temperature of the cathode rises as the discharge progresses and then drops to −11 °C. This is because the heating has just ended and the PTC heating wire is still emitting heat in the low temperature environment, causing the battery temperature to continue to rise at the end of the heating (i.e. at the start of discharge). However, when the temperature rises to -9 °C, the battery temperature drops as the battery is in a −40 °C environment. But, during the discharge process, the temperature of the cathode is above −15 °C due to the heat generated by the battery discharge. When the battery is discharged at 1C and 2C rates, the temperature rises throughout the discharge process due to the high discharge current and the massive heat generated

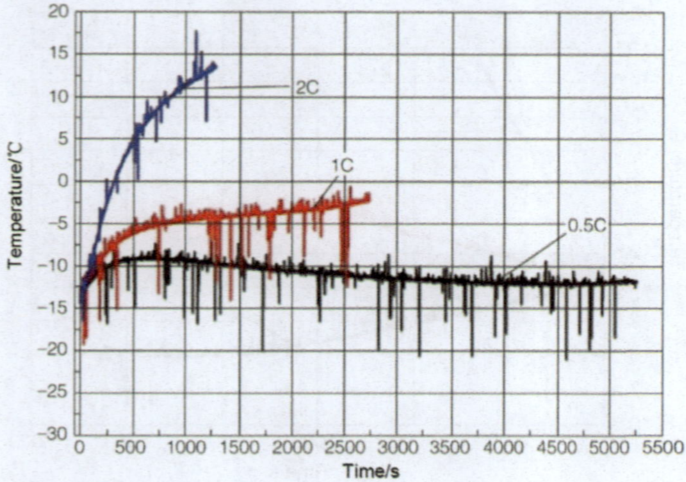

Fig. 6.20 Change of temperature at the battery cathode with the discharge process at different discharge rates

by the battery. The temperatures at the end of discharge are at −2 °C and 13 °C respectively.

The change curve of temperature at the battery anode with the discharge process at different discharge rates is shown in Fig. 6.21. As can be seen from the figure, the temperature change at the anode follows the same trend as at the cathode. As can be

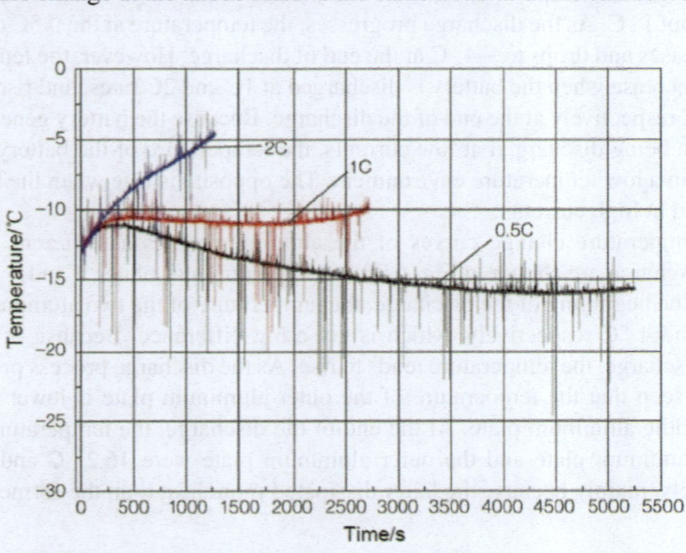

Fig. 6.21 Change of temperature at the battery anode with the discharge process at different discharge rates

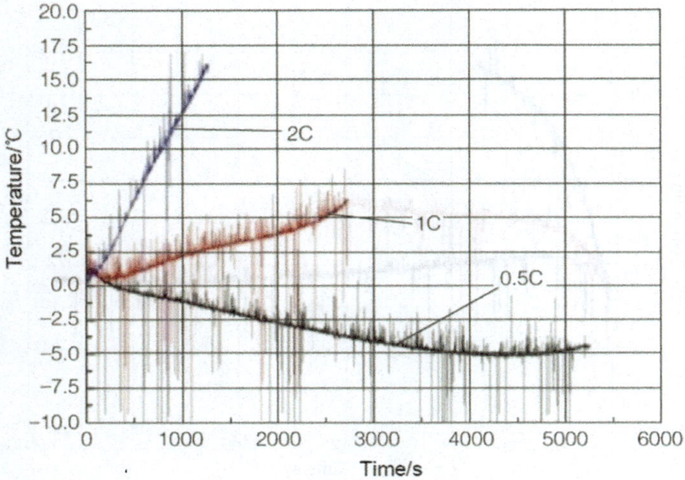

Fig. 6.22 Temperature change curve at the centre of the battery at different discharge rates (taking battery 3 as an example)

seen from Figs. 6.19 and 6.20, the temperature at the cathode is higher than that at the anode.

The temperature change curve at the centre of the battery at different discharge rates (taking battery 3 as an example) is shown in Fig. 6.22. At the beginning of the discharge, the temperature of the battery at each discharge rate is basically the same, about 1 °C. As the discharge progresses, the temperature at the 0.5C discharge rate decreases and drops to −4 °C at the end of discharge. However, the temperature tends to increase when the battery is discharged at 1C and 2C rates, and rises to 6 °C and 16 °C respectively at the end of the discharge. Because the battery generates less heat when being discharged at low currents, the temperature of the battery tends to decrease in a low temperature environment. The opposite is true when the battery is discharged at high currents.

The temperature change curves of the aluminum plates at different locations during discharge are shown in Fig. 6.23 (taking aluminum plates 3 and 4 as examples). At the beginning of the discharge, the temperature of the two aluminum plates is 2 °C and 4 °C respectively, which is not a big difference. Because of the high current discharge, the temperature tends to rise. As the discharge process progresses, it can be seen that the temperature of the outer aluminum plate is lower than that of the middle aluminum plate. At the end of the discharge, the temperatures of the middle aluminum plate and the outer aluminum plate were 16.2 °C and 10.5 °C respectively, mainly because the latter dissipated more heat than the former.

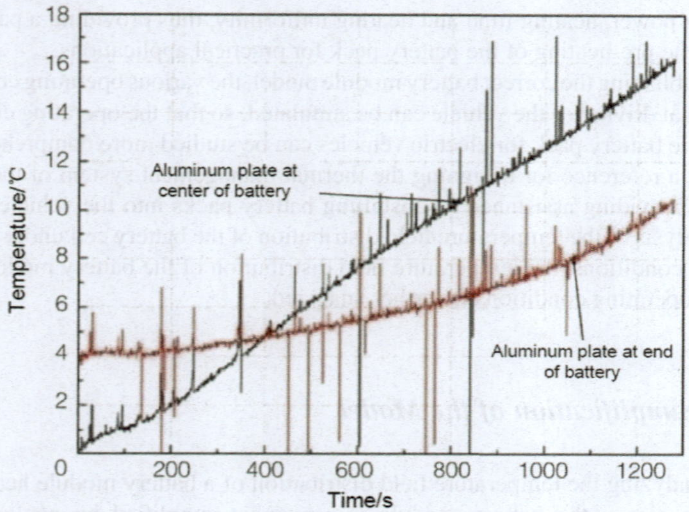

Fig. 6.23 Temperature change curve during discharge of aluminum plates at different positions

6.2 Finite Element Simulation Analysis for Heating Batteries with PTC

During the experiment, the temperature sensors could not spread over the entire battery box, and the sensors could not be buried inside the battery cells, so the measured temperature could only reflect the changes in the temperature field of the battery pack in general, and not the temperature changes inside the battery cells. Furthermore, the experiments are time and effort consuming and the measured data is only representative of the results under certain experimental conditions, so it is important to build a 3D model of the battery pack in order to simulate the temperature field of the pack using Fluent software or other computational fluid dynamics (CFD) software. The simulation must be based on the experiment, and the simulation parameters must be continuously adjusted according to the experimental results until the simulation reflects the experimental results well. The simulation model can then be extended to obtain data that cannot be collected in the experiment or to simulate the results of other non-experimental conditions. The close integration of simulation and experimentation helps to understand the changes in the temperature field of the entire battery pack during the self-heating process of the PTC heating material.

In this section, a battery module consisting of a square aluminum-plastic film battery is simulated and modeled, with the aim of analyzing the heat generation characteristics of the battery module in a low temperature environment. The aluminum plate heating method is used to analyze the effect of this heating method on the temperature field and charge/discharge performance of the battery module in terms of heating power, heating time and heating uniformity, thus providing a parametric basis for the pre-heating of the battery pack for practical applications.

By establishing the correct battery module model, the various operating conditions in the actual driving of the vehicle can be simulated, so that the operating characteristics of the battery pack for electric vehicles can be studied more comprehensively, providing a reference for designing the thermal management system of the battery pack and providing assistance for installing battery packs into the vehicles. Based on the analysis of the temperature field distribution of the battery cell under different operating conditions, the temperature field distribution of the battery module under different operating conditions is further analyzed.

6.2.1 Simplification of the Model

Before analyzing the temperature field distribution of a battery module heated with aluminum plates, the battery module is somewhat simplified by neglecting the wires between battery cells and using insulated plastic insulation sheets. A three-dimensional model of the battery module shown in Fig. 6.24 is created, which consists of 3 battery cells and 4 heating aluminum plates located between the cells and at the outermost end.

These aluminum plates have dimensions of 170 mm x 5 mm x 198 mm, which are arranged along the x-axis on both sides of the cells and are in close contact with the cells to heat the cells at low temperatures and dissipate heat at high temperatures. The

Fig. 6.24 Geometric model of the battery module

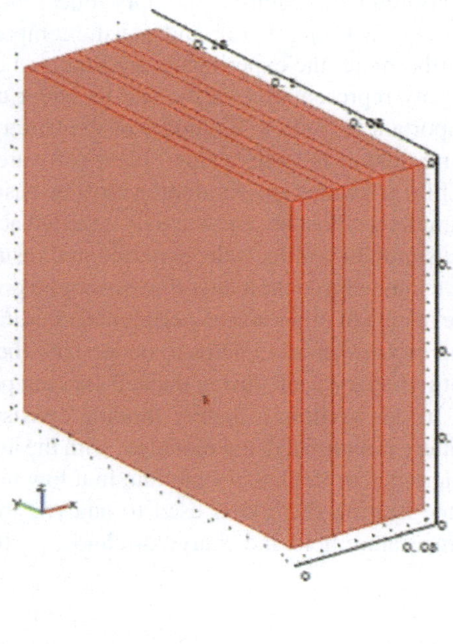

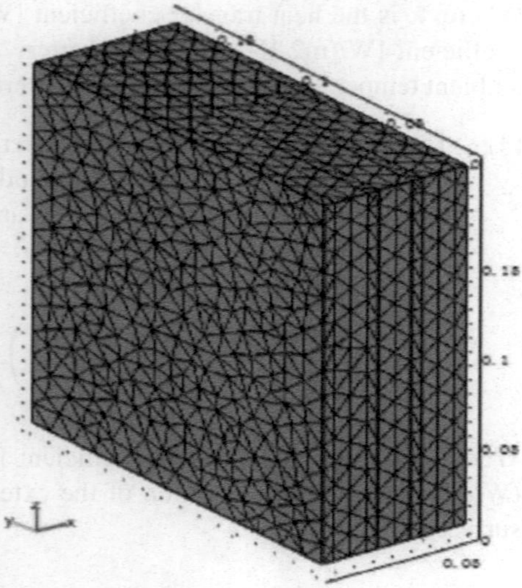

Fig. 6.25 Finite element mesh of the battery module

battery module is 64.1 mm along the x-axis, 170 mm along the y-axis and 198 mm along the z-axis. The battery cells and the heating aluminum plates are divided into hexahedral grids with guaranteed computer accuracy. The divided grids have a total of 16,742 and 24,825 degrees of freedom, and the divided grids are shown in Fig. 6.25. The material properties and heat generation rate of the 3 battery cells are the same as for the previous battery cells. The battery module consists of 3 battery cells connected in series with a capacity of 35A-h and a voltage rating of 11.1 V. The boundary conditions on the battery surface are changed when these 3 battery cells are formed into a battery module. The contact surface between the battery cell and the aluminum plate is part of the internal boundary. And, the exterior of the battery module is in contact with air.

6.2.2 *Initial and Boundary Conditions*

(1) The initial temperatures of the battery modules are selected as −40 °C, −30 °C, −20 °C and −10 °C respectively.
(2) The outer surface of the battery module is in contact with air, which belongs to the third category of boundary conditions. Namely, the boundary conditions for convective heat transfer are as follows:

$$-\lambda\left(\frac{\partial t}{\partial n}\right)_w = h\left(t_w - t_f\right) \tag{6.1}$$

Where, λ is the heat transfer coefficient [W/(m–K)]; h is the surface heat transfer coefficient [W/(m2-K)]; tw is the battery module wall temperature (K); tf is the ambient temperature (K); and n is the external normal to the heat transfer surface.

(3) The boundary conditions of the battery module cell and the heating aluminum plate are the second type of boundary conditions. Namely, the boundary conditions for the heat flux density are as follows:

$$-\lambda\left(\frac{\partial t}{\partial n}\right)_w = q_w \tag{6.2}$$

Where, λ is the heat transfer coefficient [W/(m–K)]; q_w is the heat flux density (W/m2); and n is the direction of the external normal at a point on the boundary surface.

6.2.3 *Model Validation and Analysis of Simulation Results*

1. Validation of the battery module simulation model

The battery module simulation model is validated by analyzing the change in temperature of each part of the battery module during the heating process. The battery module is preheated at −40 °C for 25 min and the results are compared to the simulation results of the battery module under the same conditions. A comparison of the simulated data with the experimental data for the centre of aluminum plate 4 at a typical location is shown in Fig. 6.26.

As can be seen from Fig. 6.26, the simulation data for the battery module heated at −40 °C for 25 min is in general agreement with the experiment data, and the temperature of the battery module reaches around 0 °C at the end of heating (t = 1500 s). Throughout the heating process, the simulated temperature is slightly lower than the experiment temperature, but the error does not exceed 2 °C. The accuracy of the battery module simulation model used in this section is adequate. Therefore, the relationship between heating power, heating time and heating termination temperature in this heating method can be simulated by the established battery module simulation model, in order to study the temperature field distribution of the battery module under different heating conditions and the heating effect of various heating methods.

2. Constant power heating at low temperatures

The distribution of the temperature field of the battery module when heated with a constant power (35 W) at low temperatures (−40 °C, −30 °C and −20 °C) is shown in Figs. 6.27, 6.28 and 6.29. Heating times are 25 min (−40 °C), 20 min (−30 °C)

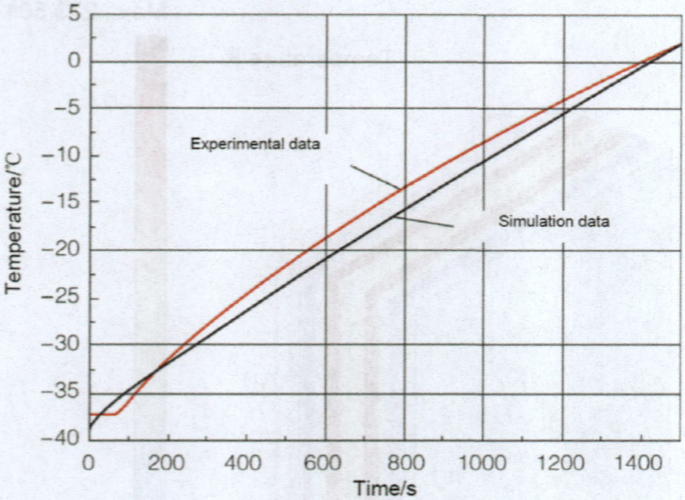

Fig. 6.26 Experiment verification of the centre of aluminum plate 4 in a typical position

and 30 min (−20 °C), step length is 1 s, and the number of steps is 1500, 1200 and 1800 respectively.

As can be seen from Fig. 6.27, the minimum temperature is −5.819 °C and the maximum temperature is 0.501 °C when the battery module is heated at −40 °C using heating aluminum plates. This means that the purpose of improving the working environment of the battery pack cannot be achieved under certain circumstances. Therefore, it is possible to achieve the desired operating temperature of the battery module by extending the heating time or increasing the heating power.

As can be seen from Figs. 6.27, 6.28 and 6.29, the minimum temperatures of the battery module after heating the heating aluminum plate at a power of 35 W for 30 min at −40 °C, −30 °C and −20 °C are -5.819 °C, 4.092 °C and 14.1 °C respectively. This represents an increase of 34.181 °C, 34.092 °C and 34.1 °C respectively compared to the initial ambient temperature. This shows that the use of heating aluminum plates to heat the battery modules enables the module temperature to rise to a higher value in a shorter period of time. Therefore, the pre-heating time of the battery pack for low temperature environments is reduced.

As also can be seen from Figs. 6.27, 6.28 and 6.29, the higher temperature areas of the battery module at the end of heating are concentrated on the two middle aluminum plates. The maximum temperatures of the battery module at each ambient temperature (−40 °C, −30 °C and −20 °C) are 0.501 °C, 9.55 °C and 19.52 °C respectively, which are 6.32 °C, 5.458 °C and 5.42 °C higher than the lowest temperatures. This is mainly due to the fact that the heating aluminum plates in the middle dissipate less heat than the plates on either side, resulting in lower temperatures of the aluminum plates on both sides. Compared to other heating methods such as heating the bottom of the

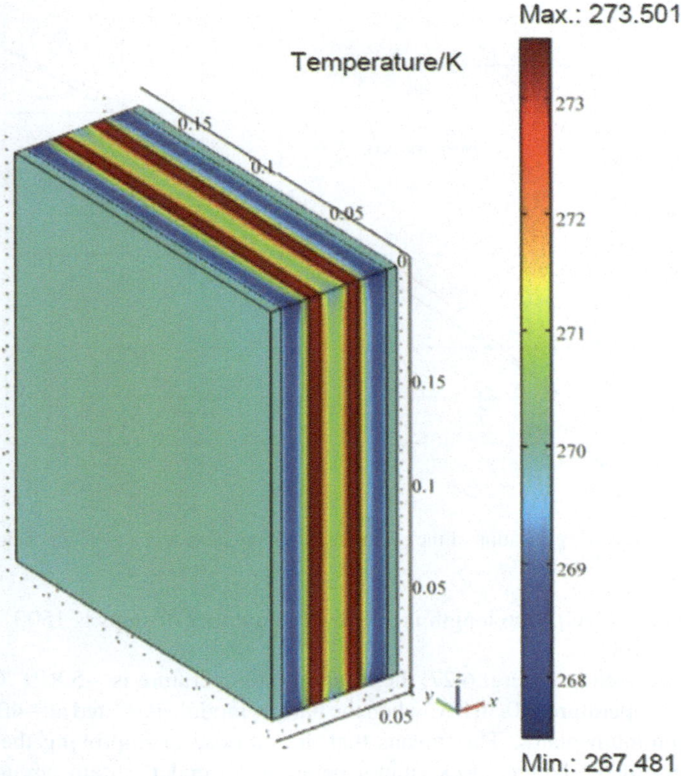

Fig. 6.27 Temperature field distribution of the battery module at an initial ambient temperature of −40 °C

pack, the addition of heating aluminum plates on both sides of the single cell is a significant improvement in terms of heating uniformity and heating time.

The use of heating aluminum plates improves the working environment of the battery modules considerably. In low-temperature environments, the temperature of the battery module can be brought up to the desired value in a relatively short period of time by means of this type of heating scheme. As a result, the performance of the battery pack in low temperature environments is improved, which can be easily realized in practical applications.

3. Heating strategies for PTC aluminum plates in low temperature environments

Through the above analysis, adding heating aluminum plates on each side of the battery cell to heat the battery module can well meet the heating needs of the battery module in terms of heating power and heating time. In order to guide the selection of the heating method and the design of the thermal management system of the battery pack in practical applications, it is investigated how to determine the heating time and heating power according to the external environment when using heated aluminum

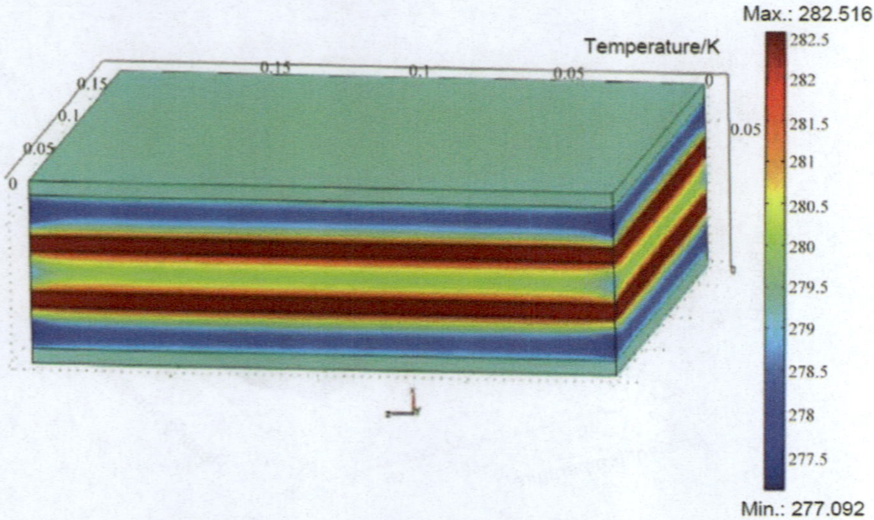

Fig. 6.28 Temperature field distribution of the battery module at an initial ambient temperature of −30 °C

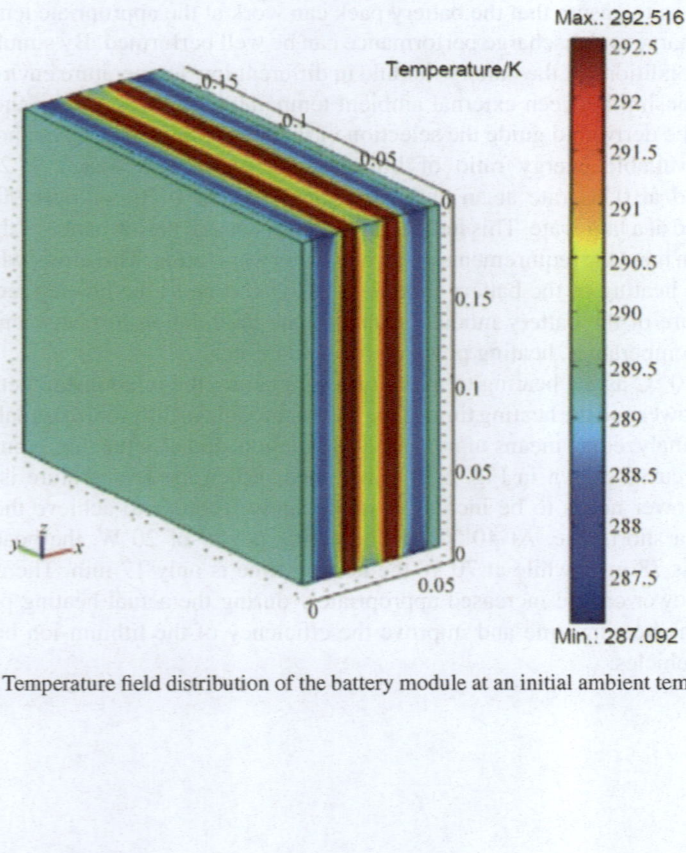

Fig. 6.29 Temperature field distribution of the battery module at an initial ambient temperature of −20 °C

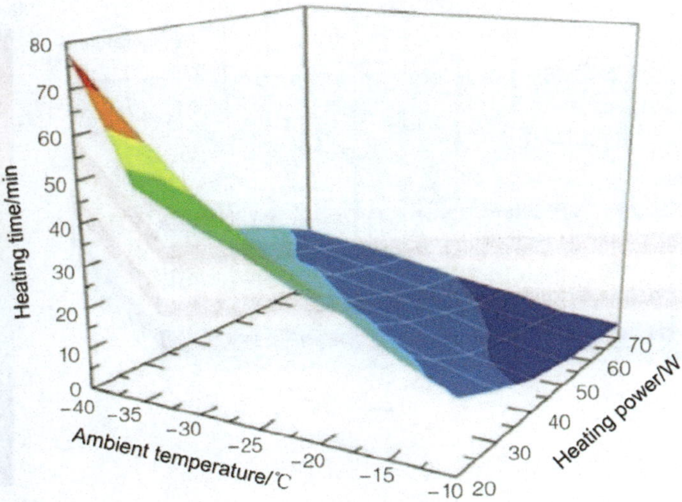

Fig. 6.30 Relationship between external ambient temperature, heating power and heating time

plates, so as to ensure that the battery pack can work at the appropriate temperature and the charge and discharge performance can be well performed. By simulating the heating conditions of the battery module in different low temperature environments, the relationship between external ambient temperature, heating power and heating time can be derived to guide the selection of the appropriate heating method.

The available energy ratio of lithium-ion batteries can reach 95.2% when discharged at 0.3C rate at an ambient temperature of 0 °C, or over 90% when discharged at a large rate. This indicates that the discharge performance of the battery at 0 °C can meet the requirements of normal battery operation. Therefore, when simulating the heating of the battery module, 0 °C is chosen as the heating termination temperature of the battery module to investigate the relationship between external ambient temperature, heating power and heating time.

Using 0 °C as the heating termination temperature, the relationship between the heating power and the heating time of the battery module in different external environments is analyzed by means of numerical simulation, and also the three-dimensional variation curve shown in Fig. 6.30 is obtained. When the temperature is low, the heating power needs to be increased significantly in order to achieve the heating target in a short time. At 40 °C and a heating power of 20 W, the heating time required is 78 min, while at 70 W the heating time is only 17 min. Therefore, the heating power can be increased appropriately during the actual heating process to shorten the heating time and improve the efficiency of the lithium-ion battery for electric vehicles.

By analyzing the variation relationship between external temperature, heating time and heating power, it can provide a reference for how to choose the heating method in practical applications and support the design of the battery thermal management system control strategy.

6.3 Study of Self-Heating Characteristics of PTC-Based Batteries

When designing a heating system for lithium-ion battery packs in electric vehicles, the source of energy for heating is an unavoidable issue. Normally this can be obtained externally, for example through a charging post when the electric vehicle is being charged. However, if there is no external electric energy, then it can only be provided by the electric vehicle itself. In the case of hybrid vehicles, the battery pack can be heated by engine coolant. However, for purely electric vehicles, the battery pack must be heated by its own energy. Based on this, Sects. 6.3 and 6.4 focus on heating the PTC with the battery pack's own power supply. This research will help to make it possible to pre-heat the battery pack of an electric vehicle by relying on the battery pack's own energy.

6.3.1 *Self-Heating Scheme and Experimental Design*

The scheme is shown in Fig. 6.31, based on which two modes of power supply can be used: externally powered and powered by the battery pack itself. The PTC resistance strip is embedded in a slotted aluminum plate. The aluminum plate is then placed between the sides of the battery cells in order to quickly transfer the heat generated by the PTC to the cells through the aluminum plate. The extra slots in the aluminum plate form air ducts for high temperature heat dissipation. This scheme allows for the

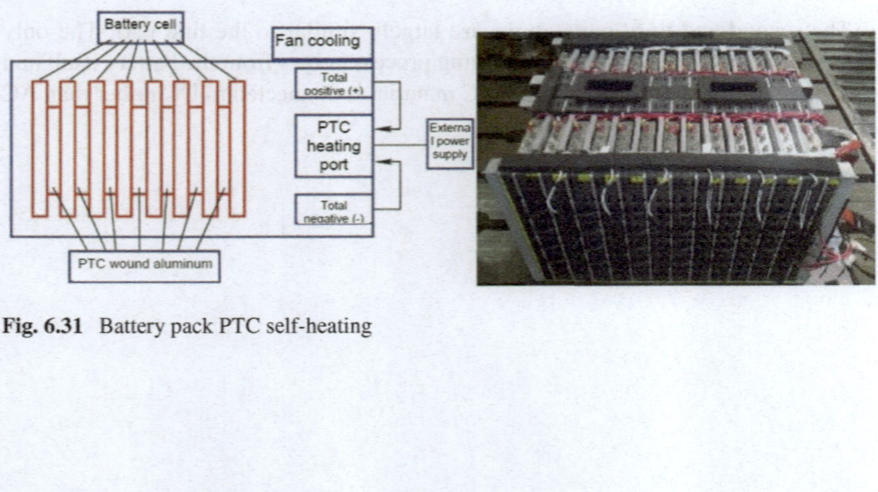

Fig. 6.31 Battery pack PTC self-heating

integration of low temperature heating and high temperature cooling of the battery pack.

The experiment consists of three parts:

① The first part is an experiment in which the battery pack is heated by external electrical energy. It is compared with the experiment in which the heating is done by its own electrical energy.
② The second part is an experiment where the battery pack is heated by its own power when fully charged (SOC = 100%).
③ The third part is an experiment where the battery pack is heated by its own power when not fully charged (SOC = approx. 60%).

The heating experiment, which relies on external electrical energy, consists of two heats.

The first heating raises the average temperature of the battery pack from −40 ~ −30 °C to −20 °C, while the second heating raises it to 0 °C again. Each heating is followed by a pulse charge and discharge experiment at different rates, which is used to study the recovery of the battery charge/discharge performance. Finally, after the second heating, a 1C constant rate discharge experiment is carried out to investigate how much capacity the battery can discharge after heating. The following steps can be followed:

① Place the battery pack in a −40 °C oven for 5 h to bring the battery pack temperature down to between −30 and 40 °C.
② Connect the PTC material to 220 V AC for heating.
③ Temporarily stop heating when the average temperature inside the battery box rises to −20 °C.
④ Let the battery pack subject to the pulse charge and discharge experiment at different rates.
⑤ Repeat the above ②.
⑥ Stop heating when the average temperature of the battery rises to 0 °C.
⑦ Repeat the above ④.
⑧ Implement 1C constant rate discharge experiment on the battery pack until the discharge cut-off voltage.

The second and third parts above are largely similar to the first part. The only difference is that the energy for the heating process comes from the battery itself and not from external 220 V AC, so the PTC material is connected to DC rather than AC current.

6.3.2 *Study on the Temperature Characteristics of Self-Heating Scheme*

1. Externally powered heating and self-heating process heating

As can be seen from the description in Sect. 6.3.1, the heating experiment is carried out twice. The first heating raises the average temperature of the battery pack from −40–−30 °C to −20 °C, while the second heating raises it from −20 °C to approx. 0 °C. The results of the two heating experiments are shown in Tables 6.6 and 6.7.

As can be seen from Tables 6.6 and 6.7:

(1) Despite the poor discharge capacity of the battery pack at low temperatures, the ability to supply power to the PTC for self-heating cannot be ignored. When the SOC is 100%, the temperature rise rate of the self-heated battery pack in the table is approximately 73% of that of the externally electrically heated battery pack, with significantly improved heating effect.

(2) The battery pack SOC also has an effect on self-heating. The higher the SOC of the battery pack, the higher the total voltage and the higher the temperature rise rate during heating.

However, the data in Tables 6.6 and 6.7 only represent the surface temperature of the battery, while the internal temperature of the battery is not available due to the inability to bury the sensor. Also, as the aluminum plate inside the battery pack is

Table 6.6 First heating

Heating conditions	Heating time/min	Heating	Temperature rise rate/(°C /min)
Externally-powered heating	31	−39.8 to −20.3 °C	0.629
Self-powered heating (SOC = 100%)	34.2	−39.4 to −20.7 °C	0.459
Self-powered heating (SOC = 60%)	43.33	−32 to −20.3 °C	0.270

Table 6.7 Second heating

Heating conditions	Heating time/min	Heating	Temperature rise rate/(°C /min)
Externally-powered heating	45	−23.2 to −0.5 °C	0.504
Self-powered heating (SOC = 100%)	48	−19.3 to −2.4 °C	0.352
Self-powered heating (SOC = 60%)	52	−19.7 to −2.7 °C	0.327

attached to the side of the battery cell, the temperature measured is likely to be that of the surface of the aluminum plate.

For external electric energy, the average temperature of the entire battery pack rises quickly because the PTC generates heat quickly due to the high and stable voltage. However, it takes time for the heat to pass from the PTC to the aluminum plate, then from the plate to the surface of the battery and finally from the surface to the inside of the battery, so the temperature inside the battery is lower than the actual average pack temperature measured.

For the self-heating of the battery pack, in addition to the heat generated by the external PTC material, heat is also generated inside the battery due to the battery discharge to PTC material, so that the internal temperature of the battery does not differ too much from the measured temperature in the case of externally-powered heating.

2. Battery pack voltage and temperature changes during self-heating

During the self-heating process of the battery pack, the temperature of the battery pack slowly increases. As the temperature rises, the voltage of the battery pack also rises. When the battery pack SOC = 100%, the changes in battery pack voltage and temperature during the two heating processes are shown in Figs. 6.32 and 6.33 respectively.

As shown in Fig. 6.32, the initial total voltage of the battery pack is 190 V and the initial average temperature is −36.4 °C. When the PTC material is connected, the total voltage quickly drops to 142 V and the internal resistance component voltage is 48 V. After 34.2 min, the heating is completed and the average temperature rises from −36.4 °C to −20.7 °C (i.e. an increase of 15.7 °C) and the total voltage rises from 142 to 172 V (i.e. an increase of 30 V), which indicates a 30 V reduction in the internal resistance component voltage.

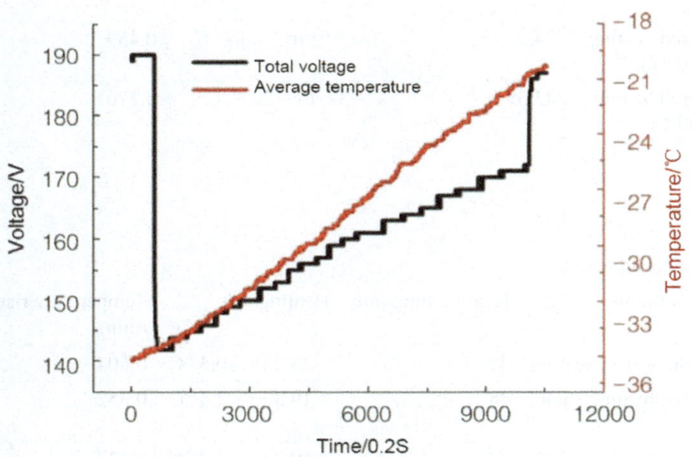

Fig. 6.32 First self-heating when the battery pack is fully charged

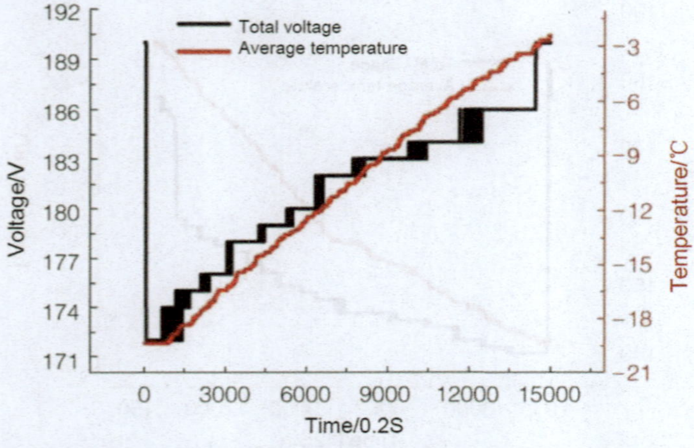

Fig. 6.33 Second self-heating when the battery pack is fully charged

Based on the same principle, the initial total voltage for this self-heating is 190 V and the initial average temperature is −19.3 °C, as shown in Fig. 6.33. While the cathode and anode of battery pack are connected to the PTC material, the total voltage drop is 172 V and the internal resistance component voltage is 18 V. The heating time is 48 min and the average temperature rises from −19.3 to −2.4 °C (i.e. an increase of 16.9 °C). The total voltage rises from 172 to 186 V, at which point the cathode and anode of battery pack are disconnected from the PTC material. At this point, the total voltage rises again to 190 V (i.e. an increase of only 4 V), which indicates that the internal resistance component voltage is only 4 V. The internal resistance decreases significantly due to the increase in temperature.

When the battery pack SOC = 60%, the changes in battery pack voltage and temperature during the two heating processes are shown in Figs. 6.34 and 6.35 respectively.

As shown in Fig. 6.34, power is drawn from the battery pack to the PTC material, which then heats the battery through the PTC material. While the battery pack is connected to the PTC material, the total battery pack voltage drops rapidly from 183 to 140 V. This indicates that the voltage loaded onto the PTC material is 140 V and that the difference between 183 and 140 V, i.e. 43 V, is divided by the internal resistance.

Due to the heating effect of the PTC material, the battery pack temperature slowly increases (see average temperature curve in Fig. 6.34). As the temperature increases, the internal resistance of the battery pack decreases, resulting in less voltage division and a slow increase in the total voltage loaded onto the PTC material (see total voltage curve in Fig. 6.34). The heating time is 43.33 min. At the end of heating, the average temperature rises from −32 °C at the time of unheating to −20.3 °C (i.e. an increase of 11.7 °C). The total voltage rises from 140 V at the beginning of the heating to 160 V (i.e. an increase of 20 V).

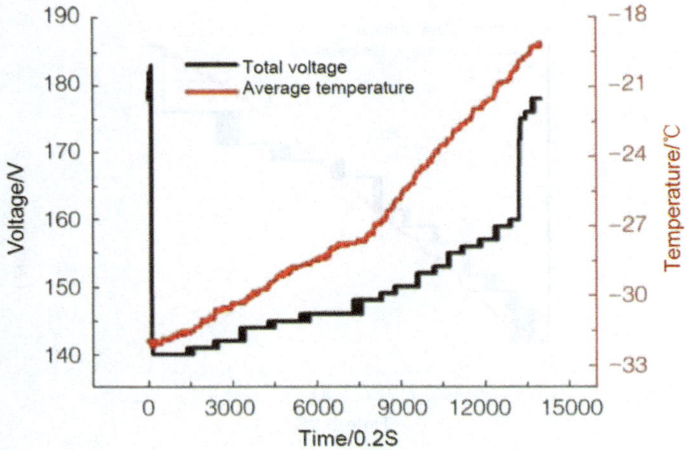

Fig. 6.34 First self-heating when the battery pack is not fully charged (SOC = 60%)

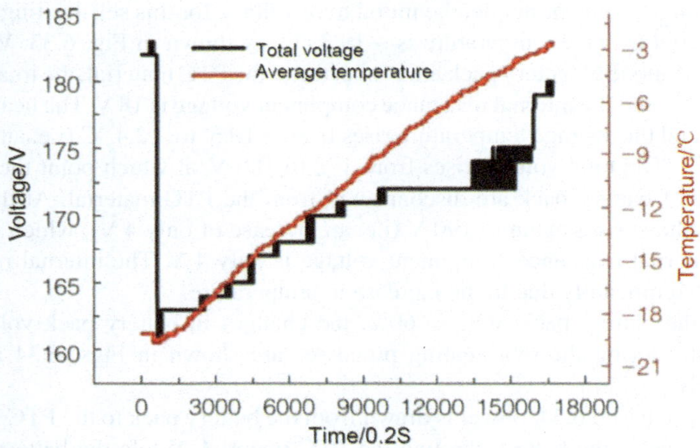

Fig. 6.35 Second self-heating when the battery pack is not fully charged (SOC = 60%)

Based on the same principle, as shown in Fig. 6.35, while the battery pack is connected to the PTC material, the total battery pack voltage drops rapidly from 182 to 161 V. This indicates that the voltage loaded onto the PTC material is 161 V and that the difference between 182 and 161 V, i.e. 21 V, is divided by the internal resistance.

Due to the heating effect of the PTC material, the battery pack temperature slowly increases (see average temperature curve in Fig. 6.35). As the temperature increases, the internal resistance of the battery pack decreases, resulting in less voltage division and a slow increase in the total voltage loaded onto the PTC material (see total voltage

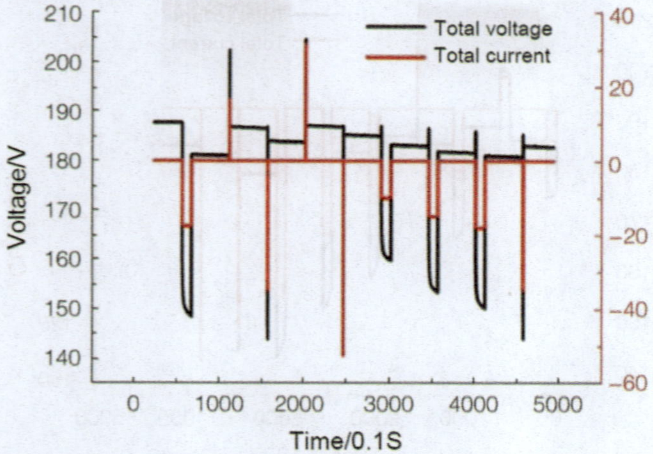

Fig. 6.36 Pulse charge and discharge experiment after the first heating using external electric energy (SOC = 100%)

curve in Fig. 6.35). The heating time is 52 min. At the end of heating, the average temperature rises from −19.7 to −2.7 °C (i.e. an increase of 17 °C). The total voltage rises from 161 V at the beginning of the heating to 175 V (i.e. an increase of 14 V).

3. Pulse charge and discharge performance of self-heated batteries

At the end of each heating, the battery pack can be subjected to a pulse charge and discharge experiment at different rates. This experiment gives an initial idea of the battery pack's discharge capacity after heating. The pulse charge and discharge curves after the first heating and after the second heating using external electric energy are shown in Figs. 6.36 and 6.37, respectively. The average battery pack temperature rises from −39.8 °C to −20.3 °C during the first heating, while it rises from −23.2 °C to 0.5 °C during the second heating.

As shown in Fig. 6.36, after the first heating using external electric energy, the battery pack can be discharged at 0.5C rate for 10 s, but cannot be discharged at 1C rate and cannot be charged at 0.5C rate. As shown in Fig. 6.37, the battery can be discharged at a 3C rate for 10 s after the second heating with external electric energy, but cannot be discharged at a 3.5C rate. For charge performance, the battery pack can be charged at 0.5C rate for 10 s, but not at 1C rate. This indicates that the discharge capacity of the battery pack has not recovered significantly after the first heating. However, after the second heating, the battery can be discharged at a 3C rate for 10 s, which indicates that the discharge performance of the battery pack has recovered considerably. For charge performance, it is difficult to recharge the battery pack as the SOC is 100%.

A pulse charge and discharge experiment is also carried out after full charge self-heating. The pulse charge and discharge curves after the first self-heating and second full charge self-heating are shown in Figs. 6.38 and 6.39 respectively. The average

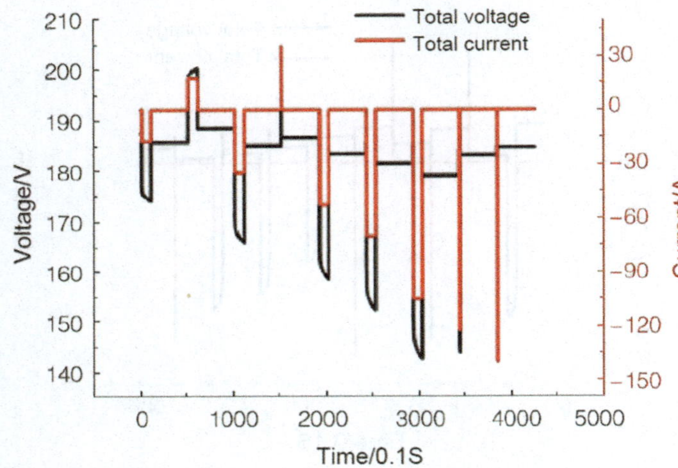

Fig. 6.37 Pulse charge and discharge experiment after the second heating using external electric energy (SOC = 100%)

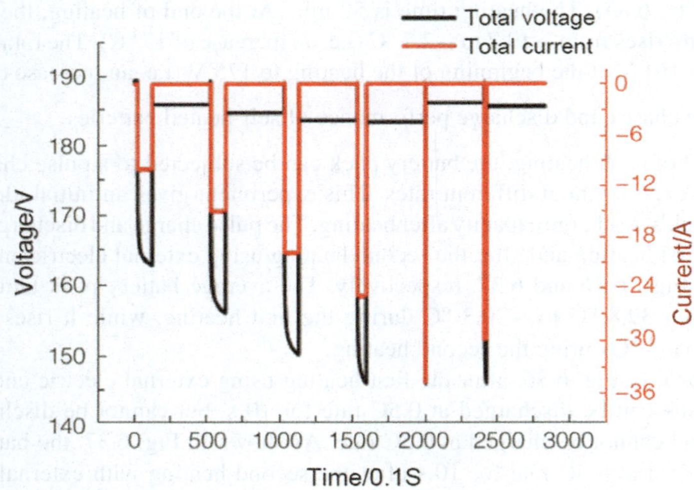

Fig. 6.38 Pulse charge and discharge experiment after the first full charge self-heating (SOC = 100%)

battery pack temperature rises from -39.4 °C to −20.7 °C during the first heating and from −19.3 to −2.4 °C during the second heating.

As can be seen from Figs. 6.38 and 6.39, when the battery SOC = 100%, after the first self-heating, the battery can be discharged at 0.57C rate for 10 s, but not at 1C rate, but after the second heating, the battery can be discharged at 3C rate. After the dynamic charge and discharge performance after full charge self-heating

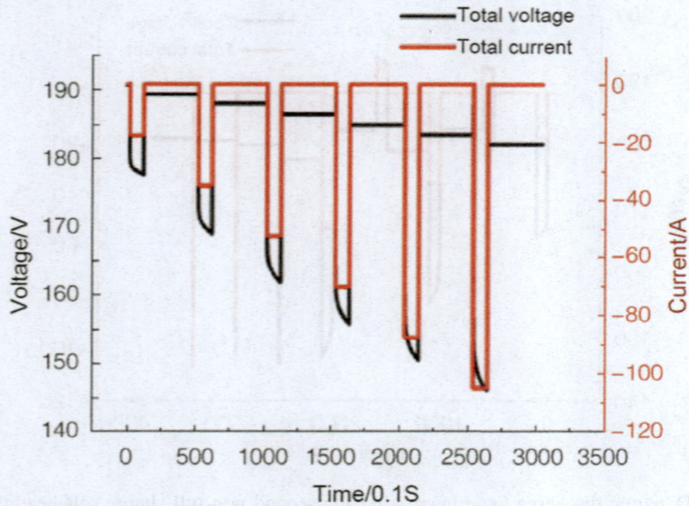

Fig. 6.39 Pulse charge and discharge experiment after the second full charge self-heating (SOC = 100%)

is compared with that after externally-powered heating, it is clear that the difference in charge and discharge capacity between the two battery packs after heating is not significant.

A pulse charge and discharge experiment is also carried out after non-full charge self-heating. The pulse charge and discharge curves after the first self-heating and second self-heating at non-full charge are shown in Figs. 6.40 and 6.41 respectively.

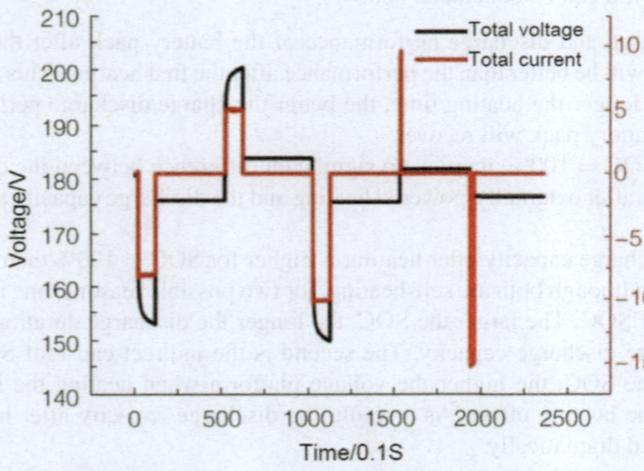

Fig. 6.40 Dynamic discharge experiment after the first non-full charge self-heating (SOC = 60%)

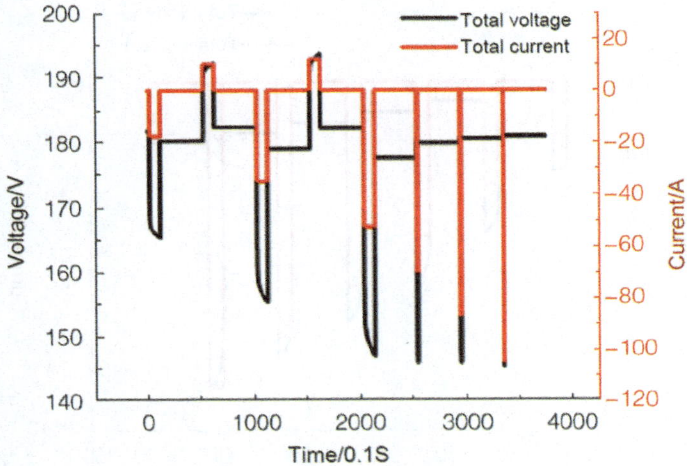

Fig. 6.41 Dynamic discharge experiment after the second non-full charge self-heating (SOC = 60%)

The average battery pack temperature rises from −32 °C to −20.3 °C during the first heating and from −19.7 °C to -2.7 °C during the second heating.

As can be seen in Figs. 6.40 and 6.41, when SOC = 60%, after the first self-heating, the battery pack can be discharged at 0.29C rate for 10 s, but not at 0.43C rate for 10 s. After the second heating, the battery pack can be discharged at 1.5C rate for 10 s, but not at 2C rate for 10 s. For charging capability, after the first heating, the battery pack can be discharged at 0.14C rate for 10 s, but not at 0.29C rate. After the second heating, the battery can be discharged at 0.34C rate for 10 s.

In summary, it can be concluded below:

(1) The charge and discharge performance of the battery pack after the second heating will be better than the performance after the first heating. This indicates that the longer the heating time, the better the charge/discharge performance of the battery pack will recover.

(2) When SOC = 100%, there is no significant difference between the discharge capacity after externally-powered heating and the discharge capacity after self-heating.

(3) The discharge capacity after heating is higher for SOC = 100% than for SOC = 60%, although both are self-heating, for two possible reasons: one is a direct result of SOC. The larger the SOC, the longer the discharge duration and the better the discharge capacity. The second is the indirect cause of SOC. The larger the SOC, the higher the voltage platform when heating the PTC, the better the heating effect. As a result, the discharge capacity after heating is improved dramatically.

4. Comparative analysis of the discharge performance of batteries with self-heating and external heating

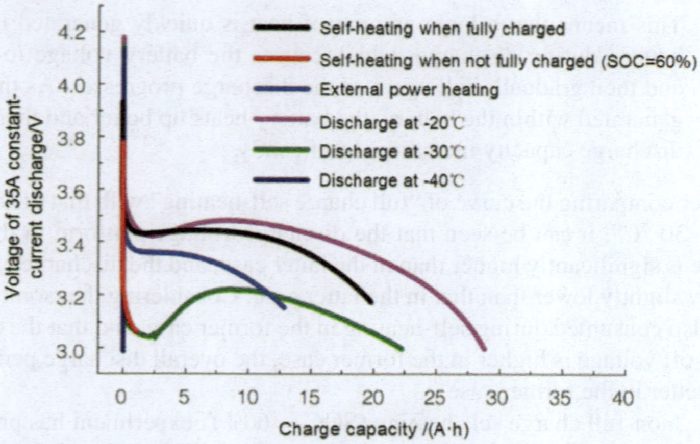

Fig. 6.42 Comparison of 1C rate discharge capacities of heated and unheated batteries

The 1C rate discharge curve after heating of the battery pack is compared with the 1C rate discharge curve of an unheated battery cell at low temperature. The results are shown in Fig. 6.42. The voltage of the battery pack is also converted into an average single voltage for comparison purposes.

The following points can be seen from Fig. 6.42:

(1) The 1C rate discharge capacity after full charge self-heating is 19.834 A-h, while the 1C rate discharge capacity after full charge externally-powered heating is 12.853 A-h. Therefore, the former is significantly greater than the latter. This is because during full charge self-heating, the battery pack needs to be discharged to the PTC material. Therefore, in addition to the heat generated by the external PTC material during the self-heating process, heat is also generated inside the battery. In contrast to externally-powered heating, i.e. the heat generated by the PTC passes through the aluminum plate and then through the battery housing, the self-heating of the battery is more effective in raising the internal temperature of the battery. Therefore, for the same SOC, self-heating of the battery is more effective than externally-powered heating. As a result, the 1C rate discharge capacity of the heated battery is also higher.

(2) After comparing the curve of "externally-powered heating "with that of "discharge at −30 °C", it can be found that the discharge capacity in the former case is smaller than that in the latter case. The reasons for this include the following two points:

① In the former case, the cut-off voltage is set high and the average single voltage reaches 3.1 V before it is cut-off. However, if the cut-off voltage is also set at 3.1 V in the latter case, there will be little or no discharge.

② In the latter case, the voltage during the discharge process changes according to the law of first falling, then rising and finally falling again.

This means that a large amount of heat is quickly generated inside the battery during discharge, which causes the battery voltage to rise first and then gradually fall again as the discharge progresses. As this heat is generated within the battery, the battery heats up better and therefore the discharge capacity increases significantly.

(3) After comparing the curve of "full charge self-heating" with that of "discharge at −30 °C", it can be seen that the discharge voltage platform in the former case is significantly higher than in the latter case, and the discharge capacity is only slightly lower than that in the latter case. Considering that some capacity is also consumed during self-heating in the former case, and that the discharge cut-off voltage is higher in the former case, the overall discharge performance is better in the former case.

(4) The "non-full charge self-heating (SOC = 60%)" experiment has proved that it's almost impossible to discharge in this case. The reasons for this include three main points:

① Since the battery itself has an SOC of only 60% and part of its capacity is consumed during the self-heating process, the battery SOC will be even smaller.

② Due to the low temperature of −40 °C, the brief heating does not allow for a large increase in the internal temperature of the battery.

③ The cut-off voltage is set too high and the 1C discharge rate is too large.

In summary, the following two points emerge from the analysis:

(1) The actual internal temperature rise of the battery during the heating process is very important. As it is not possible to bury sensors inside the battery for temperature measurement and the thermal conductivity coefficient of the battery is also small, the temperature measured on the surface is hardly reflective of the actual internal temperature. If the surface temperature is used as the internal temperature of the battery to make a prediction of the battery discharge capacity, it is prone to large errors.

(2) Self-heating of the battery is not only an emergency method when there is no external power supply, but also enables the internal heating of the battery through the heat generated by the battery itself during discharge. As a result, the discharge capacity in a short period of time after self-heating even exceeds the discharge capacity when relying solely on external electric energy for heating.

6.3.3 Battery Pack PTC Self-Heating Characteristics

In order to investigate the change in heat generation rate during self-heating of the PTC material, the heating current is monitored with an ammeter at average intervals of 1–2 min during the battery pack PTC material self-heating experiment, and the battery pack voltage is measured using the battery management system. By multiplying the current and voltage, the amount of power supplied to the PTC material by the battery pack can be calculated. Assuming that all the power is used for heat generation, the heat generation rate of the PTC material can be calculated.

For self-heating of the battery pack, in addition to the heat generation rate of the PTC material, the heat generation rate of the battery pack's own discharge process also needs to be considered. The heat generation rate of the battery during discharge can be calculated using the Bernardi battery heat generation rate model, and combined with the monitored voltage, current and temperature data.

The curves of heating power and heating current for the first and second full charge and non-full charge self-heating are shown in Figs. 6.43, 6.44, 6.45 and 6.46.

As can be seen from Figs. 6.43, 6.44, 6.45 and 6.46:

(1) During the first self-heating, the heating current tends to decrease. This is because the heating current decreases as the resistance of the PTC material increases with increasing temperature. However, during the first self-heating, although the heating current is decreasing, the heating power is increasing, because the total voltage of the battery pack is also increasing as the pack heats up. Compared to the second self-heating, the curve of the first self-heating process looks more complex.

(2) During the second self-heating, the total voltage of the battery pack has largely stabilized, although it is also increasing, as the temperature of the battery pack has increased significantly. However, the resistance of the PTC material is still increasing, so the heating current is decreasing and the heating power is also decreasing.

Taking the first full charge self-heating (see Fig. 6.43) as an example, the current is measured 19 times during the entire heating process and the corresponding total

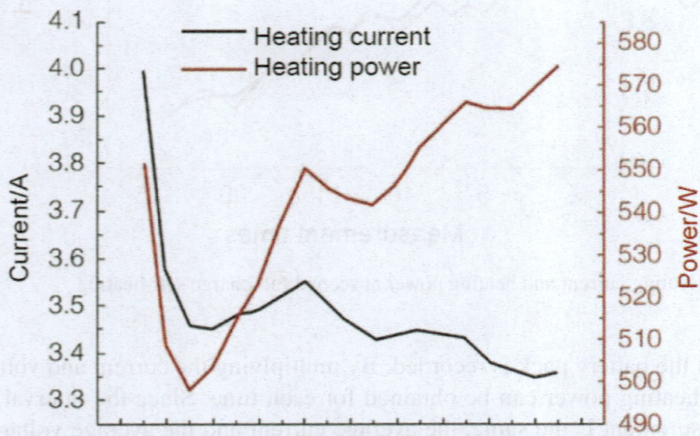

Fig. 6.43 Heating current and heating power at first full power self-heating

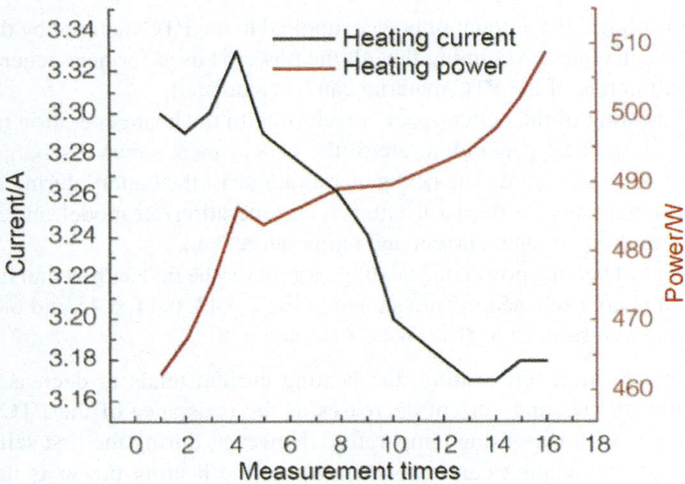

Fig. 6.44 Heating current and heating power for first non-full charge self-heating (SOC = 60%)

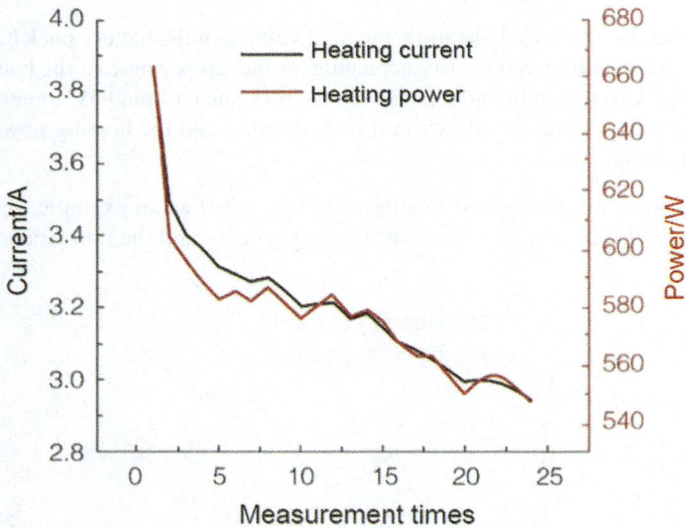

Fig. 6.45 Heating current and heating power at second full charge self-heating

voltage of the battery pack is recorded. By multiplying the current and voltage each time, the heating power can be obtained for each time. Since the interval between each measurement is the same, the average current and the average voltage can be found first, and then the average power can be derived. Finally, the average power is multiplied by the total process duration. Assuming that all the power is used to generate heat, the total amount of heat generated can be found. Based on the actual

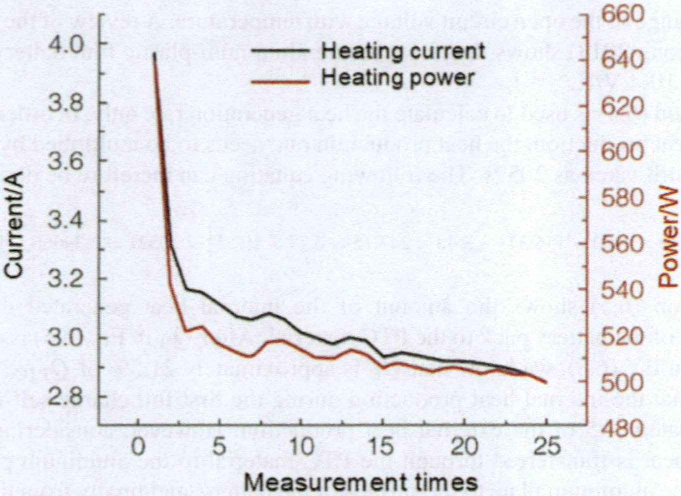

Fig. 6.46 Heating current and heating power at second non-full power self-heating (SOC = 60%)

measured data, the average of the 19 measured currents is 3.48 A and the average voltage is 156.37 V. The average heating power is therefore 544.17 W and the duration is 34.2 min (i.e. 2052s), so the heat produced by the external PTC material at the first full charge self-heating is as follows:

$$Q_{\mathrm{PTC}} = 544.17\mathrm{W} \times 2052\mathrm{s} \approx 1.116637 \times 10^{6}\mathrm{J} \tag{6.3}$$

Where, Q_{PTC} is the amount of heat produced by the PTC material at full charge self-heating.

As the battery pack is discharged to the PTC material during self-heating, the internal heat generation rate during the first full charge self-heating is also required. Since the average current from the 19 measurements is known to be 3.48 A, the formula for the Bernardi heat generation rate is derived as follows:

$$q_{\mathrm{B}} = I_{\mathrm{L}}(E_0 - U_{\mathrm{L}}) - I_{\mathrm{L}}T(\mathrm{d}E_0/\mathrm{d}T) \tag{6.4}$$

Where, IL is the average current from 19 measurements, which is 3.48 A. E0 is the initial voltage of the battery pack before it is discharged to the PTC material, which is 190 V. UL is the voltage of the battery when it is discharged to the PTC material, which is taken as the average of 19 measurements and is 156.37 V. T is the average temperature of the battery during discharge to the PTC material and is calculated on the basis of the Kelvin temperature. During the first full charge self-heating, the battery pack temperature rises from −39.4 °C to −20.7 °C. The average temperature can be taken to be approximately −30 °C, i.e. 243.15 K. dE0/dT is the rate of change of the open circuit voltage with temperature. A review of the reference (Guangchong 2013) shows that for a square aluminum-plastic film battery, dE0/dT is 2.79 × 10-4 V/K.

Equation (6.4) is used to calculate the heat generation rate only. In order to calculate the heat production, the heat production rate needs to be multiplied by the time, which is still taken as 2052s. The following equation can therefore be derived:

$$Q_B = \left[3.48 \times (190 - 156.37) - 3.48 \times 243.15 \times 2.79 \times 10^{-4}\right] \times 2052\text{J} = 2.39666 \times 10^5\text{J} \quad (6.5)$$

Equation (6.5) shows the amount of the internal heat generated during the discharge of the battery pack to the PTC material. After Q_B in Eq. (6.5) is compared to Q_{PTC} in Eq. (6.3), we know that Q_B is approximately 21.5% of Q_{PTC}. So, it can be seen that the internal heat production during the first full charge self-heating is approximately 1/5 of the external heat production. However, considering that the external heat is transferred through the PTC material to the aluminum plate, then through the aluminum plate to the surface of the battery, and finally from the surface of the battery to the internal parts of the battery, while the internal heat is generated directly inside the battery, so although the internal heat production is only 1/5 of the external heat production, the effect of the internal heat production on the heating effect of the battery cannot be ignored. In the following section, the effect of external and internal heat generation is simulated and analyzed using Fluent thermal simulation software.

6.4 Simulation Analysis of Self-Heating Based on PTC Battery

In this section, a 1/4 model of the battery box will be built in Gambit software using Fluent-based thermal simulation of the battery pack and then imported into Fluent for thermal simulation calculations. Finally, a temperature cloud can be used to show the distribution of the internal temperature field of the battery box. During the simulation, the temperature values calculated at each step will be recorded so that the temperature change at any point in the heating process of the battery pack can be described at each moment.

6.4.1 Simplification of the Model

In order to carry out a thermal simulation, the geometric model of the battery pack must first be established, which can generally be done in Gambit software. When building the geometric model, the model is simplified as below:

(1) The battery cell is regarded as an isotropic object made of a single material. Furthermore, thermophysical parameters such as thermal conductivity coefficient and specific heat capacity are equal everywhere within the battery cell and do not vary with temperature during the heating process. This assumption is mainly based on the fact that the simulation is for the whole battery pack. This assumption cannot be made if only a battery cell is simulated thermally.
(2) Objects such as connections, wires and insulating sheets inside the battery case are ignored. This assumption is made mainly to simplify the model and thus speed up the calculations. At the same time the influence of these objects on the heating and warming process of the battery is relatively small.
(3) The PTC material is ignored and the heat generation rate of the PTC material is taken directly as the heat generation rate of the aluminum plate. This assumption is made mainly due to the high thermal conductivity coefficient of the aluminum plate and the fact that the PTC material is embedded in the grooves of the aluminum plate and therefore the heat generated by the PTC material can be quickly transferred to the aluminum plate.
(4) The aluminum slots are neglected and the aluminum sheet is considered as a flat plate that is flat on both sides. This assumption is made mainly because the battery pack is heated without a fan and therefore the aluminum plate slots are not required as air ducts, and because the PTC material is also embedded in the aluminum plate slots in the battery pack. If the PTC material is considered to be part of the aluminum plate, then the aluminum plate slots can be ignored as they are filled with PTC material.
(5) As the fan is not switched on during the heating and self-heating of the PTC material, the designs of the fan and the corresponding air inlet and outlet are ignored in the model. It is assumed that the interior of the battery pack is closed during the thermal simulation.
(6) For a 1/4 model of a battery pack consisting of 12 cells, the temperature field of the entire pack can be simulated by setting up two symmetry surfaces.
(7) The specific structure of the battery box housing is ignored and the battery box housing is considered as a regular rectangular housing. This assumption simplifies the model and speeds up the calculation. Air exists between the housing and the battery cells, so the shape of the housing affects the airflow field distribution. However, as the convective heat transfer of the gas is negligible compared to the direct contact heat transfer of the aluminum plates, the specific shape of the battery box housing does not have a large effect on the temperature field distribution of the entire battery pack.

6.4.2 Establishment of the Geometric Model

Following the principles of the model simplification in the previous subsection, the final 1/4 model consists of four main components:

① 12 battery cells.
② 11 aluminum plates and a half aluminum plate next to the symmetrical surface.

Table 6.8 Geometric modeling related parameters (1/4 model)

Structure	Dimensions (L x W x H, in mm)
Single battery	180 × 14.7 × 246
Aluminum plate	170 × 5 × 198
Battery housing	220 × 262.5 × 296

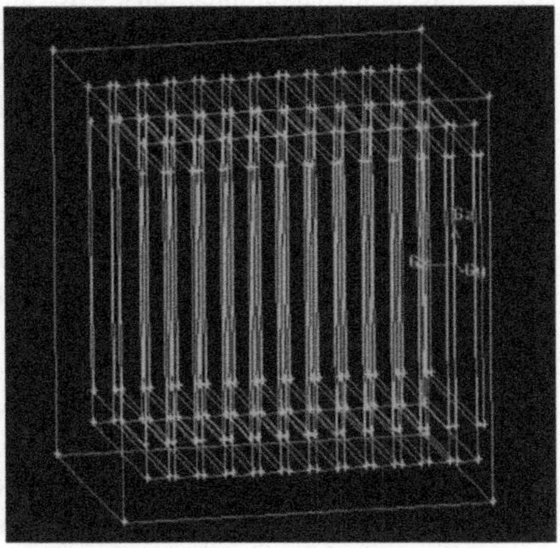

Fig. 6.47 1/4 model of a battery pack (12 cells)

③ Battery box housing.
④ Air between the housing and the aluminum plate and the battery cell.

The relevant specification parameters for modeling are shown in Table 6.8.

When building the geometric model, the aluminum plates are placed between the two battery cells, all aluminum plates and battery cells are placed inside the battery housing, the air layer is set between the battery housing and the aluminum plates and battery cells, and symmetrical surfaces are set on the right side and rear side of the entire model respectively. The geometric model completed in Gambit software is shown in Fig. 6.47.

6.4.3 Analysis of Simulation Results

When the first full charge heating is taken as an example, the heat generation rate is calculated according to its principle and the results are then imported into Fluent software for calculation. As the current is measured 19 times during the first full charge self-heating, there are 19 results for the external and internal heat generation rates. The duration of the first full charge self-heating is 2052 s, which is divided into 19 time periods, each lasting 108 s. Therefore, the value of the internal and external heating rate of the battery is changed every 108 s, and then written into the UDF program and imported into Fluent for calculation.

In the specific calculation of the heat generation rate, the volume of the aluminum plate and the battery cell needs to be taken into account. To calculate the external heat generation rate, the external heating power is divided by the total volume of all aluminum plates, as the PTC material is ignored and the external heat production is considered to be generated directly by the aluminum plates themselves. There are 48 cells in the battery box, which are arranged in two rows, each consisting of 24 cells. For these 24 cells, one aluminum plate is arranged between every two cells. Thus there are 23 aluminum plates, or 46 plates in two columns. The size of each aluminum plate is 170 mm x 5 mm x 198 mm, so the volume of each plate is 1.683 $\times$ 10^{-4}m^3 and the total volume of the 46 pieces is 7.7418 $\times$ 10^{-3}m^3.

Based on the same principle, the total volume of all the cells is also taken into account when calculating the internal heat generation rate. As the cell size is 180 mm $\times$ 14.7 mm x 246 mm, the volume of one cell is 6.5092 $\times$ 10^{-4}m^3 and the total volume of 48 cells is 3.1244 $\times$ 10^{-2}m^3.

By dividing the external heating power and internal heat generation rate from each of the 19 measurements by the corresponding volume, the heat generation rate per internal unit volume of the external aluminum plate and the battery cell can be obtained. The results are written into the UDF program and imported into Fluent for calculation. For comparison purposes, the values are first calculated for the case of combined internal and external heating, and then for the case of external heating only.

The relevant thermal physical parameters for the square aluminum-plastic film cells used in this section include:

① The average density of the batteries is 2182.7 kg/m^3.
② The specific heat capacity of the batteries is 1100 J/(K-kg).
③ The average thermal conductivity coefficient of the batteries is 0.895 W/(m–K).

As the experiment is carried out in the closed environment of the thermostat and the blower is not switched on inside the thermostat, the value of the convective heat transfer coefficient between the battery box housing and the external environment can be determined as 5 W/(m^2-K) according to engineering experience.

The experiment results are shown in Fig. 6.48, where the three curves are as follows:

① Simulated average temperature of all cells when heat is generated by external PTC material only during self-heating.
② Simulated average temperature of all cells when heat is generated internally and externally together during self-heating.
③ The average temperature of the battery pack measured in experiments.

As it is not possible to bury the sensor inside the cell during the experiment, the temperature is mainly measured on the outside of the cell.

As can be seen from Fig. 6.48, the average temperature of all the cells when heat is generated internally and externally together is a good fit to the average temperature of the battery pack measured in experiments. However, during the actual experiment, the temperature measured is only the temperature on the outside of the battery cell. The

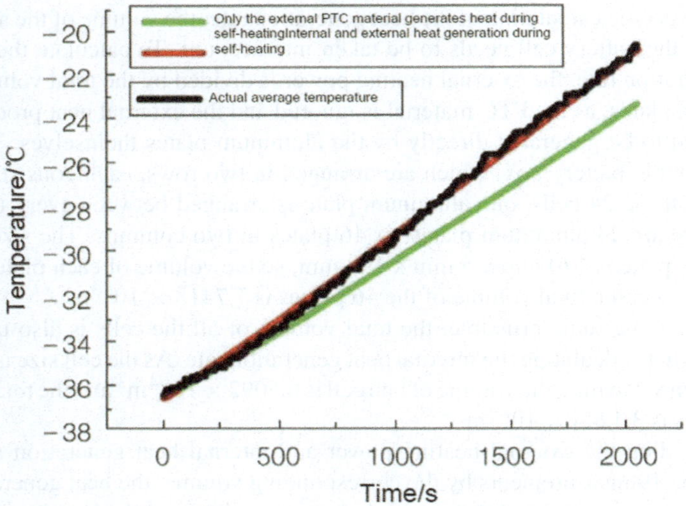

Fig. 6.48 Comparison of the simulated temperature rise at the first full charge self-heating with the actual situation

temperature measured on the outside of the cell must be higher than the temperature inside the cell because of the PTC heat generation material and the high thermal conductivity aluminum plate outside. However, the simulation results show that the average temperature of all cells is as high as the actual average temperature measured on the outside of the cell, which is due to simulation errors. During the simulations, the heat generation rate is attributed directly to the aluminum plate, which ignores the heat transfer process from the PTC material to the aluminum plate. Furthermore, the contact from the aluminum plate to the side of the battery cell is seen as a completely seamless contact, whereas in reality there is a contact thermal resistance between the aluminum plate and the side of the cell, thus resulting in a reduced heat transfer. Although the simulated the average temperature of all cells when heat is generated internally and externally together is a good fit to the actual measured average temperature on the outside of the cells within the battery pack, this is only a coincidence. In reality, the internal temperature of the cells cannot be as high as the actual measured external temperature.

One more thing can be seen in Fig. 6.48: the importance of internal heat generation. If the heat generated during the discharge to the PTC material in the self-heating of the battery pack is not taken into account, the temperature obtained in the simulation will be 2.44 °C lower than if the internal heat generation of the battery pack is taken into account. The above simulation results also show that the heat generated during the discharge to the PTC material in the self-heating of the battery pack should not be ignored.

Using the capabilities of Fluent software, the temperature distribution for each cell inside the battery pack is obtained, as shown in Fig. 6.49.

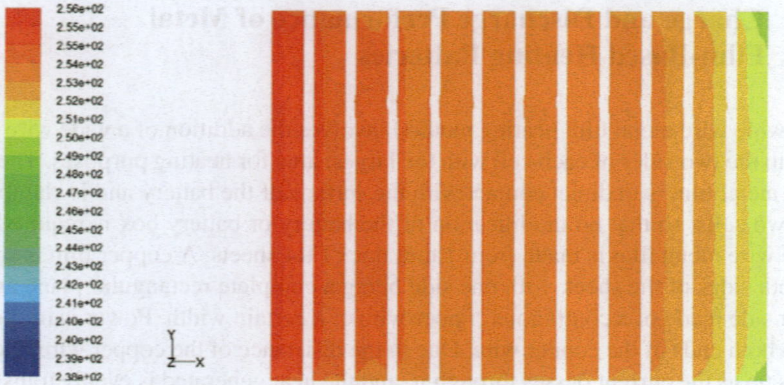

Fig. 6.49 Temperature distribution of the 12 battery cells at z = 0 (central surface) when heat is generated internally and externally together

In Fig. 6.49, the leftmost cell is the one close to the symmetrical surface and therefore has a high temperature; the rightmost cell is the one closest to the outside and therefore has a low temperature. As the above modeling takes into account air and housing, the simulated outermost layer temperature of the battery cell increases compared to the initial temperature. If air and housing are not taken into account, then the outermost temperature of the cell is set to the external ambient temperature when setting the convective heat transfer conditions. The simulated outermost layer temperature of the cell is then still the ambient temperature, which does not correspond to the actual situation. It is therefore necessary to consider the air and housing in the simulation.

By comparing the simulation results, it can be seen that the average temperature of the leftmost cell in Fig. 6.49 rises to −18.8 °C at the end of the simulation and the rightmost cell rises to −23.47 °C at the end of the simulation, i.e. a difference of 4.67 °C between the two. This shows that the temperature difference between the cells can be controlled within 5 °C at the first full charge self-heating. This is due to the fact that the outer side of the outermost battery cell is not fitted with aluminum plates of PTC material. If the aluminum plates are added, then there will be an aluminum plate on each side of any battery cell, resulting in a smaller temperature difference. Considering that the outermost aluminum plate is only in contact with the side of a single cell, a smaller aluminum plate can be added to the outermost side or a small amount of PTC heat generation material can be embedded in the aluminum plate.

6.5 Charge and Discharge Performance of Metal Film-Based Heating Batteries

The wide wire metal film heating method involves the addition of a wide wire metal film to the two sides of each cell with the largest area for heating purposes. The wide wire metal film is in direct contact with the surface of the battery and is clamped by the two cells, so that no modification of the battery or battery box is required. The wide wire metal film is made from 1 mm thick FR4 sheets. A copper film is applied to both sides of the sheet, with one side being a complete rectangular plane and the other side made of a continuous copper wire of a certain width. Power is introduced from both ends of the copper wire. Due to the resistance of the copper wire, the wire heats up as the current passes through it, and the heat generated is evenly transferred to the battery through the copper film plane, so that the battery is heated. In view of the need for heat dissipation, FR4 can be replaced by an aluminum-based material. A slotted aluminum cooling plate with a certain thickness is inserted between the two cells so that the cells can be dissipated through the slotted aluminum plate. The wide wire metal film heating method has the following main advantages:

(1) The heating device is simple, easy to install, easy to realize, reliable and has a wide working temperature range.
(2) The heating power of the heating unit can be flexibly adjusted to meet the heating requirements under different circumstances.
(3) Direct contact with and clamping by the battery cell reduces heating power losses.
(4) As there is a wide wire metal film on both sides of each battery cell, the whole battery pack is heated evenly, thus avoiding the effect of uneven heating on the consistency of the battery pack.
(5) Adding an aluminum cooling plate between the battery cells can dissipate heat from the battery pack.

In order to analyze the heating effect, a wide wire metal film is used to heat the square aluminum-plastic film cell. The wide wire metal film is powered by an external DC supply. Three square aluminum-plastic film battery cells are connected in series to form a lithium-ion battery pack.

A wide wire metal film is applied to both sides of each cell with the largest area and the 3 cells are laminated together as shown in Fig. 6.50. In order to match the heating experiment results to the vehicle lithium-ion battery pack, three battery packs fitted with a wide wire metal film are loaded into the battery box which is placed in a thermostat as shown in Fig. 6.51.

In order to study the heating effect of wide wire metal films on lithium-ion battery packs at low temperatures, the packs are placed in a −40 °C thermostat. The standing time is extended from 5 to 8 h because the battery box has a certain insulation effect. After the standing time, the wide wire metal film is connected to the power supply to heat the battery pack. The wide wire metal film can be powered by an external

Fig. 6.50 3 batteries in series

Fig. 6.51 Lithium-ion battery pack box

power supply or by the battery pack itself. In this section, heating experiments are carried out for each of the two power supply methods.

6.5.1 Constant Current Charge and Discharge Performance After Externally-Powered Heating in Low Temperature Environment

1. The 1C rate constant current discharge of the battery pack after 15 min of heating at different heating powers

The curves for the 1C rate constant current discharge of the battery pack are shown in Fig. 6.52 after 15 min of heating with 240 W, 120 W and 90 W at an ambient temperature of −40 °C. The figure shows that as the heating power increases, there is a large difference between the initial and mid-term discharge voltages. For example, the discharge voltage of the battery pack after heating with 240 W power is on average 0.53 V higher than after heating with 90 W, with a maximum voltage difference of 1.38 V. It is important to note that the amount of heating power has a small effect on

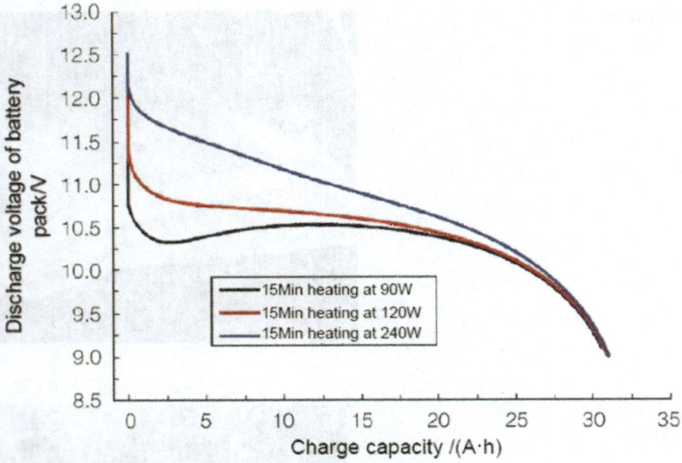

Fig. 6.52 1C rate constant current discharge of the battery pack after heating at different heating powers

the discharge capacity of the battery. For example, the discharge capacity of a battery heated at 90 W is 30.547A-h and that of a battery heated at 240 W is 30.997A-h, i.e. a difference of only 0.45A-h. This proves that under the condition of same heating time, increasing the heating power can raise the discharge voltage and discharge power of the battery pack, but cannot significantly enhance the discharge capacity of the battery.

The curves for 1C rate constant current discharge of the battery pack after 15 min of heating with 3 powers are shown in Fig. 6.53. Also, the curves of constant current discharge of the unheated cells at 1C rate at low temperatures can be seen therein. The battery pack consists of 3 cells. During discharge, the discharge voltage of the 3 cells is not identical. For comparison with the cells, the average voltage of the 3 cells is taken as the discharge voltage of the battery pack. After heating the battery pack with 90 W power for 15 min, in the early stage of discharge, the average discharge voltage of the battery pack is close to that of the battery cell at −20 °C; in the middle and late stage of discharge, the average discharge voltage of the battery pack is higher than that of the battery cell at −20 °C and close to that of the battery cell at −10 °C, and the discharge capacity of the battery pack is almost equal to that of the battery cell at −10 °C. This means that although the external heating stops, some of the heat generated by the discharge of the battery pack is exchanged with the −40 °C environment and the remaining heat continues to heat the battery, causing the temperature of the battery pack to continue to rise from −20 °C. After heating the battery pack with 120 W power for 15 min, in the early stage of discharge, the average discharge voltage of the battery pack is slightly lower than that of the battery cell at −10 °C; in the middle and late stage of discharge, the discharge curve of the battery pack gradually coincides with that of the battery cell at −10 °C. However, there are no cases where the discharge voltage of the battery pack after 90 W power heating

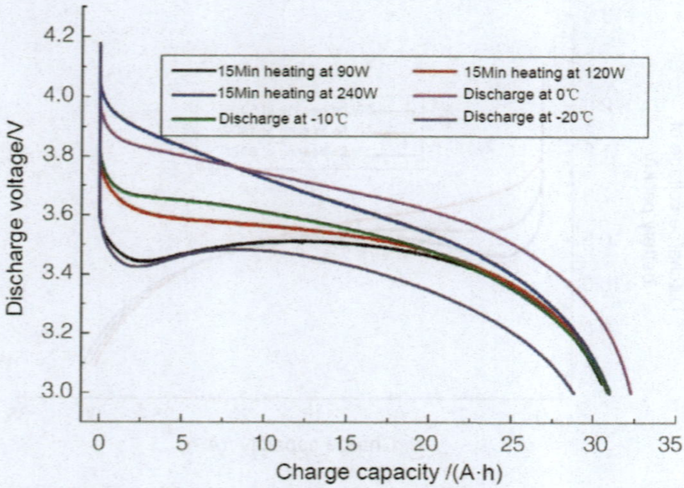

Fig. 6.53 1C rate constant current discharge of battery cell

significantly exceeds the discharge voltage of the battery cell at −20 °C. This is because the temperature of the battery pack is already close to −10 °C after heating. Although the battery pack generates a lot of heat during the discharge process, this heat can only maintain the pack temperature around −10 °C at the external ambient temperature of −40 °C and cannot increase the battery temperature any further. The entire discharge process is therefore close to that at −10 °C. After heating the battery pack with 240 W for 15 min, the average discharge voltage of the battery pack at the beginning of the discharge is slightly higher than that of the cells at 0 °C. This indicates that the temperature of the battery pack has been raised above 0 °C by the heating. However, in the middle and late stages of discharge, the average discharge voltage of the battery pack is lower than that of the cell at 0 °C and the final discharge capacity is lower than that of the cell at 0 °C. This is also due to the fact that the heat generated by the discharge of the battery pack is not sufficient to maintain the battery temperature at 0 °C after the heating has stopped.

2. 1C rate constant current discharge performance of battery packs under the conditions of different heating powers and heating times

The above analysis shows that raising the heating power can increase the battery discharge voltage when the heating time is the same. This section analyses the recovery of the battery pack's discharge performance when different heating powers and heating times are used. The curves of 1C rate constant current discharge of the battery pack under the conditions of different heating powers and heating times at an ambient temperature of −40 °C are shown in Fig. 6.54. The initial discharge voltage with 60 W heating power, extended heating time to 30 min and 1C rate constant current is higher than the discharge voltage with 90 W heating power and 15 min heating, but the difference in discharge capacity is small. The initial discharge voltage

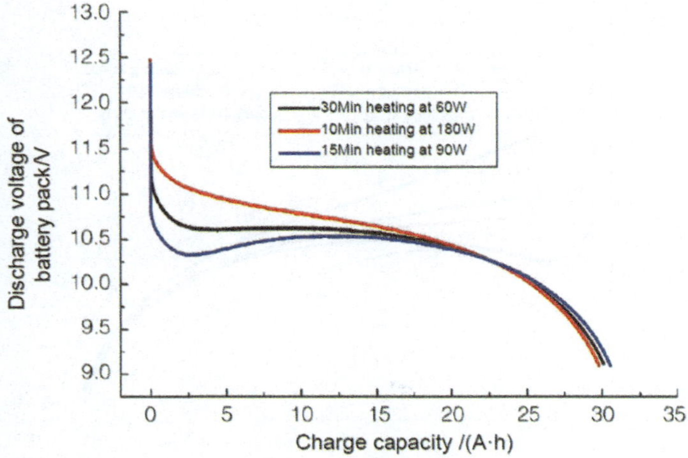

Fig. 6.54 1C rate constant current discharge of the battery pack after heating at different heating powers and heating times

with 180 W heating power, extended heating time to 10 min and 1C rate constant current is higher than the discharge voltage with 90 W heating power and 15 min heating, but the difference in discharge capacity is also small. Therefore, different heating powers and heating times can be selected according to the actual situation, so that the battery pack heating system can achieve a better heating effect.

3. 1C constant current charge performance of heated battery packs at extremely low temperatures

The 1C rate constant current charging curve for the battery pack after being heated with 240 W for 15 min at −40 °C and the 1C rate constant current charging curve of the unheated battery cell at different temperatures are shown in Fig. 6.55. By heating the battery pack, its charge performance is significantly improved and its temperature is raised to between 0 and 10 °C. As the charging and heating of the battery pack can be carried out using an external power supply, the heating time and the uniformity of the heat applied to the battery should be considered.

6.5.2 Pulse Charge and Discharge Performance After Externally-Powered Heating in Low Temperature Environment

According to the results of the above-mentioned low-temperature heating experiments on the battery packs, the charge and discharge performance of the heated packs was substantially improved. Since the 1C constant current charge/discharge

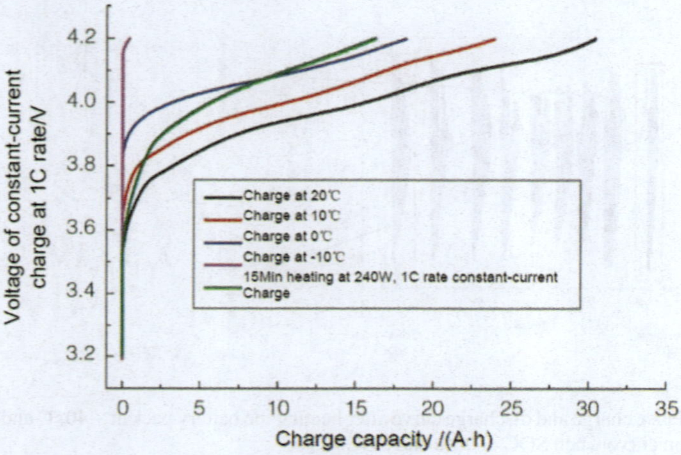

Fig. 6.55 1C rate constant current charge of battery cell

experiment is carried out on the heated battery pack, the maximum charge/discharge power that can be achieved when the battery is heated at very low temperatures is not known. Therefore, the pulse charge and discharge experimental study is done on the heated battery pack in this section.

1. Pulse charge and discharge experiment on heated battery pack at −40 °C

The battery pack is fully charged at a constant current and voltage of 1/3C at room temperature, then left to stand for 8 h in a −40 °C thermostat, and finally heated for 15 min at 120 W. After heating, the battery pack is subjected to pulse charge and discharge with a maximum discharge current of 280 A and a maximum charging current of 210 A. The results are shown in Fig. 6.56. As can be seen from the figure, the discharge performance of the battery pack is significantly improved after heating. Initially, the discharge current can reach a maximum of 210A. As the pulse charge and discharge proceeds, the battery pack can be discharged at 280A. The charge performance of the battery is relatively poor. When the SOC is greater than 50%, the charging current cannot exceed 50A. As the capacity decreases, the charging current gradually increases, eventually reaching 210 A.

2. Comparison of pulse experiments on the heated and unheated battery packs at −20 °C

The comparison results of pulse experiments on the heated and unheated battery packs at −20 °C are shown in Figs. 6.57 and 6.58. Figure 6.57 shows the pulse curve of an unpreheated battery pack. Figure 6.58 shows the pulse charge and discharge

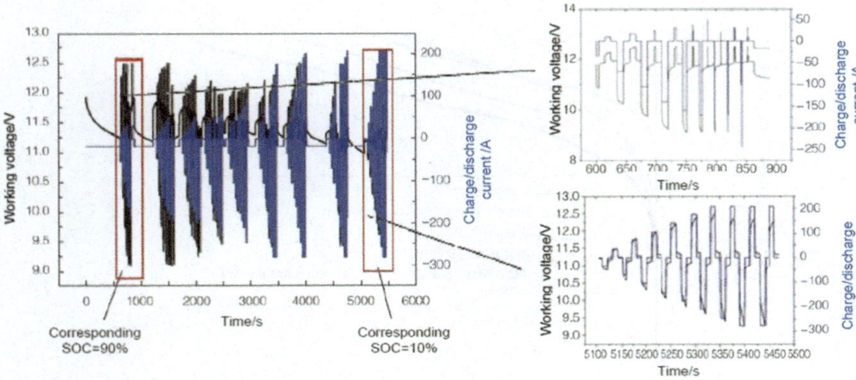

Fig. 6.56 Pulse charge and discharge curve after heating the battery pack at −40 °C and pulse local amplification curve when SOC = 90% and SOC = 10%

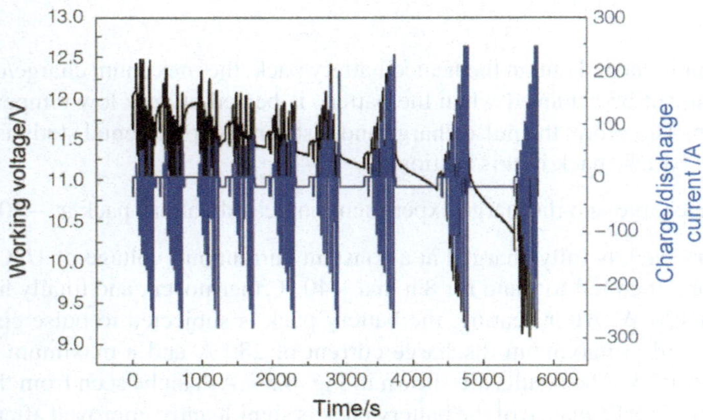

Fig. 6.57 Pulse charge and discharge curve for unheated battery pack at −20 °C

curve of the battery pack after a 15 min preheating at 120 W. As can be seen from the figure, the discharge performance of the battery pack is significantly improved after the pre-heating, with a significant increase in discharge voltage. The pulse charge and discharge curves for the pre-charged/unpreheated battery pack when the SOC is 90% are shown in Figs. 6.59 and 6.60 respectively. The pre-heated battery pack can be discharged at a rate of 8C. The unpreheated battery pack can only be discharged at a rate of around 2C, with the discharge voltage close to the cut-off voltage. This shows that preheating can significantly improve the low temperature performance of the battery pack.

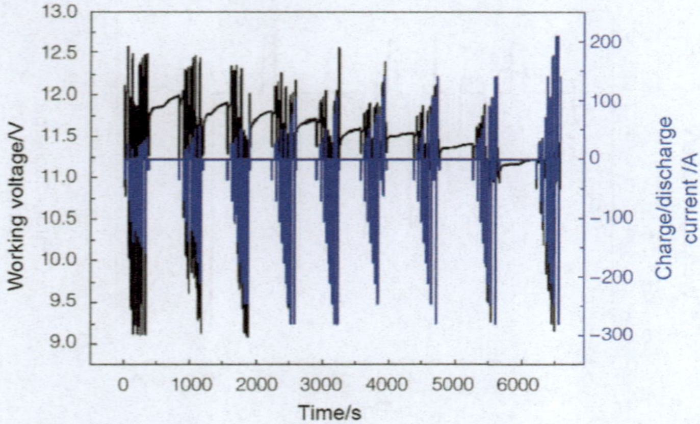

Fig. 6.58 Pulse charge and discharge curve of a preheated battery pack at −20 °C

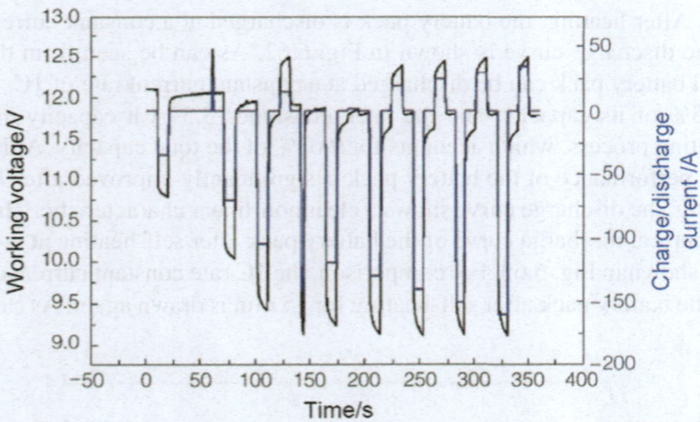

Fig. 6.59 Pulse curve for unheated battery pack at −20 °C when SOC = 90%

6.5.3 *Charge and Discharge Characteristics of Low Temperature Self-Heating Batteries*

The ability of the battery pack itself to act as a heating source to power a wide wire metal film to restore discharge performance at low temperatures is of great importance for specific applications in electric vehicles.

The battery pack is fully charged at a constant current and voltage of 1/3C at room temperature, and then left to stand for 8 h in a −40 °C thermostat. At the end of the above standing period, the cathode and anode of battery pack are connected to the wide wire metal film to power it for 15 min. During the heating process, the discharge voltage and current of the battery pack are on the rise, as shown in

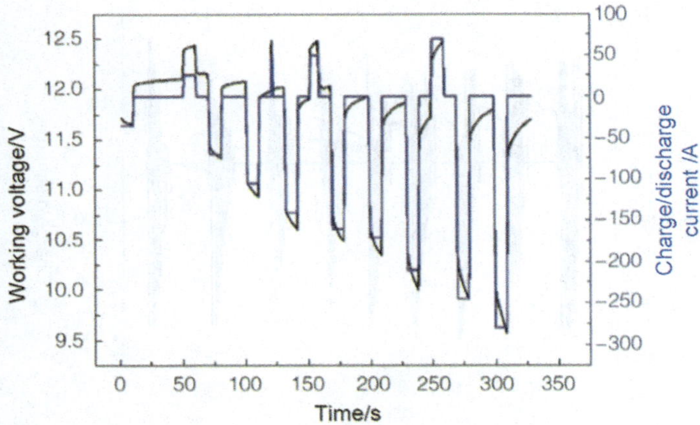

Fig. 6.60 Pulse curve for heated battery pack at −20 °C when SOC = 90%

Fig. 6.61. After heating, the battery pack is discharged at a constant current rate of 1C and the discharge curve is shown in Fig. 6.62. As can be seen from the figure, the heated battery pack can be discharged at a constant current rate of 1C, releasing around 75% of its capacity, and the pack consumes 3.37A-h capacity during the entire heating process, which accounts for 9.63% of the total capacity. Although the discharge performance of the battery pack is significantly improved after 15 min of self-heating, the discharge curve shows a clear non-linear characteristic. The 1C rate constant current discharge curve of the battery pack after self-heating at −40 °C for 25 min is shown in Fig. 6.63. For comparison, the 1C rate constant current discharge curve of the battery pack after self-heating for 15 min is drawn again. As can be seen

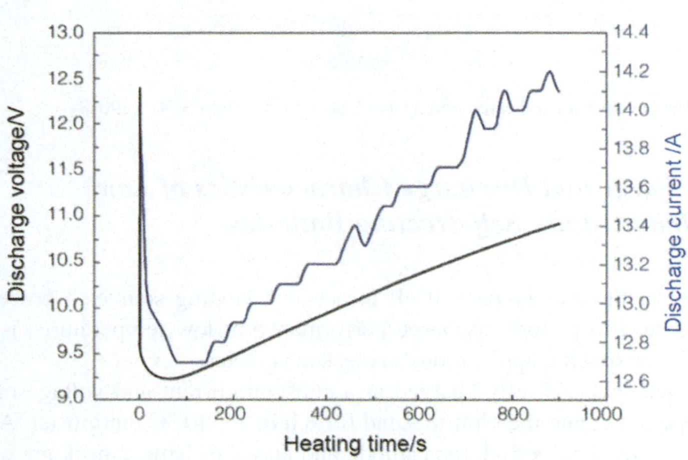

Fig. 6.61 Battery pack voltage and current variation curves during self-heating

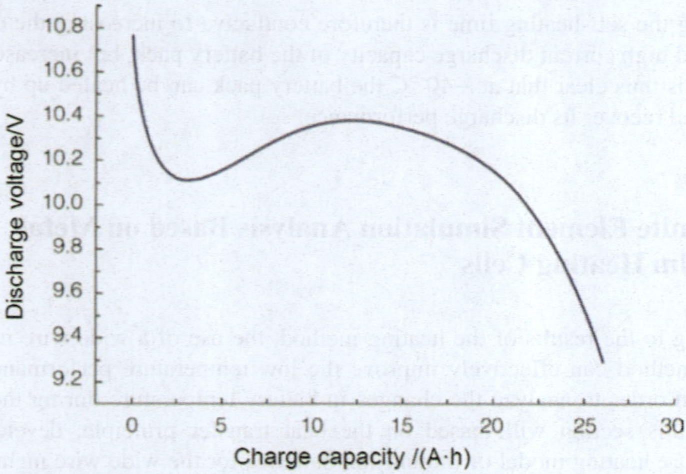

Fig. 6.62 1C rate discharge curve of a battery pack after 15 min self-heating

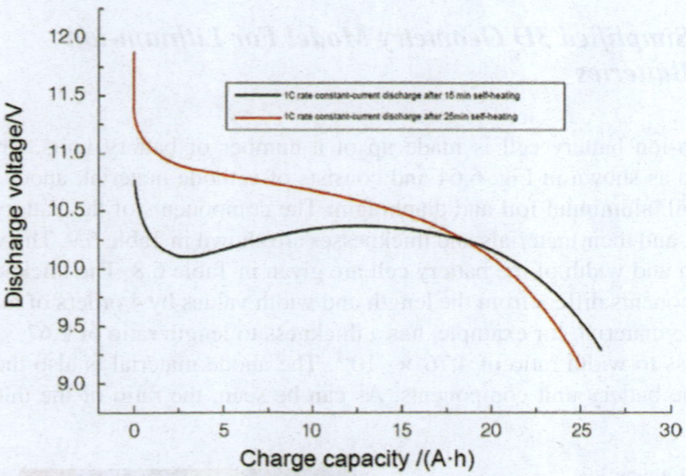

Fig. 6.63 1C rate constant current discharge curve after 15 min and 25 min self-heating of the battery pack

from the figure, the discharge voltage of the battery pack increases significantly after the self-heating time is extended. In the early and middle stages of discharge, the voltage change is relatively gentle, with a maximum increase of 0.82 V and an average increase of 0.34 V. As far as the battery packs of the whole locomotive are concerned, the maximum instantaneous power increase is 3903 W and the average power increase is 1618 W. Due to the longer self-heating time, the battery capacity consumed by heating increases and the capacity released by the battery pack decreases by 3.37%.

Extending the self-heating time is therefore conducive to increasing the discharge power and high current discharge capacity of the battery pack, but increases energy losses. It is thus clear that at −40 °C the battery pack can be heated up by its own energy and recover its discharge performance.

6.6 Finite Element Simulation Analysis Based on Metal Film Heating Cells

According to the results of the heating method, the use of a wide wire metal film heating method can effectively improve the low temperature performance of the battery. In order to analyze the changes in battery temperature during the heating process, this section will, based on the heat transfer principle, develop a low temperature heating model of lithium-ion batteries for the wide wire metal heating method.

6.6.1 Simplified 3D Geometry Model For Lithium-ion Batteries

A lithium-ion battery cell is made up of a number of battery units. The unit is structured as shown in Fig. 6.64 and consists of cathode material, anode material, copper foil, aluminum foil and diaphragm. The components of the battery unit are very thin, and their materials and thicknesses are shown in Table 6.9. The values for the length and width of the battery cell are given in Table 6.8. The thickness of the cell components differs from the length and width values by 4 orders of magnitude. The anode material, for example, has a thickness to length ratio of 2.67×10^{-4} and a thickness to width ratio of 4.76×10^{-4}. The anode material is also the thickest among the battery unit components. As can be seen, the ratio of the thickness to

Fig. 6.64 Lithium-ion battery unit structure

Table 6.9 Materials and thicknesses of the battery unit components

Composition	Material	Thickness/μm
Cathode material	$LiMn_2O_4$	80
Aluminum foil	Al	20
Anode material	Natural graphite	55
Copper foil	Cu	10
Diaphragm	PVDF	40
Shell	Aluminum laminated film	145

the length and width of the components of the battery unit is very small. If a three-dimensional model is built strictly in accordance with the structure of the battery cell, the simulation calculations for even a single battery cell are quite large. In order to reduce the computational effort of the model and to make the calculation of the heating model feasible, in this section, the battery cell is considered as a whole when building the battery model, and the internal structure of the battery cell is ignored. As the cathode and anode lugs of the battery are only 0.15 mm thick aluminum and copper sheets respectively, and the heating model is more concerned with the overall temperature and internal temperature field distribution of the battery, the cathode and anode lugs and some details that do not affect the overall temperature and internal temperature field distribution of the battery have been removed from the model. The geometric model of the battery cell is shown in Fig. 6.65.

Fig. 6.65 Geometric model of the battery cell

6.6.2 Experiment Method for Obtaining the Specific Heat Capacity of Lithium-ion Batteries

The specific heat of the battery is obtained by both theoretical calculations and experimental measurements. The experimental measurement method is based on Eq. (6.6), where the battery is heated by placing it in an adiabatic environment. The specific heat of a battery can be calculated by measuring the temperature rise of the battery and the energy used to heat it.

$$c_{\mathrm{p}} = \frac{\Delta Q}{\Delta T m} \tag{6.6}$$

where, cp is the specific heat capacity of the battery [J/(kg-K)]; ΔQ is the energy used for heating (J); ΔT is the temperature rise of the battery (K); and m is the mass of the battery (kg).

The specific heat of the battery cell is measured using an accelerated rate calorimeter (ARC). Calorimetry measures the specific heat of a substance using the flow of heating → monitoring → following. In an adiabatic chamber with a heating source, a fixed power heating unit is used to heat the battery cells, as shown in Fig. 6.66. A battery cell with a heater and temperature sensor installed is placed in the adiabatic chamber so that temperature changes can be measured by a temperature sensor attached to the cell surface, as shown in Fig. 6.67. The ambient temperature inside the adiabatic chamber is then tracked by the heating source.

In this process, since the power of the heating device is constant, the heat Q_{gen} generated in a certain time can be calculated by Eq. (6.7) as:

$$Q_{\mathrm{gen}} = P\Delta t \tag{6.7}$$

The heat (ΔQab) absorbed by the cell can be obtained from Eq. (6.6) as:

$$\Delta Q_{\mathrm{ab}} = c_{\mathrm{p}} m \Delta T \tag{6.8}$$

Fig. 6.66 Mounting heating plates on a battery cell

Fig. 6.67 Mounting heating plates on a battery cell

Since the chamber is adiabatic, the heat generated is equal to the heat absorbed, i.e.:

$$c_p m \Delta T = P \Delta t \Rightarrow c_p = \frac{P \Delta t}{m \Delta T} \tag{6.9}$$

The specific heat of the cell at different temperatures can be obtained by Eqs. (6.9). The curve of temperature variation with time during the heating of the battery cell can be obtained from the experiment to find the temperature gradient at each point. The specific heat values of the cell at different temperatures can then be obtained by fitting the experimental data, as shown in Figs. 6.68 and 6.69.

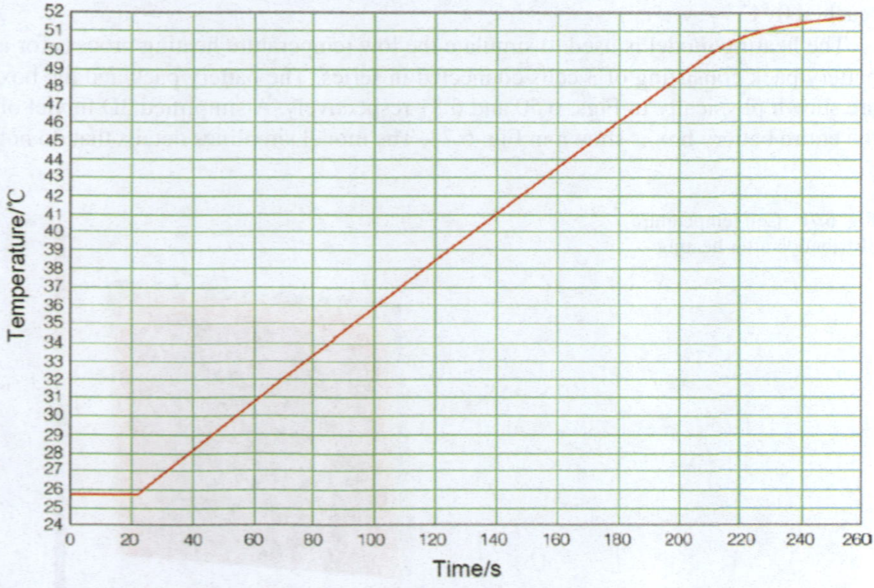

Fig. 6.68 Variation of cell temperature with heating time

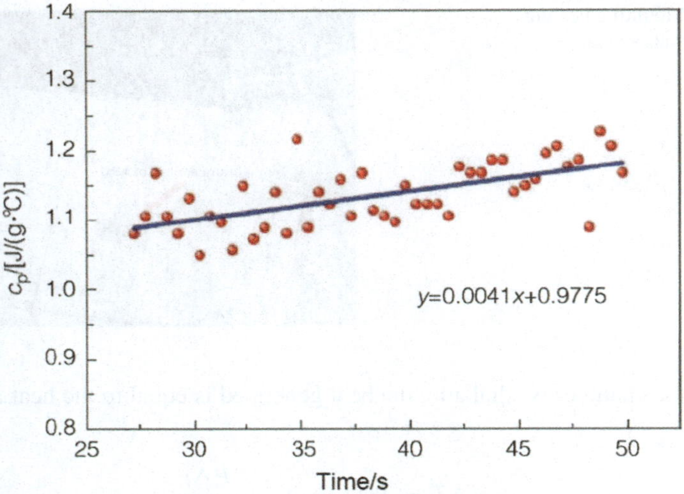

Fig. 6.69 Fitted curve for specific heat of the cell

The heating of battery cell is simulated using the established heating model. The temperature distribution of the cell after being heated with 60 W power for 15 min at −40 °C is shown in Fig. 6.70. A cross-section of the cell temperature distribution is shown in Fig. 6.71. The temperature change curve at the centre of the cell during heating is shown in Fig. 6.72. After 15 min of heating, the central temperature has reached 0 °C.

The heating model is used to simulate the low temperature heating process for a battery pack consisting of 3 cells connected in series. The battery pack and the box are shown physically in Figs. 6.50 and 6.51 respectively. A simplified 3D model of the entire battery box is shown in Fig. 6.73. The model simplifies details that do not

Fig. 6.70 Cell temperature distribution after heating

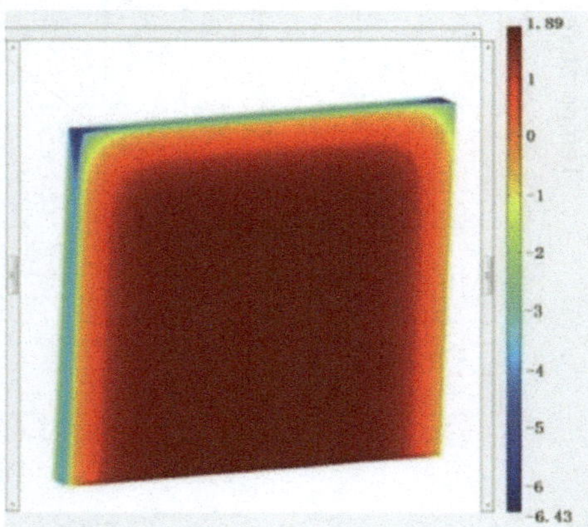

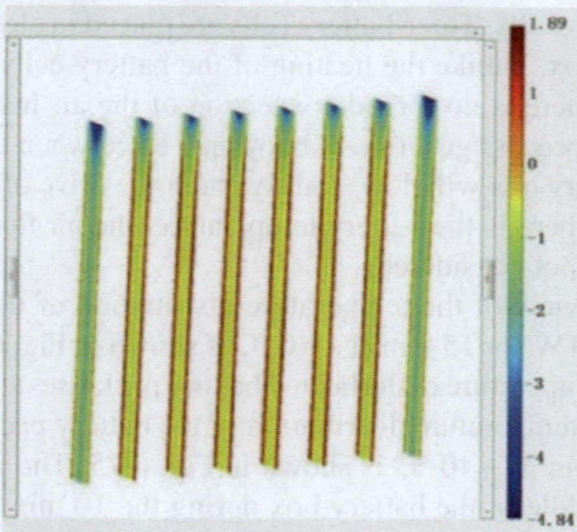

Fig. 6.71 Cell temperature profile after heating

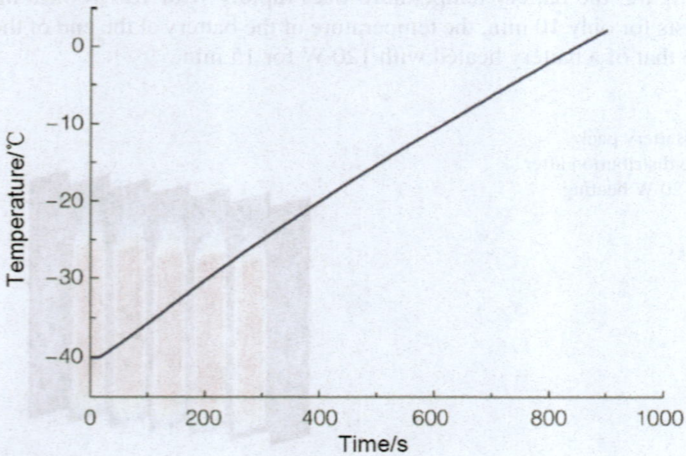

Fig. 6.72 Temperature variation curve at the centre of the cell

Fig. 6.73 Simplified 3D model of the battery box

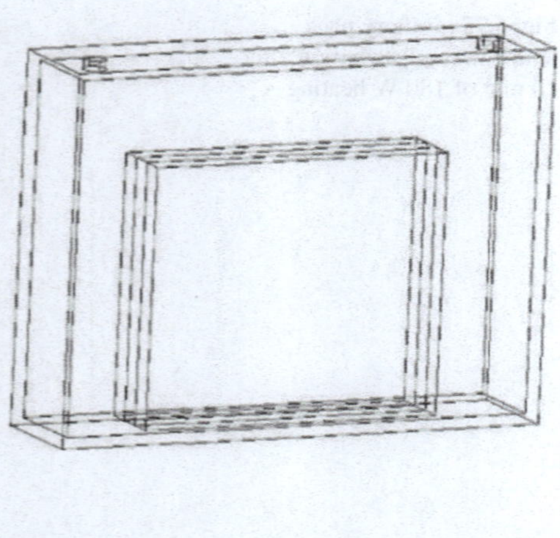

affect the calculation results. Three battery cells are placed in close proximity to each other in the battery box. Unlike the heating of the battery cells, during the heating of the battery pack, there is no forced convection of the air inside the battery box; however, the heating process generates a buoyancy force when the air is heated, and the air inside the battery box will flow slightly under the drive of the buoyancy force. As the main concern here is the battery temperature, the air flow inside the battery box due to heating is not considered.

A cross-sectional view of the temperature distribution of the battery pack after being heated with 120 W for 15 min at −40 °C is shown in Fig. 6.74. As can be seen from the figure, the temperature of the heated battery pack rises significantly. A cross-sectional view of the temperature distribution of the battery pack after being heated with 180 W for 10 min at −40 °C is shown in Fig. 6.75. The average temperature of the cells in the middle of the battery box during the 10 min heating with 180 W power is compared to the average temperature of the cells in the middle of the box during the 15 min heating with 120 W power, as shown in Fig. 6.76. As can be seen from the figure, the battery temperature rises rapidly with 180 W heating. As the heating lasts for only 10 min, the temperature of the battery at the end of the heating is close to that of a battery heated with 120 W for 15 min.

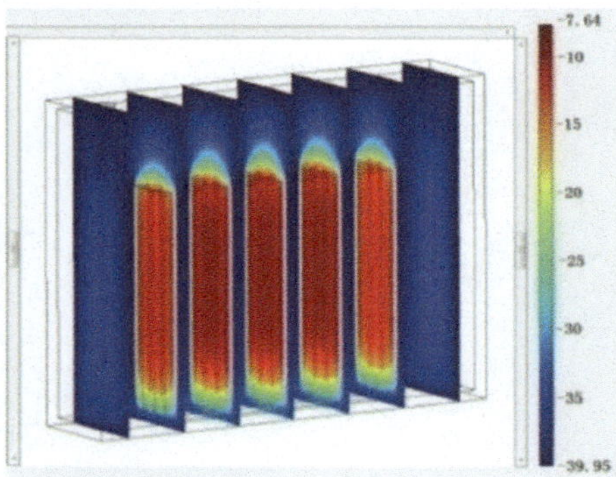

Fig. 6.74 Battery pack temperature distribution after 15 min of 120 W heating

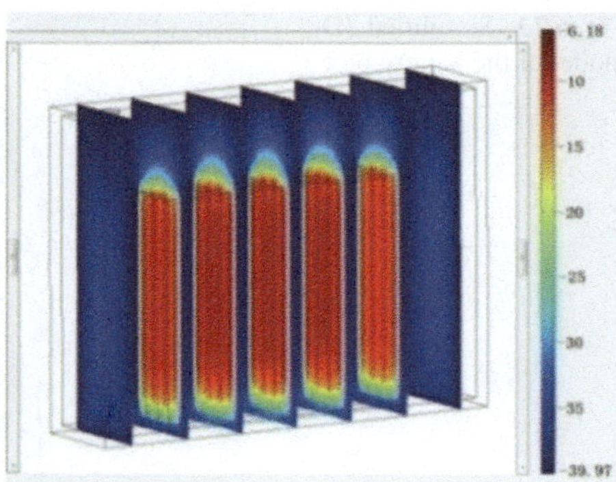

Fig. 6.75 Battery pack temperature distribution after 10 min of 180 W heating

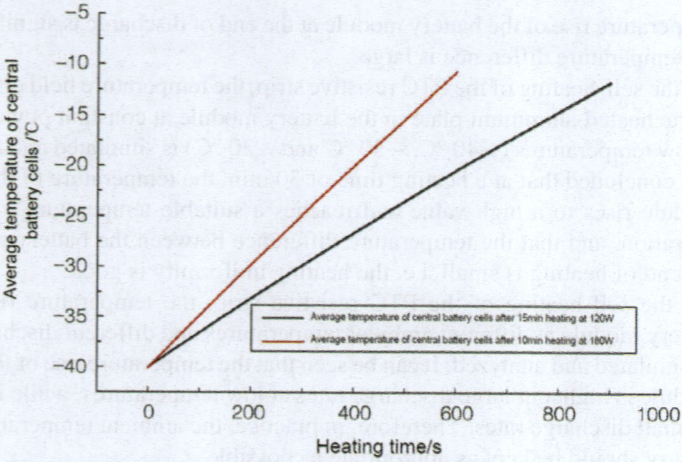

Fig. 6.76 Average cell temperature variation curve at different heating powers

The battery low-temperature heating model enables the calculation of the battery temperature field during the heating process to simulate the temperature distribution of the battery.

Summary

This chapter presents a detailed experimental and simulation analysis of the heating of lithium-ion battery packs at low temperatures by PTC resistive bands, both in terms of external heating and self-heating. This chapter also provides a detailed analysis of the wide wire metal film heating method in conjunction with the experiments and simulations. The main conclusions are as follows:

(1) For the external heating of the PTC resistive strip, the discharge performance of the battery is significantly improved when discharged at different rates after the heating is completed. According to the comparison with the results of unheated batteries discharged at 0.3C rate, it can be seen that the discharge capacity and discharge energy of the heated batteries are much higher. For example, at −40 °C, the discharge capacity and discharge energy of an unheated battery discharged at 0.3C rate are 21.9A-h and 212.13 W-h respectively, but after heating for 25 min they increase to 24.11A-h and 263.58 W-h respectively. This shows that it is necessary to preheat the battery at low temperature environments.

(2) For the external heating of the PTC resistive strip, it can be seen that the large rate discharge is beneficial to the discharge performance of the battery by analyzing the large rate current discharge of the heated battery. When the battery is discharged at low temperature, the temperature rise of the battery module is small and the temperature difference of the battery module is also small when discharging at a small rate; when discharging at a large rate, the

temperature rise of the battery module at the end of discharge is significant and the temperature difference is large.

(3) For the self-heating of the PTC resistive strip, the temperature field distribution of the heated aluminum plate in the battery module at constant power (35 W) at low temperatures (−40 °C, −30 °C and −20 °C) is simulated and analyzed. It is concluded that at a heating time of 30 min, the temperature of the battery module rises to a high value and reaches a suitable temperature for battery operation, and that the temperature difference between the battery module at the end of heating is small, i.e. the heating uniformity is good.

(4) For the self-heating of the PTC resistive strip, the temperature rise of the battery module at different ambient temperatures and different discharge rates is simulated and analyzed. It can be seen that the temperature rise of the battery module is higher at large discharge rates at low temperatures, while it is lower at small discharge rates. Therefore, in practice, the ambient temperature of the battery should be kept as appropriate as possible.

(5) For the wide-wire metal film heating method, a series of heating experiments (using different heating times and heating powers) are carried out on lithium-ion battery packs at low temperatures, and static and dynamic charge and discharge experiments are carried out on the heated packs. The results show that the wide wire metal film heating method can significantly improve the low temperature charge and discharge performance of the battery pack.

(6) For the wide-wire metal film heating method, the battery pack is used as a heating source to power the wide-wire metal film for self-heating at −40 °C, and the heated battery pack is discharged at a constant 1C rate. The results show that the lithium-ion battery pack is capable of self-heating and restoring discharge performance at −40 °C using the wide-wire metal film heating method.

Reference

Guangchong F (2013) Modeling and experimental research on low temperature thermal management of lithium-ion batteries for vehicles. Beijing University of Technology, Beijing

Chapter 7
Internal Heating of Lithium-ion Batteries Based on Sinusoidal Alternating Current

The main method currently used in the research of battery heating methods is external heating, i.e. the battery is heated by the heat generated by an external heat generating device. This method is simple and easy to use, but it is time consuming as the heat needs to be transferred slowly from the outside to the inside of the battery, and it is likely that only the surface of the battery has been heated and there is no certainty that the inside of the battery has been heated in a short time. In order to improve the heating efficiency of the battery and to compensate for the disadvantages of external heating, the battery can be heated quickly by means of internal heating. In this chapter, the control strategy for sinusoidal alternating current heating of lithium-ion batteries is proposed and verified based on the principle of sinusoidal alternating current heating of lithium-ion batteries, and combined with experimental and simulation analysis of sinusoidal alternating current heating of batteries.

7.1 Principle of Sinusoidal Alternating Current Heating of Lithium-ion Batteries

The internal heating of the battery can be carried out in three ways:

(1) The heating device is buried directly inside the battery, but this affects the internal structure of the battery and is less feasible.
(2) The battery itself generates heat from charge and discharge, but the heating time is longer.
(3) AC is applied to the cathode and anode of battery, which offers advantages in terms of battery life and heating efficiency.

By applying a periodic alternating current to the cathode and anode of battery, the battery voltage rises or falls periodically around the electric potential platform and the battery is in a state of alternating charge and discharge. During this process,

J. Li, *Modeling and Simulation of Lithium-ion Power Battery Thermal Management*, Key Technologies on New Energy Vehicles,
https://doi.org/10.1007/978-981-19-0844-6_7

Joule heat is generated in the real part of the battery AC impedance, which results in internal heating of the battery. The current of a sinusoidal alternating current varies according to Eq. (7.1).

$$i(t) = A\sin(2\pi ft + \phi) \tag{7.1}$$

where, A is the current amplitude of the sinusoidal alternating current; f is the frequency of the sinusoidal alternating current; and ϕ is the initial phase.

The internal heat generation rate of the battery when heated by sinusoidal alternating current is as follows:

$$q = \left(\frac{A}{\sqrt{2}}\right)^2 Z_{\mathrm{Re}} \tag{7.2}$$

where, Z_{Re} is the value of real part of the battery AC impedance, and q is proportional to the quadratic current amplitude and the real part of battery internal impedance. The internal impedance of the battery is related to the temperature of the battery and the frequency of the alternating current. Generally speaking, the lower the temperature, the larger the real part of the battery AC impedance.

7.2 Electro-Thermal Coupling Model for Sinusoidal Alternating Current Heating of Batteries

When modeling the sinusoidal alternating current thermal characteristics of lithium-ion batteries, it is first necessary to model the battery equivalent circuit. The model should be able to reflect the internal impedance variation of the battery and the model parameters can be obtained by parameter identification, which is the first step in modeling the electro-thermal coupling of sinusoidal alternating current.

7.2.1 Equivalent Circuit Model for AC Heating

The main types of lithium-ion batteries used in vehicles include lithium iron phosphate, ternary NCM (lithium nickel cobalt manganate), lithium manganate, etc. The EIS figures of different types and models of batteries may vary. In order to provide an accurate description of the electrode reaction characteristics inside the cell, different equivalent circuit models need to be chosen for modelling. Several commonly used equivalent circuit models are shown in Fig. 7.1.

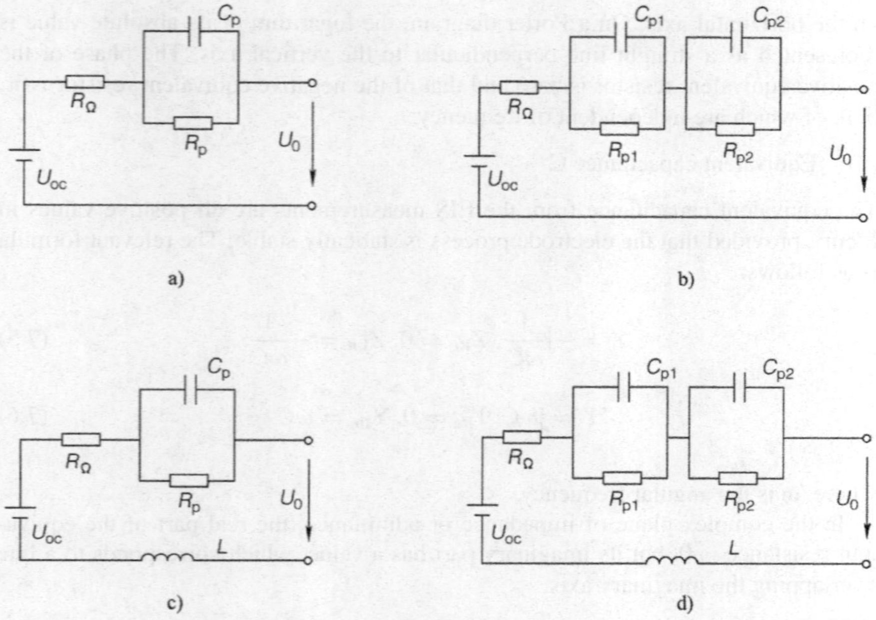

Fig. 7.1 Several commonly used equivalent circuit models, R_Ω—ohmic resistance, R_p—polarization impedance, C_p—polarization capacitance, L—high frequency response inductive resistance, R_{p1}—charge transfer impedance, C_{p1}—electric double layer capacitor, R_{p2}—Warburg diffusion impedance, C_{p2}—diffusion capacitance. R_Ω R_p C_p R_{p1} C_{p1} R_{p2} C_{p2}

(1) Equivalent resistance R

The equivalent resistance R in an electrochemical reaction is a signed quantity, with positive values representing resistance and negative values representing reactance. The value is related to the electrode area and is in $\Omega \cdot cm^2$. The relevant formula is as follows:

$$Z = R = Z_{\text{Re}},\ Z_{\text{Im}} = 0 \tag{7.3}$$

where, Z is the impedance; R is the equivalent resistance; Z_{Re} is the real part of the impedance; Z_{Im} is the imaginary part of the impedance.

$$Y = \frac{1}{R} = Y_{\text{Re}},\ Y_{\text{Im}} = 0 \tag{7.4}$$

where, Y is the admittance; Y_{Re} is the real part of the admittance; and Y_{Im} is the imaginary part of the admittance.

In the complex plane of impedance or admittance, the imaginary part of the equivalent resistance is 0, but its real part has a value, which corresponds to a point on the horizontal axis. On a Porter diagram, the logarithm of its absolute value is represented as a straight line perpendicular to the vertical axis. The phase of the positive equivalent resistor is zero and that of the negative equivalent resistor is π, both of which are independent of frequency.

(2) Equivalent capacitance C

The equivalent capacitance from the EIS measurements are all positive values in F/cm^2, provided that the electrode process is statically stable. The relevant formula is as follows:

$$Z = -\mathrm{j}\frac{1}{\omega C}, Z_{\mathrm{Re}} = 0, Z_{\mathrm{Im}} = -\frac{1}{\omega C} \tag{7.5}$$

$$Y = \mathrm{j}\omega C, Y_{\mathrm{Re}} = 0, Y_{\mathrm{Im}} = \omega C \tag{7.6}$$

where, ω is the angular frequency.

In the complex plane of impedance or admittance, the real part of the equivalent resistance is 0, but its imaginary part has a value, which corresponds to a line overlapping the imaginary axis.

(3) Equivalent inductance L

The equivalent inductance from the EIS measurements are all positive values in H·cm^2, provided that the electrode process is statically stable. The relevant formula is as follows:

$$Z = \mathrm{j}\omega L, Z_{\mathrm{Re}} = 0, Z_{\mathrm{Im}} = \omega L \tag{7.7}$$

$$Y = -\frac{\mathrm{j}}{\omega L}, Y_{\mathrm{Re}} = 0, Y_{\mathrm{Im}} = -\frac{1}{\omega L} \tag{7.8}$$

In the complex plane of impedance or admittance, the equivalent inductance is represented as a straight line coinciding with the negative half-axis of the imaginary axis.

(4) Constant phase angle element CPE

Due to the extremely complex electrode reactions, as well as the porous and rough nature and the adsorption phenomenon of the electrode surfaces, the purely capacitive part of the equivalent circuit deviates, i.e. the "dispersion effect" occurs. Therefore, the constant phase angle element CPE is used to represent the capacitance component in order to obtain a better fitting effect. The equation is as follows:

$$Z_{CPE} = \frac{1}{Y_0}(\mathrm{j}\omega)^{-n} \tag{7.9}$$

$$Z_{\mathrm{Re}} = \frac{\omega^{-n}}{Y_0}\cos\frac{n\pi}{2},\ Z_{\mathrm{Im}} = -\frac{\omega^{-n}}{Y_0}\sin\frac{n\pi}{2} \tag{7.10}$$

where, the CPE element contains two parameters, Y_0 and n. Y_0 represents the generalized capacitance in $1/(\Omega \cdot \mathrm{cm}^2 \cdot \mathrm{s}^n)$. Since it represents the effect when the equivalent capacitance deviates, it also always takes a positive value. The relationship of their phases is as follows:

$$\tan\varphi = \tan\frac{n\pi}{2},\ \varphi = \frac{n\pi}{2} \tag{7.11}$$

where, n is the "dispersion index". When $n = 0$, CPE is the resistance; when $n = 1$, CPE is the capacitance; when $n = -1$, CPE is the inductance; in particular, when $n = 0.5$, CPE is the Warburg impedance. When there is a dispersion effect on the electrode surface, the n value is always in the range of 0.5 ~ 1.

Various complex equivalent circuits can be composed of the basic equivalent components mentioned above in series or parallel. The various equivalent components reflect the different characteristics of the battery during the electrochemical reaction. The corresponding frequency response of the individual components of the battery impedance can be obtained from the EIS test.

In order to reduce the computational effort while maintaining the high accuracy of the model, an equivalent circuit model based on the frequency domain has been developed as shown in Fig. 7.2. The model describes the voltage-current characteristics of the battery at low temperatures. In the figure, R_Ω is the ohmic resistance; R_{ct} is the polarization resistance; C_{dl} is the polarization capacitance, also known as the electric double layer capacitor; R_{SEI} is the resistance of the SEI film; C_{SEI} is the capacitance of the SEI film; and L is the inductance, which reflects the high frequency excitation response of the battery. The Warburg diffusion resistance is very small and negligible in the frequency range studied here (f > 0.1 Hz). In the process of sinusoidal alternating current heating, both the total impedance and the real part of its impedance are important components in the calculation of the heat generation rate.

According to the above equivalent circuit model, the real part of the impedance can be expressed as follows:

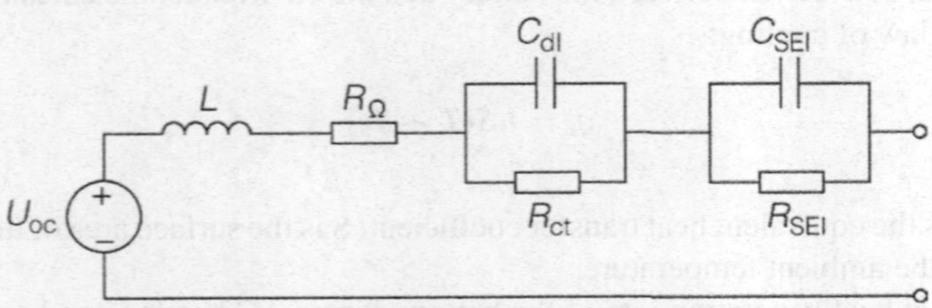

Fig. 7.2 Equivalent circuit model for AC heating

$$Z_{\mathrm{Re}}(T,f)=R_{\Omega}(T)+\frac{R_{\mathrm{ct}}(T)}{1+(2\pi f)^2R_{\mathrm{ct}}^2(T)C_{\mathrm{dl}}^2(T)}+\frac{R_{\mathrm{SEI}}(T)}{1+(2\pi f)^2R_{\mathrm{SEI}}^2(T)C_{\mathrm{SEI}}^2(T)} \tag{7.12}$$

The expression for the total impedance is as follows:

$$\begin{aligned}Z(T,f)=&\left[R_{\Omega}+\frac{R_{\mathrm{ct}}}{1+(2\pi f)^2R_{\mathrm{ct}}^2C_{\mathrm{dl}}^2}+\frac{R_{\mathrm{SEI}}}{1+(2\pi f)^2R_{\mathrm{SEI}}^2C_{\mathrm{SEI}}^2}\right]\\&+\mathrm{j}\left[2\pi fL-\frac{2\pi fR_{\mathrm{ct}}^2C_{\mathrm{dl}}}{1+(2\pi f)^2R_{\mathrm{ct}}^2C_{\mathrm{dl}}^2}-\frac{2\pi fR_{\mathrm{SEI}}^2C_{\mathrm{SEI}}^2}{1+(2\pi f)^2R_{\mathrm{SEI}}^2C_{\mathrm{SEI}}^2}\right]\end{aligned} \tag{7.13}$$

where, f is the current frequency; R_{Ω}, R_{ct}, R_{SEI}, C_{dl}, C_{SEI} and L are all closely related to temperature and current frequency.

7.2.2 *Thermal Model of AC Heated Lithium-ion Batteries*

A battery module consisting of three 18,650/2.15Ah ternary batteries connected in parallel is chosen for the study. Due to the good thermal conductivity of each component layer of this type of battery, the heat flux in the winding layer is continuous in both the radial and axial directions. The temperature difference between the interior and the surface of the battery is also not significant when a large rate current excitation is applied to it. When modeling battery AC heated lithium-ion batteries, the internal heat transfer within the battery geometry is ignored and the battery is considered as a whole. According to the principle of energy conservation, the energy balance equation inside the battery can be expressed as:

$$mC\frac{\partial T}{\partial t}=q-q_{\mathrm{t}} \tag{7.14}$$

where, m is the mass of the battery; C is the specific heat capacity; t is time; T is the temperature of the battery; q_t is the heat transfer between the battery and its surroundings; and q is the heat generation rate of the battery.

The heat convection between the battery and the environment is calculated using Newton's law of cooling:

$$q_{\mathrm{t}}=hS(T-T_{\mathrm{a}}) \tag{7.15}$$

where, h is the equivalent heat transfer coefficient; S is the surface area of the battery; and T_a is the ambient temperature.

The total heat generation rate of the battery during AC heating can be expressed in two main parts as follows:

$$q_q = q_{rev} + q_{irr} \tag{7.16}$$

where, q_{rev} is reversible heat generation and q_{irr} is irreversible heat generation. The irreversible heat generation includes ohmic heat generation and polarization heat generation, while the reversible heat generation refers to the heat generation of electrochemical reactions. According to the Bernardi cell heat generation rate equation, the reversible electrochemical reaction heat generation can be expressed as:

$$q_{rev} = -I_L T \frac{\mathrm{d}E_0}{\mathrm{d}T} \tag{7.17}$$

where, I_L is the current; E_0 is the open-circuit voltage (OCV).

In one cycle of AC heating, the algebraic sum of reversible heat generation from electrochemical reactions is zero and therefore negligible.

When the current applied to the cathode and anode of battery varies according to the following sinusoidal pattern:

$$i(t) = A\sin(2\pi f t + \phi) \tag{7.18}$$

The heat generation rate inside the battery can be calculated by the following equation:

$$q = q_{irr} = \left(\frac{A}{\sqrt{2}}\right)^2 Z_{Re}(T, f) \tag{7.19}$$

Taking into account the constraints on the safe operating voltage of the battery, the maximum permissible current amplitude can be expressed as:

$$A_{max} = \min\left(\frac{U_{max} - U_{oc}}{|Z|}, \frac{U_{oc} - U_{min}}{|Z|}\right) \tag{7.20}$$

where, U_{max} is the upper voltage limit, taken as 4.2 V; U_{min} is the lower voltage limit, taken as 2.8 V; |Z| is the mode of the total impedance, which can be expressed as:

$$|Z(T, f)| = \sqrt{Z_{Re}^2(T, f) + Z_{Im}^2(T, f)} \tag{7.21}$$

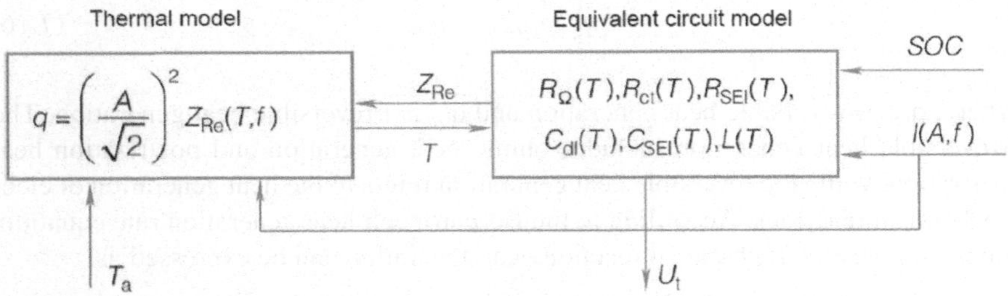

Fig. 7.3 Coupling mechanism of the electro-thermal coupling model

7.2.3 *Electro-Thermal Coupling Mechanism for AC Heated Batteries*

An equivalent circuit model for describing the voltage-current characteristics of lithium-ion batteries has been developed based on the electrochemical reaction principle. And, a thermal model of the lithium-ion battery has been developed based on the principle of energy conservation to describe its heating characteristics. Afterwards, the equivalent circuit model is combined with the thermal model to develop a coupled electrical-thermal model to comprehensively describe the electrical-thermal performance of the lithium-ion battery during sinusoidal alternating current heating, as shown in Fig. 7.3.

Based on the equivalent circuit model and parameters such as the SOC and current of the battery, the parameters of the components of the equivalent circuit model of the battery, such as resistance, capacitance and inductance, can be calculated. In turn, the terminal voltage, the total impedance and the real part of the impedance of the battery can be found. They are then substituted into the thermal model to calculate the heat generation rate during sinusoidal alternating current heating. On the other hand, the temperature in the thermal model affects the individual parameters in the equivalent circuit model, which in turn affects the terminal voltage of the battery.

7.3 Sinusoidal Alternating Current Heating Experiment and Model Validation of Lithium Ion Batteries

7.3.1 *Establishment of Experiment Platform*

The technical parameters of the selected cylindrical 18,650 ternary lithium-ion battery cells and the module consisting of three cells connected in parallel are shown in Table 7.1.

The experimental platform and equipment built for sinusoidal alternating current heating at low temperatures are shown in Figs. 7.4, 7.5 and 7.6. A bipolar power

Table 7.1 Parameters of battery cells and battery modules

Parameters	Value
Battery type	18,650 cylinder
Cathode material	Li (NiCoMn) O_2
Anode material	Graphite
Rated capacity of cell	2.15A·h
Rated voltage	3.6 V
Charge and discharge cutoff voltage	4.2/2.8 V
Mass of cell	45.0 g
Surface area of battery	$4.26 \times 10^{-3} m^2$
Rated capacity of module	6.45A·h
SOC (state of charge)	20%, 50%, 80%

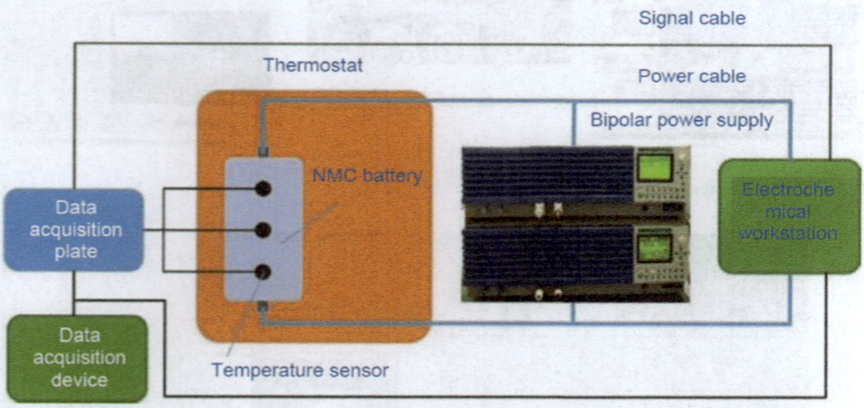

Fig. 7.4 Experiment platform

supply is used to provide a sinusoidal alternating current with an output voltage range of ± 20 V and an output current range of ±40 A. A thermostat is used to provide low temperature ambient conditions for the battery with a range of −55–50 °C. The electrochemical workstation is used to measure the electrochemical impedance spectrum of the battery under different conditions. A data acquisition instrument is used to collect data on the voltage, current and temperature of the battery. An oscilloscope is used to observe the battery voltage waveform to keep it within safe limits. In addition, a class A Pt100 chip temperature sensor is used to measure the surface temperature of the battery.

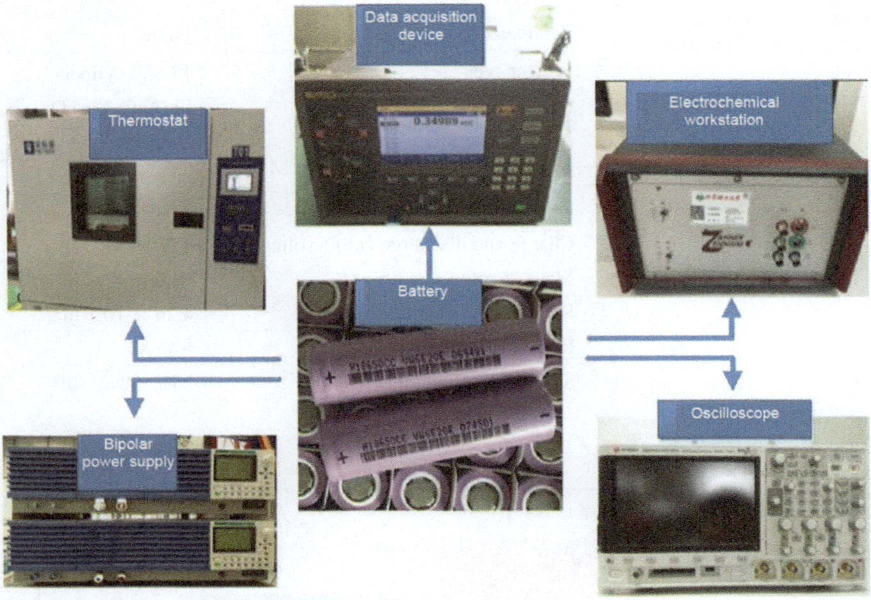

Fig. 7.5 Experiment equipment

Fig. 7.6 Experiment site

7.3.2 *Battery Impedance Characteristics at Different Temperature and SOC*

The impedance tester applies a voltage disturbance signal with an amplitude of 5 mV and a frequency range of 0.001–10000 Hz, varying sinusoidally, to the cathode and anode of battery pack, in order to test the feedback signal of the battery pack after receiving the disturbance signal. From this, the values of the real and imaginary parts of the AC impedance of the battery pack at different frequencies can be determined. To prevent the impedance test and sinusoidal alternating current heating from interfering with each other, the two should be alternated during the experiment, and the sweep range of the impedance tester should be minimized to shorten the test time and thus reduce the impact of the test on the heating process.

In the AC impedance experiment, three 18,650 battery modules connected in parallel are used as test subjects. In the experiments, measurements have been done at 0 °C, − 20 °C and − 40 °C when SOC is 80% and 50% respectively. That's to say, six test cycles have been done.

During the impedance experiment, the battery is simultaneously heated by AC current for 30 min at a time. The impedance test is carried out once before the heating starts. After the heating has started, the impedance is tested every 5 min. The impedance is therefore tested 7 times during a single heating session. To reduce the time required for each measurement, the impedance tester can be set to a minimum sweep frequency of 1 Hz, so that a measurement takes only 3 min. The amplitude and frequency of the AC heating is 2C and 8 Hz at 0 °C and −20 °C respectively. At −40 °C, a sinusoidal alternating current of 1C/8 Hz is used for heating, as the 2C amplitude would put the battery module voltage out of normal range.

After the 6 rounds of experiments have been completed, sampling points within 100 Hz are extracted for each round of experiment. Within 100 Hz, the impedance tester selects 64 frequencies (minimum 1 Hz and maximum 97.217 Hz) to measure the AC impedance of the battery. By connecting the values of the real part of the AC impedance obtained from these 64 sampling points into a curve, the variation of the real part of the AC impedance with frequency up to 100 Hz can be observed. Following this plotting method, a curve of the real part of the battery AC impedance as a function of frequency up to 100 Hz for each of the seven measurements in the six rounds of experiments has been produced, as shown in Figs. 7.7, 7.8, 7.9, 7.10, 7.11 and 7.12.

The following points can be summarized from Figs. 7.7 , 7.8, 7.9, 7.10, 7.11 and 7.12:

(1) The lower the temperature, the higher the value of the real part of the AC impedance, provided that both frequency and battery SOC are the same.
(2) The lower the frequency, the higher the value of the real part of the AC impedance in the range 0–100 Hz, provided that both temperature and battery SOC are the same.
(3) At −40°C, the change in the value of the real part of the impedance after heating is obvious.

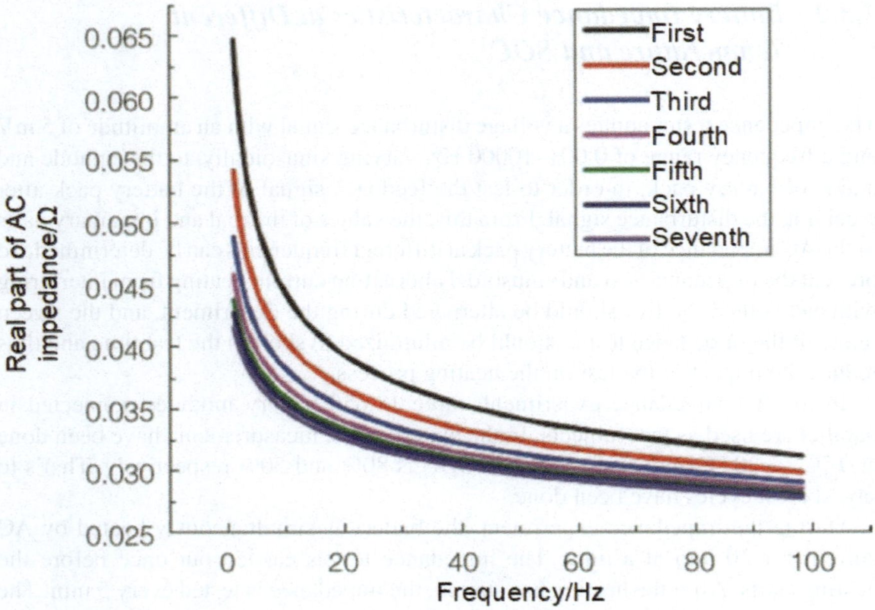

Fig. 7.7 Variation of the real part of the AC impedance during sinusoidal alternating current heating at 0 °C (SOC = 80%)

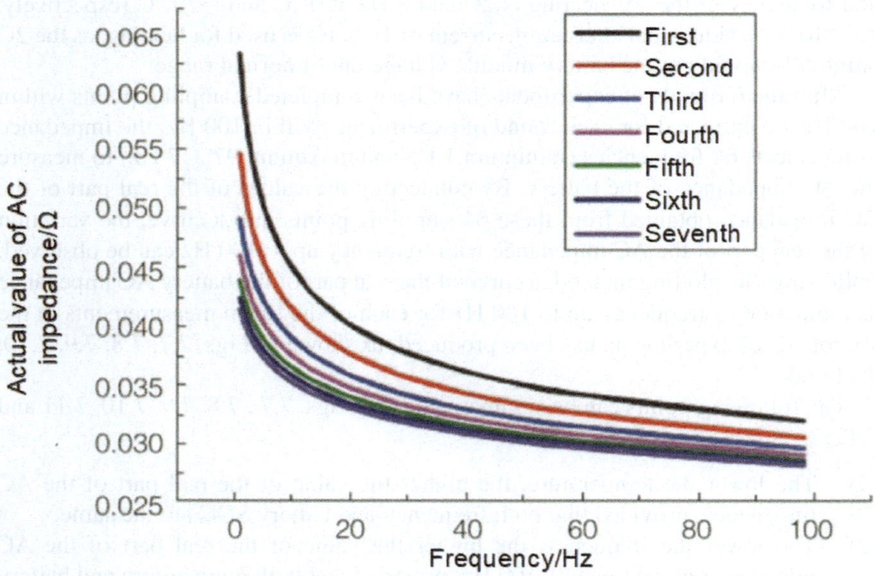

Fig. 7.8 Variation of the real part of the AC impedance during sinusoidal alternating current heating at 0 °C (SOC = 50%)

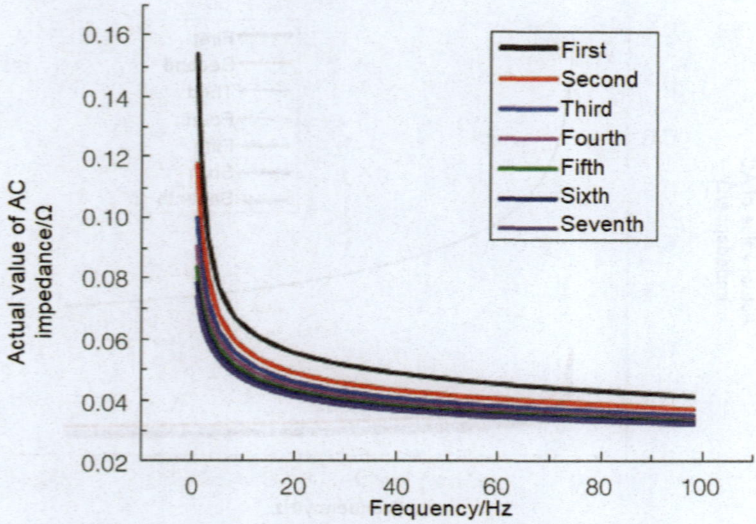

Fig. 7.9 Variation of the real part of the AC impedance during sinusoidal alternating current heating at −20 °C (SOC = 80%)

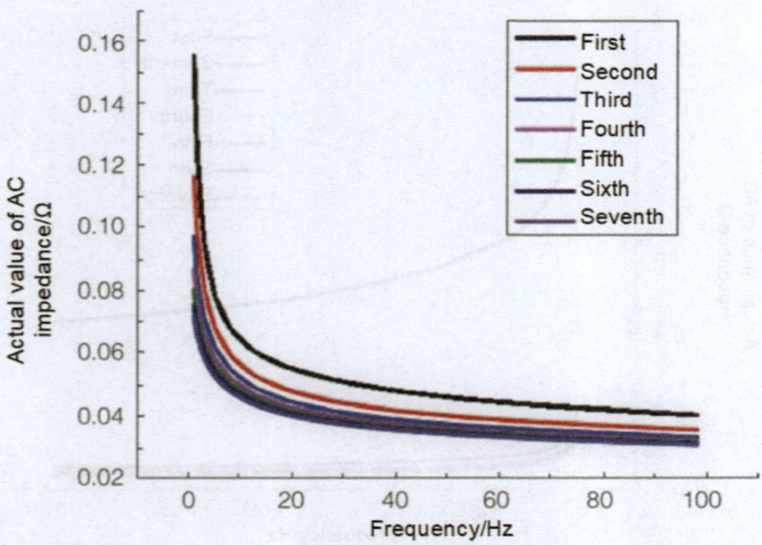

Fig. 7.10 Variation of the real part of the AC impedance during sinusoidal alternating current heating at −20 °C (SOC = 50%)

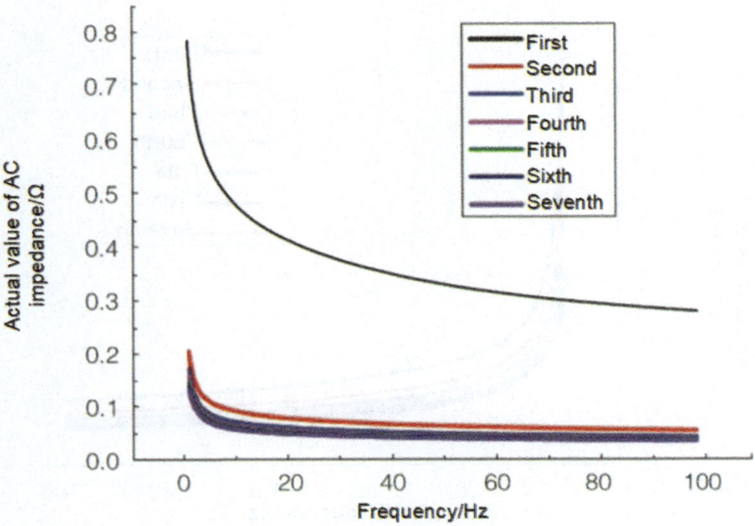

Fig. 7.11 Variation of the real part of the AC impedance during sinusoidal alternating current heating at −40 °C (SOC = 80%)

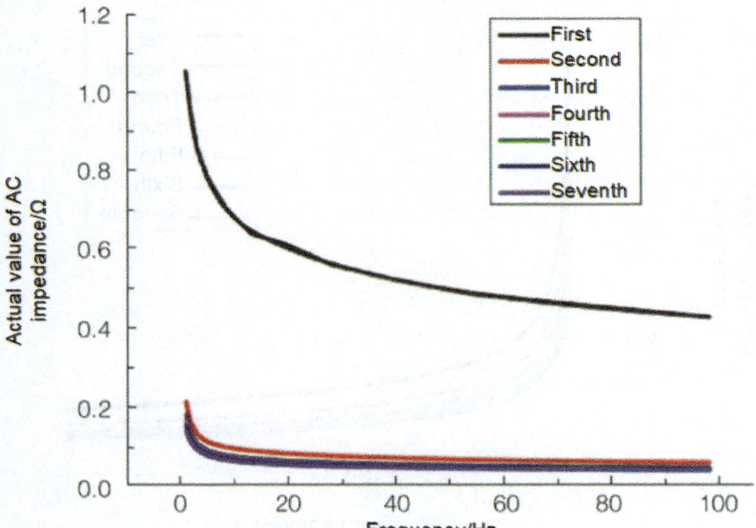

Fig. 7.12 Variation of the real part of the AC impedance during sinusoidal alternating current heating at −40 °C (SOC = 50%)

7.3.3 Verification of Equivalent Circuit Models for Sinusoidal Alternating Current Heating

1. Principle of EIS test of model parameters

Electrochemical impedance spectroscopy (EIS) is a method of measuring the impedance spectrum of an electrode system when it is perturbed by a sinusoidal quantity. This method abstracts the electrode process as an equivalent circuit consisting of electrical components such as resistors, capacitors and inductors connected in series and parallel. It investigates the principle of the electrode system by solving for the parameters of the circuit elements based on impedance mapping. The output of a linear and stable electrochemical system is a sinusoidal response Y when a sinusoidal quantity U with an angular frequency ω is fed into the system. If the ratio of the response to the input is a function G, then we have:

$$Y = G(\omega)U \tag{7.22}$$

If U is the sinusoidal current excitation and Y is the voltage response, then G is the impedance of the system, symbolized as Z. If U is the sinusoidal voltage excitation and Y is the current response, then G is the admittance of the system, symbolized as Y. Z and Y are collectively referred to as the admittance, and Y = 1/Z. In EIS figure, Z_{Re} and Z_{Im} are mostly used as horizontal and vertical coordinates.

$$Z(\omega) = Z_{Re} - jZ_{Im} \tag{7.23}$$

A model of the electrochemical reaction is shown in Fig. 7.13. In the figure, the electric double layer capacitor C_d is generated by the inactive ions of the electrolyte, which does not undergo a chemical reaction except the charge of distribution charge; the internal resistance R_Ω refers to the internal resistance of the electrode to the electrolyte; the Faraday impedance Z_f is generated by the active ions of the electrolyte, which undergoes a redox reaction and a transfer of charge.

The corresponding circuit model is shown in Fig. 7.14a. The Faraday process can be subdivided into matter transfer and charge transfer processes, corresponding to the Warburg impedance and charge transfer impedance, respectively, as shown in Fig. 7.14b. The expressions for the real and imaginary parts of the impedance can be obtained from Fig. 7.14 as follows:

$$Z_{Re} = R_\Omega + \frac{R_{ct} + \sigma\omega^{-1/2}}{(C_d\sigma\omega^{-1/2} + 1)^2 + \omega^2 C_d^2 (R_{ct} + \sigma\omega^{-1/2})^2} \tag{7.24}$$

$$Z_{Im} = \frac{\omega C_d (R_{ct} + \sigma\omega^{-1/2})^2 + \sigma\omega^{-1/2}(\omega^{1/2} C_d \sigma + 1)}{(C_d\sigma\omega^{-1/2} + 1)^2 + \omega^2 C_d^2 (R_{ct} + \sigma\omega^{-1/2})^2} \tag{7.25}$$

where, σ is the coefficient related to the transfer of substances.

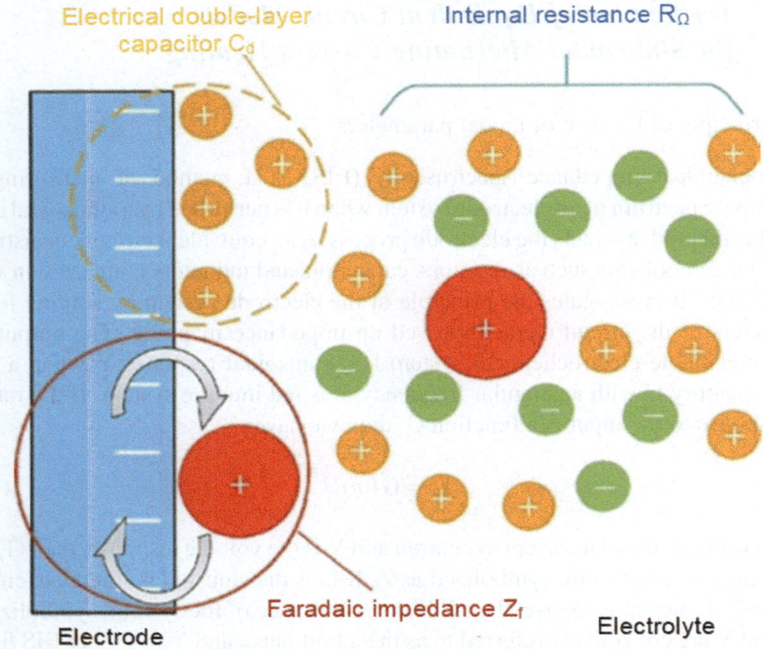

Fig. 7.13 Electrochemical reaction model

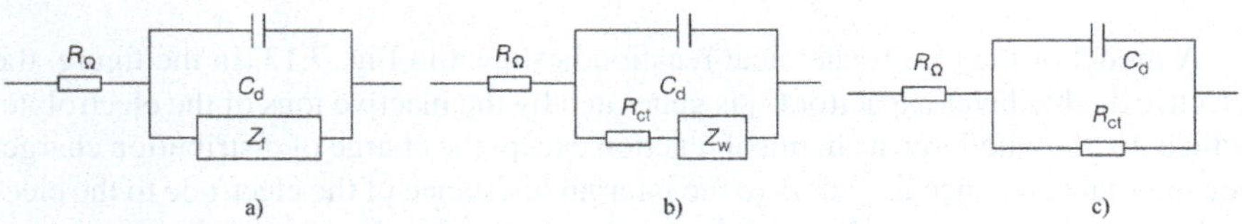

Fig. 7.14 Corresponding circuit model

When ω tends to zero (low frequency), the relationship between Z_{Re} and Z_{Im} can be simplified to Eq. (7.26). The EIS figure is a straight line with a slope of 1. It intersects the Z-axis at the point $(R_\Omega + R_{ct} - 2\sigma^2 C_d, 0)$:

$$Z_{Im} = Z_{Re} - R_\Omega - R_{ct} + 2\sigma^2 C_d \tag{7.26}$$

When ω is very large (high frequency), the period of signal change is very short, so it is too late for material transfer to occur. At this point, the effect of the Warburg impedance is not reflected. The equivalent circuit model can be simplified to Fig. 7.14c, from which the relationship between the real and imaginary parts of the impedance can be obtained, as shown in Eq. (7.27). This is represented in the

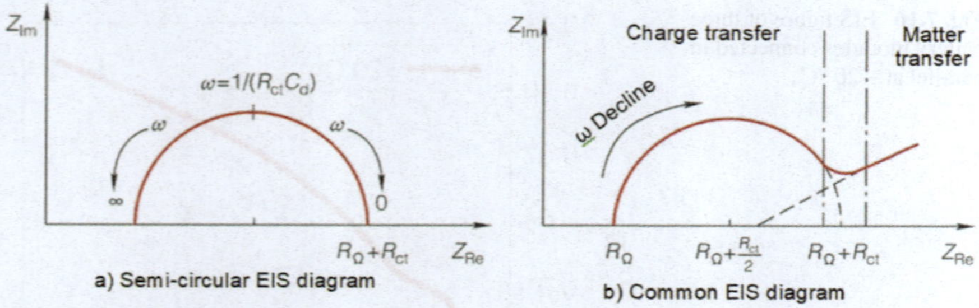

Fig. 7.15 Two EIS figures

EIS figure as a semicircle with $R_\Omega + \frac{R_{ct}}{2}$ as the centre and $\frac{R_{ct}}{2}$ as the radius, as shown in Fig. 7.15a.

$$(Z_{Re} - R_\Omega - \frac{R_{ct}}{2})^2 + Z_{Im}^2 = \left(\frac{R_{ct}}{2}\right)^2 \tag{7.27}$$

Based on the above analysis, it can be seen that in the low frequency region, matter transfer is dominant and in the high frequency region, charge transfer is dominant, as shown in Fig. 7.15b. In addition, multiple equivalent circuits can be used for fitting analysis of a given EIS figure.

2. EIS results for lithium-ion batteries at different temperatures and SOCs

In order to investigate the impedance characteristics of lithium-ion batteries during electrochemical reactions at low temperatures, electrochemical impedance spectroscopy (EIS) is used to obtain the AC impedance of the batteries at different temperatures and different SOC states to provide a basis for the identification of the parameters of the equivalent circuit model established in Chap. 2 and the calculation of the heat generation rate of the sinusoidal alternating current heating process. The EIS figure of three 18,650 NCM lithium-ion batteries connected in parallel at −20 °C is shown in Fig. 7.16. In the Nyquist figure, the sections from the left to the right correspond to the impedance values from high to low frequency respectively. As can be seen from the figure, the EIS curve consists of an approximate straight line, an approximate semicircular arc and an approximate diagonal line. In the figure, the straight line indicates the presence of inductance in the battery. This hysteresis current due to the effect of inductance is different from the induction current, which is related to the inherent properties of the electrodes, such as porosity, surface roughness and other factors. It also indicates that the battery is a viscous system. It can then be deduced that the equivalent circuit model has inductors in series and that the effect is more pronounced in the high frequency region. At the point in the figure where the high and mid frequency regions meet is the ohmic resistance. The imaginary part of the impedance at this point has a value of zero, and the horizontal coordinate of this point's intersection with the real axis is the value of the approximate ohmic

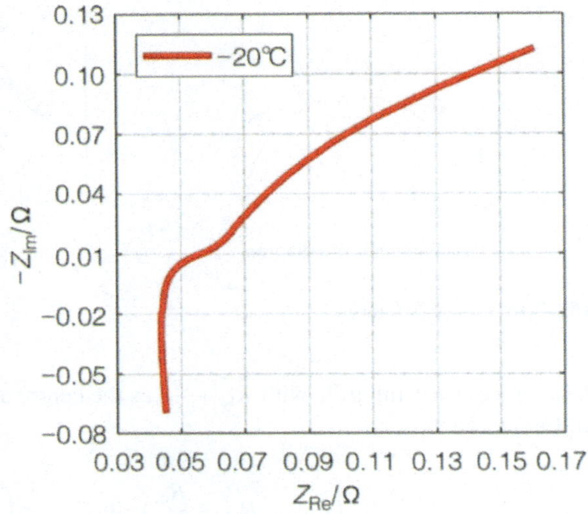

Fig. 7.16 EIS figure of three battery modules connected in parallel at −20 °C

resistance. This value is related to the transport of lithium ions in the active material of the cathode and anode, the diaphragm and the electrolyte.

The shape of the mid-frequency region of the EIS figure appears as an approximate semicircle. It is related to the charge transfer process and corresponds in the equivalent circuit model to the parallel part of the polarization resistor and the polarization capacitor. The polarization resistor is also known as the charge transfer resistor and the polarization capacitor is the electric double layer capacitor described above. It can be inferred from the slightly flattened shape of the semicircle that the polarized capacitor is not an ideal pure capacitor. The low frequency region appears as an approximate diagonal line, which is a reflection of the Warburg diffusion impedance. The low frequency region corresponds to the solid phase diffusion of lithium ions in the active material of the cathode and anode, which is caused by the concentration difference. Warburg impedance is more pronounced at very low frequencies.

It can also be seen from Fig. 7.16 that the intersection of the flat circular arc in the mid-frequency region and the sloping line in the low-frequency region is not sufficiently clear. This indicates that both electrochemical polarization and differential concentration polarization are occurring within the battery and that the total impedance is the result of both working together. In particular, in the low frequency region, the concentration polarization is dominant, which is reflected in the diagonal lines of the Warburg impedance. In the high frequency region, however, the electrochemical polarization is dominant, which is reflected in the flat arc of the electric double layer capacitor. Thus, during the electrode reaction, a charge transfer reaction occurs first, resulting in electrochemical polarization. This is followed by a lithium ion diffusion reaction, which produces a concentration polarization.

As the battery module connected in series or in parallel always serves as the basic unit of the lithium-ion batteries used in electric vehicles, in this chapter, a battery module consisting of three NCM18650 batteries connected in parallel is chosen as the object of study for EIS test and sinusoidal alternating current heating experiment. Nine battery cells with good consistency are selected and numbered from 1 to 9. Before conducting the EIS test, in order to activate the active material inside the new battery and improve its performance, the battery is first charged and discharged 3 times at the nominal rate using a KIKUSUI PBZ20-40 bipolar power supply, and then the rate capacity of the battery will be determined. Afterwards, the SOC of batteries 1 to 3 is adjusted to 80% and the batteries are connected in parallel to form module 1; the SOC of batteries 4 to 6 is adjusted to 50% and the batteries are connected in parallel to form module 2; the SOC of batteries 7 to 9 is adjusted to 20% and the batteries are connected in parallel to form module 3. One temperature measurement point is placed in the middle of the surface of each battery, and then a Pt100 chip temperature sensor is attached to the temperature measurement point to measure the surface temperature of the battery.

The Zahner Zennium electrochemical workstation shown in Fig. 7.5 is used to perform the EIS test on the battery module. For this purpose, the voltage control mode is chosen, the excitation voltage is set to 5 mV, the frequency range of the sweep is set to 10–1 ~ 104 Hz, and the number of samples is set to 147. The temperature range of the measurement is −25 to 25 °C. The EIS test is carried out at 5 °C intervals and the battery module is placed in a Ykytech thermostat before each measurement. The battery pack is held at the set temperature for at least 4 h to equalize the temperature of the battery pack. Six tests are done at each temperature and at each SOC to reduce errors and improve data reliability.

The EIS figure for a battery module with SOC = 20% at different temperatures is shown in Fig. 7.17. As can be seen from the figure, the shape of the electrochemical impedance spectrum of the battery module changes significantly as the temperature decreases, with both the real and imaginary parts of the battery impedance becoming correspondingly larger. Unlike the impedance spectrum at room temperature (25 °C), the shape of the arcs and slopes of the impedance spectrum at low temperatures changes. For example, the characteristic frequencies corresponding to both the diffusion reaction process and charge transfer processes gradually decrease at low temperatures; at the same time, the charge transfer impedance becomes progressively larger, causing the corresponding semicircle to become flatter.

In terms of the individual frequency bands, the high-frequency part has a smaller degree of variation compared to the low-frequency part. This is due to the fact that the low frequency part mainly characterizes the diffusion process of lithium ions in the active material of the cathode and anode, i.e. the lower solid phase diffusion coefficient of lithium ions in the active material is the main reason for the higher impedance at low temperatures. In addition, low temperatures also affect the ionic conductivity of the electrolyte and the impedance of the SEI membrane. The EIS figure at −15, −10 and −5 °C with different SOCs is shown in Fig. 7.18. As can be seen from the figure, the impedance spectrum value rises slightly as the SOC of the battery increases at the same temperature and frequency. However, overall, the impedance is negligibly affected by SOC.

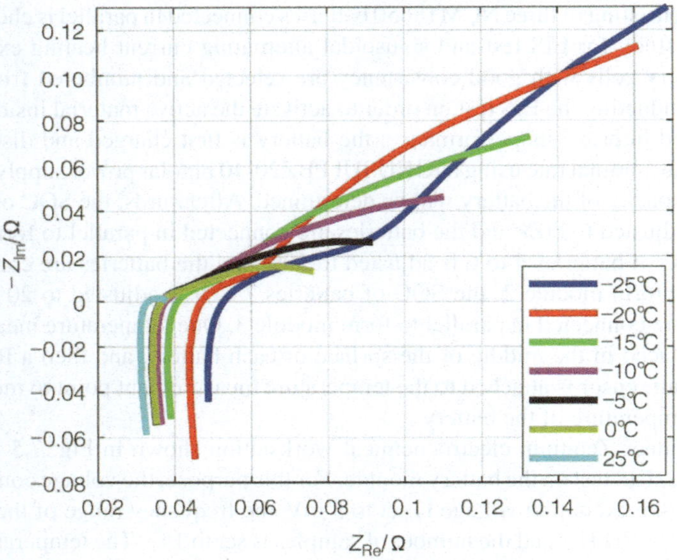

Fig. 7.17 EIS figure at different temperatures

3. Verification of equivalent circuit models

The equivalent circuit model of the battery can be abstracted from the electrochemical impedance spectrum. The shapes of the different frequency bands in the EIS curve correspond to the combination of different components or various components connected in series and parallel in an equivalent circuit. So, once the EIS data has been obtained, the corresponding equivalent circuit model can be created. Afterwards, a fitting tool or some optimization algorithm is used to fit or identify the specific parameters of the individual circuit components. Commonly used fitting tools include ZView software and common algorithms include least squares, genetic algorithms and artificial neural network algorithms.

Based on the measured EIS figure and the above theory, a second order RLC equivalent circuit model can be developed. A genetic algorithm can also be used to identify the parameters of the components in the equivalent circuit model, the results of which are shown in Fig. 7.19.

As can be seen from the figure, R_{Ω}, R_{ct} and R_{SEI} all decrease with increasing temperature, but R_{ct} decreases much faster than R_{Ω} and R_{SEI}; and R_{ct} dominates the total low temperature impedance of the battery. This is because low temperatures lead to an increase in electrolyte viscosity, which slows down the transport of lithium ions. This is consistent with the findings of reference 6.

The model is validated by substituting the results of the parameter identification into the equivalent circuit model. The impedance module-frequency curve and phase-frequency curve at −20 and −10 °C are shown in Fig. 7.20. The errors between

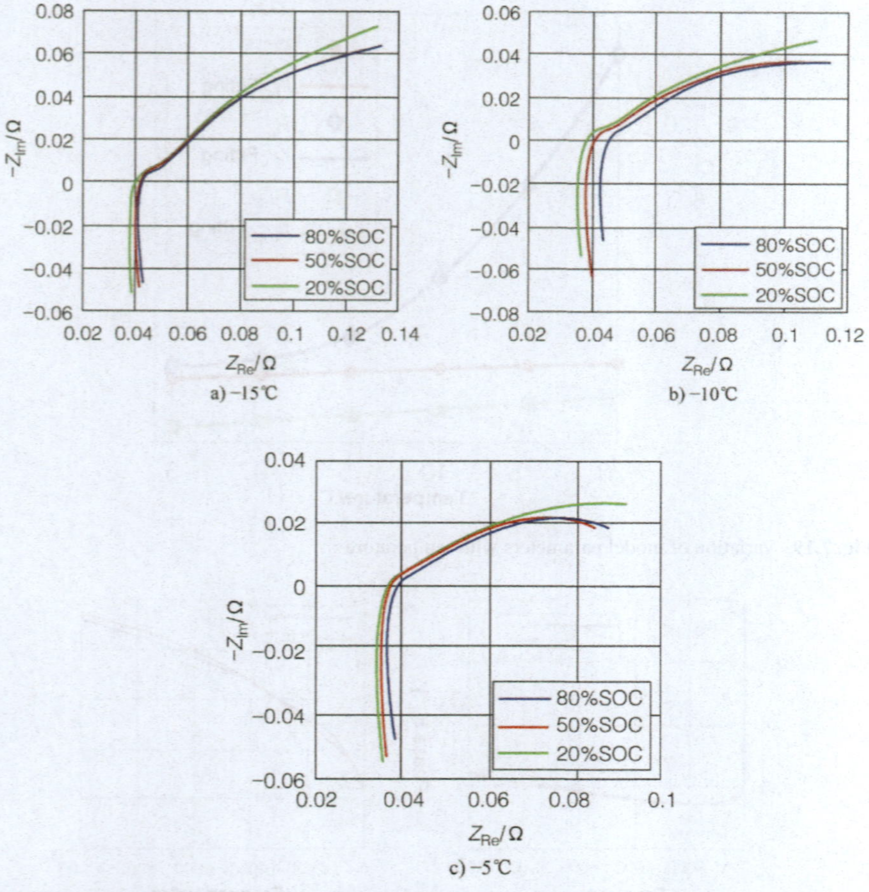

Fig. 7.18 EIS figure at different temperatures and SOCs

the model simulation and the experimental measurements in the figure are shown in Table 7.2.

As can be seen in Fig. 7.20, the model simulation and experimental measurement curves of the impedance module at −20 °C are very similar to each other, and their variation patterns are basically the same, with a maximum error of 0.0116Ω, a mean error of 0.0020Ω and a root mean square error (RMSE) of 0.0017Ω. The simulation and experimental curves for the phase at −20 °C coincide well with each other, with a maximum error of 6.0281°, a mean error of 2.2416° and an RMSE of 1.1731°. The RMSEs of the impedance module and phase at −10 °C are both lower than at −20 °C, which are 0.0011Ω and 1.0524° respectively. Overall, the accuracy of the equivalent circuit model parameter identification is high, with the error within the acceptable range.

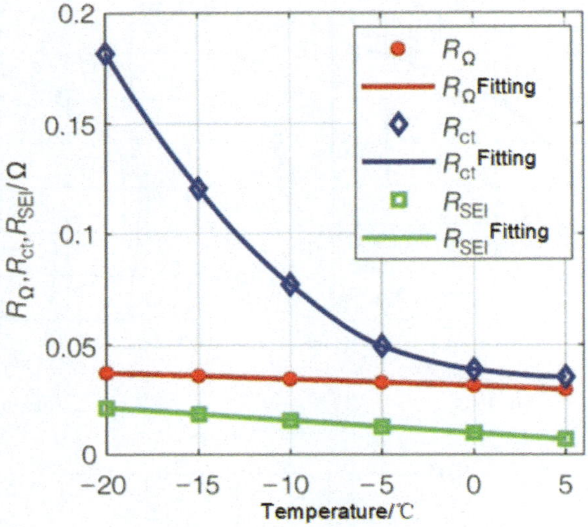

Fig. 7.19 Variation of model parameters with temperature

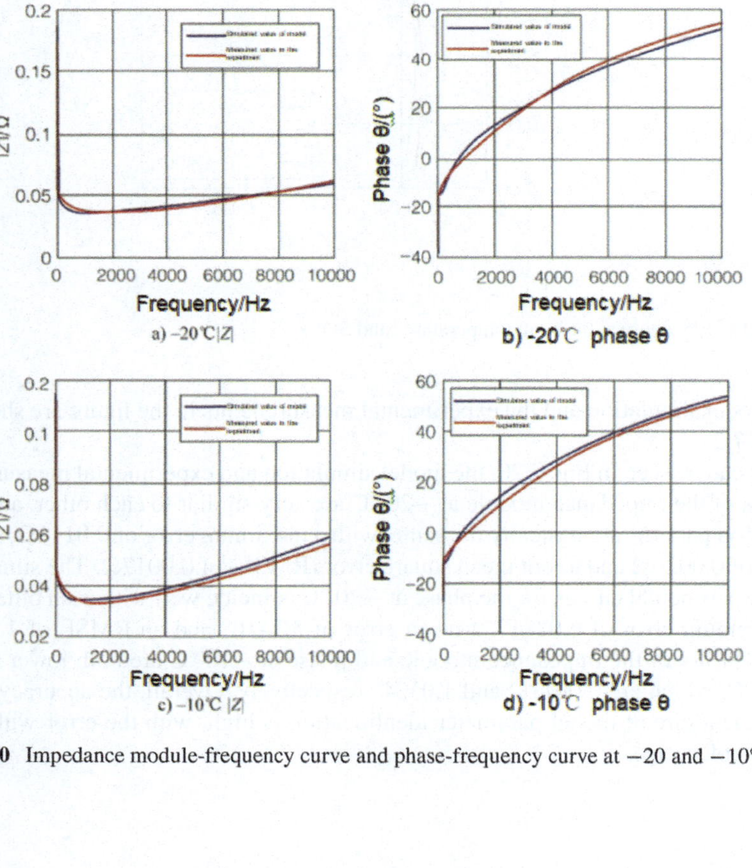

Fig. 7.20 Impedance module-frequency curve and phase-frequency curve at −20 and −10°C

Table 7.2 Results of the accuracy evaluation of the equivalent circuit model

Evaluation parameter	|Z| (−20 °C)	Θ (−20 °C)	|Z| (−10 °C)	Θ (−10 °C)
Maximum error	0.0116 Ω	6.0281°	0.0057 Ω	6.6913°
Average error	0.0020 Ω	2.2416°	0.0017 Ω	2.1736°
RMSE	0.0017 Ω	1.1731°	0.0011 Ω	1.0524°

7.3.4 *Experimental Validation and Analysis of the Electro-Thermal Coupling Model*

Based on the theory of the electro-thermal coupling model, the MATLAB program is written based on a genetic algorithm, and the thermophysical parameters of the battery are identified. Therefore, it can be known that the convective heat transfer surface transfer coefficient h of the battery is 9.93 W/(m^2·K), the specific heat capacity c of the battery is 996.65 J/(kg·K), and the general value of h for natural convection of air is 10–20 W/(m^2·K). The battery modules in this paper are first encapsulated in a foam insulated box filled tightly with foam sheets and then placed in a thermostat. As a result, the battery module is better insulated, resulting in a slightly smaller value for h. The identification of h is therefore judged to be reasonable. The value of the specific heat capacity c varies with the temperature of the battery, but not by much. The experimentally measured specific heat capacities of the batteries in Ref. 7 are shown in Table 7.3, from which it can be seen that the c values at low temperatures identified in this paper are close to the experimentally measured values and are within the reasonable range.

By substituting the identified circuit parameters and thermophysical parameters into the electro-thermal coupling model, the heat generation rate and temperature rise characteristics of the battery can be simulated. The temperature of the thermostat is set to −20 °C and the 3 battery modules are placed in the thermostat for 5 h so that the internal temperature of the battery is the same as the ambient temperature. The parameters of the bipolar power supply are set to a sinusoidal alternating current output mode with a given amplitude and frequency. The output key is pressed to heat the battery module, and the change in surface temperature and voltage of each individual cell is observed and collected using a data acquisition instrument. It is sufficient to repeat the process when applying a sinusoidal alternating current with different parameters. In this paper, 1C (6.45A)/1000 Hz and 2C (12.9A)/1500 Hz sinusoidal alternating currents are selected for the heating study, and the experimental and simulation results are shown in Fig. 7.21.

Table 7.3 Specific heat capacity of batteries at different temperatures

Temperature (°C)	−20	−10	0	10	20
Specific heat capacity/[J/ (kg·K)]	990.5	1007.1	1035.7	1051.3	1067.9

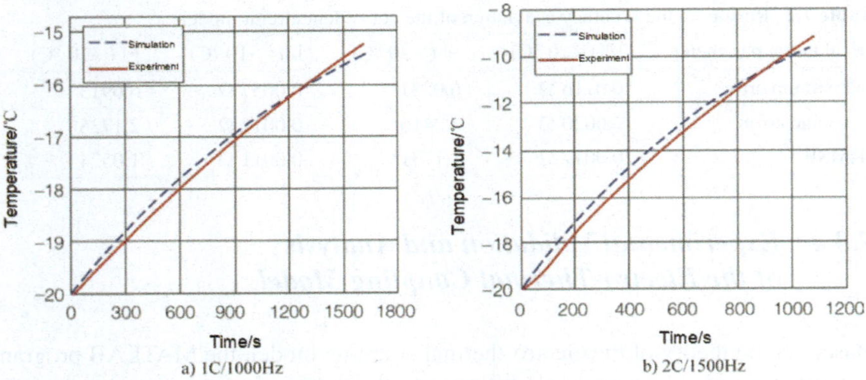

Fig. 7.21 Battery temperature rise curves at 1C/1000 Hz and 2C/1500 Hz

As can be seen in Fig. 7.21, when a sinusoidal alternating current is applied at a frequency of 1000 Hz and an amplitude of 1C, the battery warms from −20 to −15.39 °C (experiment) and −15.45 °C (simulation) in 1600 s; when a sinusoidal alternating current is applied at a frequency of 1500 Hz and an amplitude of 2C, the battery warms up from 20 to −9.84 °C (experiment) and −9.93 °C (simulation) in 1000 s. The temperature rise curves obtained from the simulation are very similar to the experimentally measured temperature rise curves and the trend is relatively consistent. The battery temperature calculated by the model simulation reflects the actual battery surface temperature very well, with a maximum temperature difference of 0.55 °C and an average temperature difference of 0.36 °C. The root mean square error (RMSE) between the simulation results and the actual temperature is shown in Table 7.4. The root mean square error for a 1.5C/1500 Hz sinusoidal alternating current is 0.1730 °C. However, for a 1C/1000 Hz sinusoidal alternating current, this error is reduced to 0.1367 °C. In particular, the root mean square error for both is less than 0.2 °C, which is very satisfactory. The electrical-thermal coupling model is therefore accurate.

Table 7.4 Results of the accuracy evaluation of the electrical-thermal coupling model

Parameters	1C/1000 Hz	1.5C/1500 Hz
Maximum error/°C	0.2891	0.5499
Average error/°C	0.1951	0.3574
RMSE/°C	0.1367	0.1730

7.4 Analysis of the Heating Effect of AC Frequency and Amplitude on Batteries

The main factors affecting the heat generation rate include the amplitude A and frequency f of the applied sinusoidal alternating current, so the next step is to investigate the effect of these two variables on the heating effect of the battery separately using the control variable method.

Firstly, to investigate the effect of the current amplitude A on the heat generation rate, a constant current frequency (e.g. 1500 Hz) and different current amplitudes (e.g. 1C (6.45A), 1.5C (9.675A) and 2C (12.9A)) are used for heating. Similarly, before heating, the battery is placed in a −20 °C thermostat for 5 h to equalize the internal and external temperature of the battery. Each of the 3 battery modules is heated. As an example, the temperature rise of the battery module at SOC = 20% is plotted in Fig. 7.22a.

From the figure, we can find that when the battery is heated at 1C rate, the temperature rise curve of the battery is relatively gentle, the battery heats up to −16.5°C at 1080 s (18 min), and in other words, the battery heats up slowly; when the battery is heated at 1.5C rate, the temperature rise curve of the battery becomes steeper than when the battery is heated at 1C rate, the battery heats up to −13.8°C at 1080 s, and in other words, the battery heats up faster; when the battery is heated at 2C

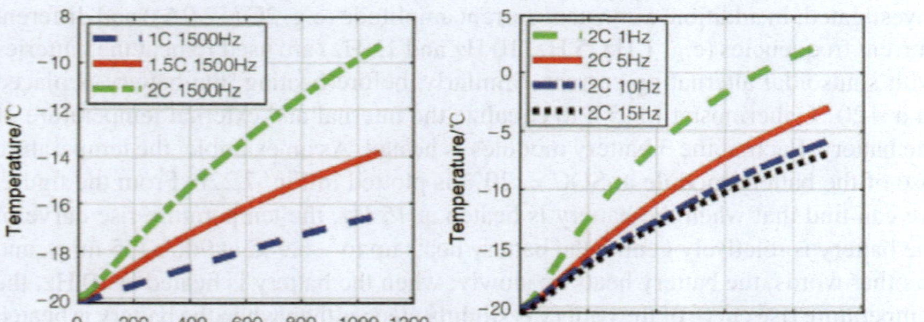

Fig. 7.22 Temperature rise curve of a battery at different current amplitudes and frequencies

rate, the temperature rise curve steepens further, the battery heats up to −9.2 °C at 1080 s (i.e. a 10.8 °C increase in temperature), and in other words, the warming effect has become noticeably better. By comparing the three curves, it is easy to see that the higher the amplitude of the applied sinusoidal alternating current at the same frequency, the better the heat generation effect.

At the same time, the terminal voltage signal of the battery is measured with an oscilloscope during the heating process using sinusoidal alternating current. Using the heating process using 2C/1500 Hz sinusoidal alternating current as an example, the terminal voltage waveform of the battery is shown in Fig. 7.22c. As can be seen from the figure, the voltage has a peak-to-peak value of 1.09 V and a peak value of 545 mV. At this point the OCV of the battery is 3.57 V. From this, the terminal voltage of the battery can be calculated to be 3.57 V + 0.545 V = 4.115 V < 4.2 V. This means that it is within the safe voltage range and is acceptable.

However, the current amplitude is not as high as it could be, as excessive current amplitude can cause the battery voltage to exceed the safety threshold. In other words, an overvoltage or undervoltage can occur. As we know, the overvoltage or undervoltage of battery is very dangerous because it triggers side reactions within the battery and leads to the decomposition of the SEI film, the reaction of the anode with the electrolyte, the reaction of the cathode with the electrolyte, and the decomposition reaction of the electrolyte itself, which in turn leads to a rapid build-up of heat inside the battery and even thermal runaway when large amounts of gas is being generated.

The effect of the current frequency f on the heating effect of the battery is then investigated. In addition, a constant current amplitude (e.g. 2C (12.9A)) and different current frequencies (e.g. 1 Hz, 5 Hz, 10 Hz and 15 Hz) are used to heat the batteries with sinusoidal alternating current. Similarly, before heating, the battery is placed in a −20 °C thermostat for 5 h to equalize the internal and external temperature of the battery. Each of the 3 battery modules is heated. As an example, the temperature rise of the battery module at SOC = 20% is plotted in Fig. 7.22b. From the figure, we can find that when the battery is heated at 15 Hz, the temperature rise curve of the battery is relatively gentle, the battery heats up to −6.9°C at 900 s (15 min), and in other words, the battery heats up slowly; when the battery is heated at 10 Hz, the temperature rise curve of the battery is slightly steeper than when the battery is heated at 15 Hz (but in general, the change is not significant), the battery heats up to − 5.8°C at 900 s, and in other words, the battery heats up more quickly; when the battery is heated at 5 Hz, the temperature rise curve is significantly steeper, the battery heats up to −3 °C at 900 s, and in other words, the battery heats up much faster significantly; when the battery is heated at 1 Hz, the temperature rise curve steepens even further, the battery heats up to + 2.7 °C at 900 s (i.e. a 22.7 °C increase in temperature), and in other words, the rate of temperature rise has increased significantly. By comparing the four curves, it can be concluded that the lower the frequency of the applied sinusoidal alternating current at the same value, the better the heat generation effect of the battery.

It is worth noting that when the frequency of the heating current is low (e.g. 1 Hz, 5 Hz, 10 Hz) and the current amplitude is high (e.g. 2C (12.9A)), the terminal voltage of the battery can be significantly out of the safe voltage range. Using an oscilloscope, the terminal voltage signals of the batteries during the heating process of each of the above groups are measured. When the heating process using 2C/10 Hz sinusoidal alternating current is taken as an example, the terminal voltage waveform of the battery is shown in Fig. 7.22c. As can be seen from the figure, the voltage has a peak-to-peak value of 2.09 V and a peak value of 1.045 V. At this point the OCV of the battery is 3.57 V. From this, the terminal voltage of the battery can be calculated to be 3.57 V + 1.045 V = 4.615 V > 4.2 V. This means that it obviously exceeds the safe voltage range and is therefore extremely dangerous.

Furthermore, the lower the frequency of the current at the same current amplitude, the greater the exceedance of the safe voltage (i.e. the more severe overvoltage). The total impedance at low and medium frequencies increases as the frequency decreases. According to Ohm's law, the overvoltage becomes more severe as the frequency decreases. Therefore, although the heating effect of low-frequency large-amplitude current is better, it can have a negative impact on the structure and safety of the battery and can even lead to safety incidents such as thermal runaway. This is unacceptable for its application in electric vehicles. Therefore, when selecting the parameters for the sinusoidal alternating current, the limits of the safe operating voltage of the battery need to be taken into account.

By comparing Fig. 7.22a with Fig. 7.22b, it can be seen that the effect of current amplitude on the heating effect is significantly greater than the effect of current frequency on the heating effect. This is because the heat generation rate of the battery, q, is proportional to the frequency, f, and the quadratic value of the current amplitude, A. Therefore, for improving the heat generation effect, increasing the current amplitude contributes much more to increasing the heat generation rate than decreasing the current frequency.

7.5 Mechanistic Analysis of the Effect of AC Heating on Battery Life

7.5.1 *Effect of Low Temperature Polarization Voltage and Low Temperature Lithium Ion Deposition*

The value of the battery polarization voltage when the battery is heated at low temperature is greater than that when the battery is heated at room temperature. The current rate should be selected such that the polarization voltage is stable and the terminal voltage of the battery does not exceed the upper and lower limits of the safe operating voltage, so as to prevent lithium ion deposition due to polarization and to reduce the adverse effects on battery capacity and life.

When a sinusoidal alternating current excitation is applied to the battery, the electrode reaction process of the lithium ion meets the requirement of the Butler-Volmer Equation:

$$j^{\text{Li}} = i_0\left[\exp\left(\frac{\alpha_a F}{RT}\eta\right) - \exp\left(\frac{\alpha_c F}{RT}\eta\right)\right] \tag{7.28}$$

To prevent the deposition of lithium ions on the anode surface, there exists:

$$\phi_s - \phi_l > E_{\text{Li+/Li}} \tag{7.29}$$

The overpotential of the electrode reaction can be defined as follows:

$$\eta = \phi_s - \phi_l - U_e \tag{7.30}$$

where, Ue is the open circuit voltage of the electrode (V); ϕ_s is the solid phase potential (V); ϕ_l is the liquid phase potential (V); $E_{Li+/Li}$ is the decomposition potential of the lithium ion, generally taken as 0 V.

The voltage of the solid–liquid phase is influenced by the ambient temperature and the SOC. Its computational formula is as follows:

$$U = f(\text{SOC}) + \frac{\partial U_{eq}}{\partial T}(T - T_{ref}) \tag{7.31}$$

where, U_{eq} is the voltage of the battery in equilibrium (V); T_{ref} is the reference temperature (K).

At low temperatures, lithium ions move slower in the electrolyte and the resistance to embedding and disembedding at the cathode and anode increases significantly, which makes the electrochemical impedance much greater than it is at room temperature. The lower the temperature, the slower the movement of lithium ions in the electrode active material and the electrolyte. Moreover, the embedded lithium impedance of the anode is greater than the de-lithium impedance of the cathode, which makes the concentration polarization more severe. As a result, lithium ion deposition is likely to occur at low temperatures. Christian et al. found that when charging the battery at −2 °C with a current greater than 0.5C, the amount of lithium ion deposited became significantly higher. For example, the lithium ion deposition at 0.5C is about 5.5% of the capacity, while the lithium ion deposition at 1C is 9% of the capacity (Von Lüders et al. 2017).

Veronika Zinth et al. (2014) studied the deposition of lithium ions in NCM18650 lithium-ion batteries at −20 °C. They found that when the battery was charged at low temperatures, the kinetic conditions deteriorated, which resulted in reduced charge capacity, slower lithium embedding in the graphite cathode, and consequentially deposition of lithium ions on the cathode surface. Although some of the deposited lithium metal can still be re-embedded in the graphite after the battery has been left for a period of time, the short shelf time in practice does not allow the entire lithium metal to be re-embedded in the graphite, so lithium ions are still deposited on the surface of the anode.

When a low frequency sinusoidal alternating current is applied to the battery at low temperatures, lithium precipitation can occur near the anode. The accumulation of lithium ions can lead to the formation of lithium dendrites, which, when they grow longer, can puncture the diaphragm and cause a short circuit inside the battery or even a thermal runaway and other accidents. In contrast, when applying medium and high frequency sinusoidal alternating currents to the battery, no lithium deposits are produced (Wang et al. 2013). This makes them more suitable for battery heating.

7.5.2 Principle of Electrode Reaction During Low Temperature AC Heating

The sinusoidal alternating current heating method offers a new option for improving the low-temperature performance of lithium-ion batteries, as it allows for rapid and uniform warming of the battery. Sinusoidal alternating current heating allows for alternating lithium ion de-lithiation and lithium embedding within the cathode and anode materials. The lithium ions that are disembeded from the cathode during each cycle are embedded in the anode, so that lithium deposition is less likely to occur. However, the electrochemical reaction is a complex process and there can be a mismatch between de-lithiation and lithium embedding in the same cycle.

The principle of the electrode reaction when the battery is excited by different currents at low temperatures is shown in Fig. 7.23. Figure 7.23a illustrates the lithium precipitation process at the battery anode during low temperature DC charging. The

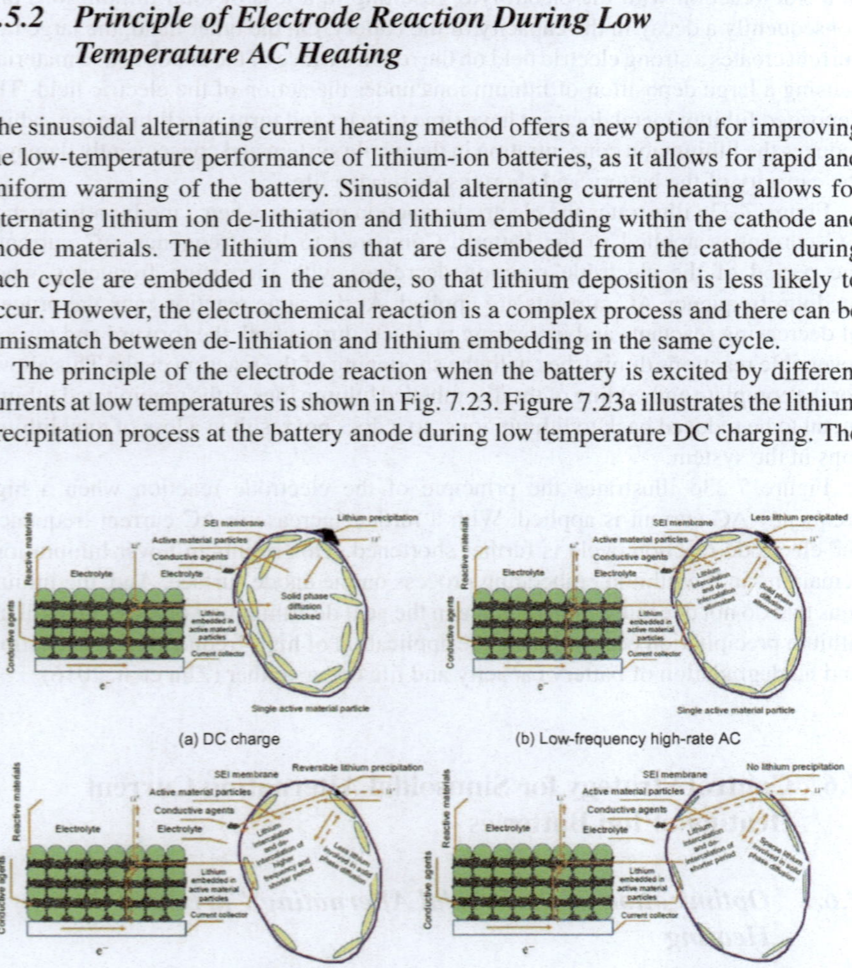

Fig. 7.23 Electrode reaction principle model

solid phase diffusion of lithium ions within the anode active material is slow and their embedding at the anode is significantly slower than their de-lithiuming at the cathode. As a result, the remaining lithium ions are not embedded in the anode in time for each cycle and precipitate out as lithium metal on the anode surface.

Figure 7.23b illustrates the process of applying a low frequency, high rate AC current to a battery at low temperatures. In this case, even though de-lithiation and lithium embedding alternate between the anode and cathode in a single cycle, the low frequency of the AC current and the complex nature of the electrode reactions within the battery cause some of the precipitated lithium metal to be reacted away in a side-reaction with the electrolyte, resulting in a loss of total lithium ions and consequently a decay in the capacity of the battery. On the other hand, the large rate current creates a strong electric field on the rough surface of the anode active material, causing a large deposition of lithium ions under the action of the electric field. The deposited lithium metal does not have time to react and turns into lithium ion, which reduces the lithium ion concentration in the whole system and consequently damages the capacity of the battery and shortens its service life.

Figure 7.23c illustrates the electrode reaction process when a medium frequency AC current is applied to the battery. Compared to low frequency AC currents, the period of the electrode reaction decreases with increasing frequency when medium frequency AC currents are applied. At the same reaction rate, the process of decreasing reactants and increasing products during both the forward and reverse reversible reactions diminishes with the shortening of the reaction cycle. This allows for the complete embedding of the disembeded lithium ions or the precipitated lithium metal to be reduced back to lithium ions, so it does not result in a loss of total lithium ions in the system.

Figure 7.23d illustrates the principle of the electrode reaction when a high frequency AC current is applied. With a further increase in AC current frequency, the electrode reaction cycle is further shortened. This results in fewer lithium ions remaining in the lithium embedding process on the anode surface. And, the lithium ions that do not deposit will participate in the next de-lithiuming process. Overall, no lithium precipitation occurs during the application of high-frequency AC excitation, and no degradation of battery capacity and life occurs either (Zhu et al. 2016).

7.6 Control Strategy for Sinusoidal Alternating Current Heating of Ion Batteries

7.6.1 Optimization of Sinusoidal Alternating Current Heating

As there is very little change in ambient temperature and heat dissipation conditions during internal heating, the most effective way to accelerate the temperature rise of a battery is to increase the internal heat generation rate of the battery when a sinusoidal alternating current is applied. Based on the electro-thermal coupling model of the battery presented above, the heat generation rate of the battery is mainly influenced by the amplitude and frequency of the sinusoidal alternating current. The amplitude of the sinusoidal alternating current has a direct and significant effect on the heat generation rate of the battery, while the frequency has an indirect effect on the heat generation rate of the battery by changing the impedance of the battery. The choice of these key parameters therefore determines the performance of the heating strategy. Currently, some heating strategies exist to achieve a better heat generation effect of the battery by adjusting the current amplitude at a suitable fixed frequency. However, these methods ignore the influence of the battery temperature on the appropriate heating frequency. As it only uses the method of adjusting the current amplitude, the rate of temperature rise of the battery is not satisfactory. In order to further improve the effect of sinusoidal alternating current heating, the amplitude and frequency of the sinusoidal alternating current are used as variables in the proposed optimization strategy. These two variables are optimized and updated as the battery temperature rises.

In order to heat the battery to a certain temperature in the shortest possible time, the amplitude and frequency of the sinusoidal alternating current needs to be adjusted in time to maximize the heat generated by the battery. At the same time, it is important to ensure that the terminal voltage of the battery is within the safe operating range during the heating process in order to avoid damage to the battery structure and to reduce the negative impact on the capacity and life of the battery. The choice of current amplitude and frequency can therefore be equated to a constrained non-linear optimization problem.

Based on the electro-thermal coupling model of the battery, the parameters of the equivalent circuit model can be obtained by interpolation for a given battery temperature. The battery impedance in the heat generation rate expression can then be considered as a function of the current frequency, and the battery impedance can be calculated from Eq. (7.32). The objective function can be expressed as follows:

$$J = \max_{A,f}\left[\left(\frac{A}{\sqrt{2}}\right)^2 \cdot Z_{\mathrm{Re}}(T, f)\right] \tag{7.32}$$

The amplitude and frequency of the sinusoidal alternating current are used as optimization variables, which are constantly optimized and updated during the heating process. When the SOC is given, the OCV is obtained. Alternatively, the maximum voltage across the battery impedance when a sinusoidal alternating current is applied at a given temperature can be calculated from Eq. (7.33).

$$U_{Z,\max} = A|Z(T, f)| \tag{7.33}$$

where, $|Z(\mathrm{T}, \mathrm{f})|$ is the module of the total impedance, which can be calculated by Eq. (7.34); $\mathrm{U}_{\mathrm{Z,\,max}}$ is the absolute value of the maximum voltage across the impedance.

$$|Z(T, f)| = \sqrt{Z_{\mathrm{Re}}^2(T, f) + Z_{\mathrm{Im}}^2(T, f)} \tag{7.34}$$

The terminal voltage of the battery is therefore a superposition of the open circuit voltage and the voltage at the ends of the impedance, which needs to be maintained within a certain range throughout the heating process. The optimization constraint can then be expressed as follows:

$$\begin{cases} C_1 : U_{\mathrm{oc}} + U_{Z,\max} = U_{\mathrm{oc}} + A|Z(T, f)| \leq U_{\max} \\ C_2 : U_{\mathrm{oc}} - U_{Z,\max} = U_{\mathrm{oc}} - A|Z(T, f)| \geq U_{\min} \end{cases} \tag{7.35}$$

where, U_{max} and U_{min} are the upper and lower limits of the terminal voltage, respectively.

The changes in the battery SOC are ignored during the heating process, so that the open circuit voltage can be regarded as a constant value. In addition, as the battery temperature rises during the heating process, the amplitude or heat generation rate of the sinusoidal alternating current that the battery can accept will gradually increase. As a result, only the terminal voltage constraint at the current moment needs to be satisfied in each optimization, which therefore simplifies the optimization problem. The optimized heating strategy is solved by MATALB using the SQP algorithm.

The solution of optimization problems is often time consuming. Real-time optimization can be limited by the computational power of the device and is therefore difficult to implement in practice. Therefore, the heating strategy optimization should consist of two steps, i.e. the creation of the optimization strategy in the offline simulation and the online execution of the optimization strategy.

1. Creation of the optimization strategy in the offline simulation

The framework of creating the optimization strategy in the offline simulation is shown in Fig. 7.24. The first step is to obtain information on the ambient temperature, the initial temperature of the battery, the parameters of the electro-thermal coupling model and the terminal voltage constraints of the battery. Once the initial temperature of the battery is given, the corresponding model parameters are obtained by looking up the table. It is then determined whether the heating strategy (i.e. the amplitude and frequency of the current) needs to be optimized and updated. Given the regulation capacity and accuracy of the device, it is not necessary to optimize and update the strategy at every time step. The amplitude and frequency of the sinusoidal alternating current can be optimized and updated after a certain time interval (e.g. 30 s, 60 s) or temperature increments (e.g. 0.5 °C, 1 °C). Next, based on the electric-thermal coupling model of the battery, the heat generation rate and temperature of the battery can be calculated according to the amplitude and frequency of the sinusoidal alternating current. Afterwards it is determined whether the temperature of the battery has reached the heating termination temperature. If not, the model parameters are updated by interpolation based on the updated battery temperature, and the above steps are repeated until the battery temperature reaches the heating

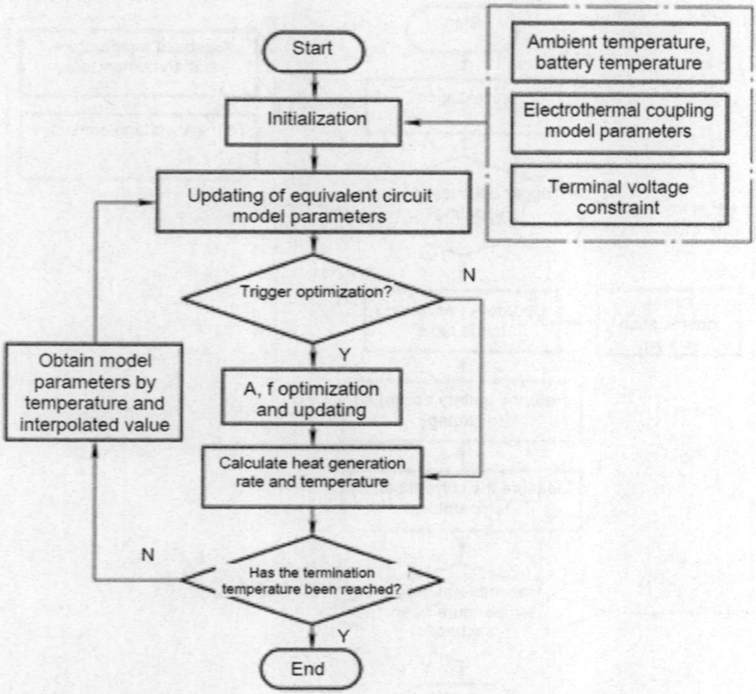

Fig. 7.24 Framework of creating the optimization strategy in the offline simulation

termination temperature. Otherwise the application of sinusoidal alternating current excitation is stopped to end the preheating process.

During the offline simulation, an optimized heating control strategy for the battery at different temperatures is established, thus providing data to support the application of the online implementation of the optimization strategy.

2. Online implementation of the optimization strategy

The framework of online implementation of the optimization strategy is shown in Fig. 7.25. First, the ambient temperature, battery temperature and voltage constraints need to be initialized first. Then, based on the set conditions, it is determined whether the heating strategy needs to be optimized and updated. When an update is required, the corresponding optimized current amplitude and frequency can be obtained according to the results obtained during the offline simulation, and based on the current temperature of the battery. Otherwise, the amplitude and frequency of the current remains unchanged. Next, the temperature of the battery is measured after heating and it is determined whether the battery temperature has reached the heating termination temperature. If not, repeat the above steps until the battery temperature reaches the heating termination temperature; otherwise the heating stops.

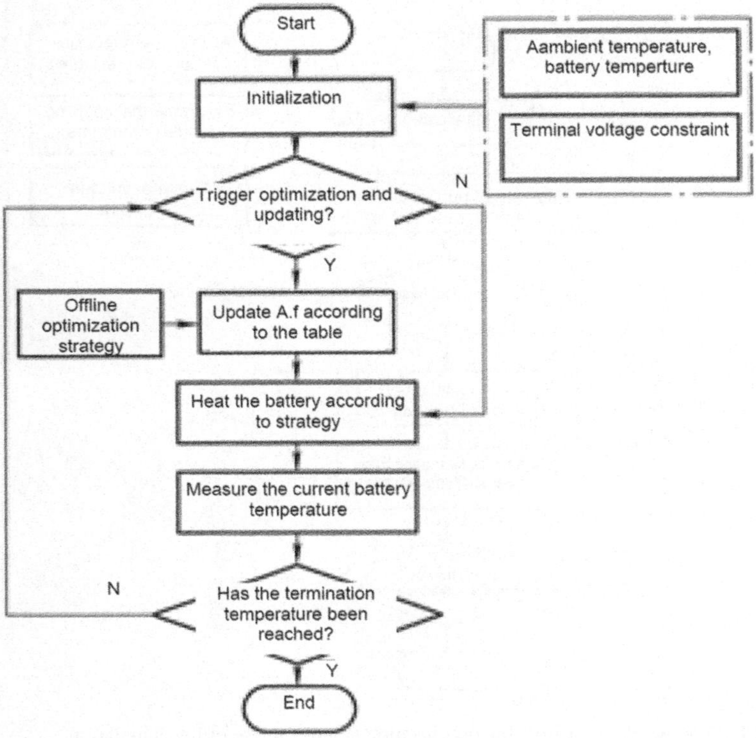

Fig. 7.25 Framework of online implementing the optimization strategy

7.6.2 *Basic Theory of the SQP Optimization Algorithm*

In simple terms, solving an optimization problem is solving for the extreme value of a function within a given set (Li and Tong 2005; Lai and He 2007). Basically, the mathematical model for all optimization problems can be expressed in the following form:

$$\min f(x) \\ \text{s.t. } x \in K \tag{7.36}$$

where, x is the decision variable; K is the feasible domain; and f(x) is a real-valued number defined on the set K.

The iterative method is a common tool for solving optimization algorithms. The procedure is shown below: the first point $x_0 \in R^n$. The point range $\{x_k\}$ is obtained according to some method. If $\{x_k\}$ is infinite, the last point is the desired one. If $\{x_k\}$ is infinite, its limit point is the optimal solution to the nonlinear programming problem being solved. In general, the iteration point $\{x_k\}$ can steadily approach the domain of local minimal point x^* and converge rapidly to x^*. This is a common characteristic of good algorithms. The iteration stops when the set termination condition is reached. The steps of the optimization algorithm are as follows:

The initial solution x_0 is given.

Step 1: Determine the search direction d_k, which is the descent direction of the objective function $f(x)$ at x_k.

Step 2: Determine the search step size α_k such that the value of $f(x)$ decreases.

Step 3: Let $x_{k+1}=x_k+\alpha_k d_k$. When x_{k+1} reaches the termination criterion, the iterative process is terminated. At this point, x_{k+1} is the approximate optimal solution; otherwise, return to Step 1. The flow is shown in Fig. 7.26.

The sequential quadratic programming (SQP) algorithm, an effective method for solving non-linear programming problems, has been more widely used in engineering practice. It offers both the advantages of global convergence and the superlinear convergence speed. The basic idea is to transform the non-linear programming problem to be solved into solving an approximate quadratic programming sub-problem. At each iteration step, the search direction d_k is determined with the help of the solution of the subproblem, the search step α_k is determined according to the requirement of reducing the value function, a new point is calculated according to the iterative equation, then the approximate solution of the subproblem continues to be solved at this new point, and finally the solution of the objective is approximated by the individual solutions obtained in the course of continuous iterations. Therefore, this optimization algorithm is called a sequential quadratic programming algorithm. The principle of the SQP algorithm is described as follows:

Non-linear programming problems to be solved:

$$\begin{aligned} &\min f(x) \\ &\text{s.t. } h_i(x) = 0, (i = 1, \ldots\ldots, l) \\ &\qquad g_i(x) \geq 0, (i = 1, \ldots\ldots, m) \end{aligned} \tag{7.37}$$

At each iteration point x_k, the quadratic programming subproblem is obtained by solving and transforming it, the solution of which is the search direction (descent direction) d_k. According to the iterative equation:

$$x_{k+1} = x_k + \alpha_k d_k \tag{7.38}$$

Determine the search step size α_k, calculate a new point x_{k+1}, and use each x_{k+1} to continuously approximate the optimal solution to the desired nonlinear programming problem. The SQP algorithm consists of three main components as follows:

(1) Constructing and computing quadratic programming subproblems

Solve for the objective function, constraints and derivatives at the initial point and at each new point, which are subsequently transformed into a quadratic programming subproblem.

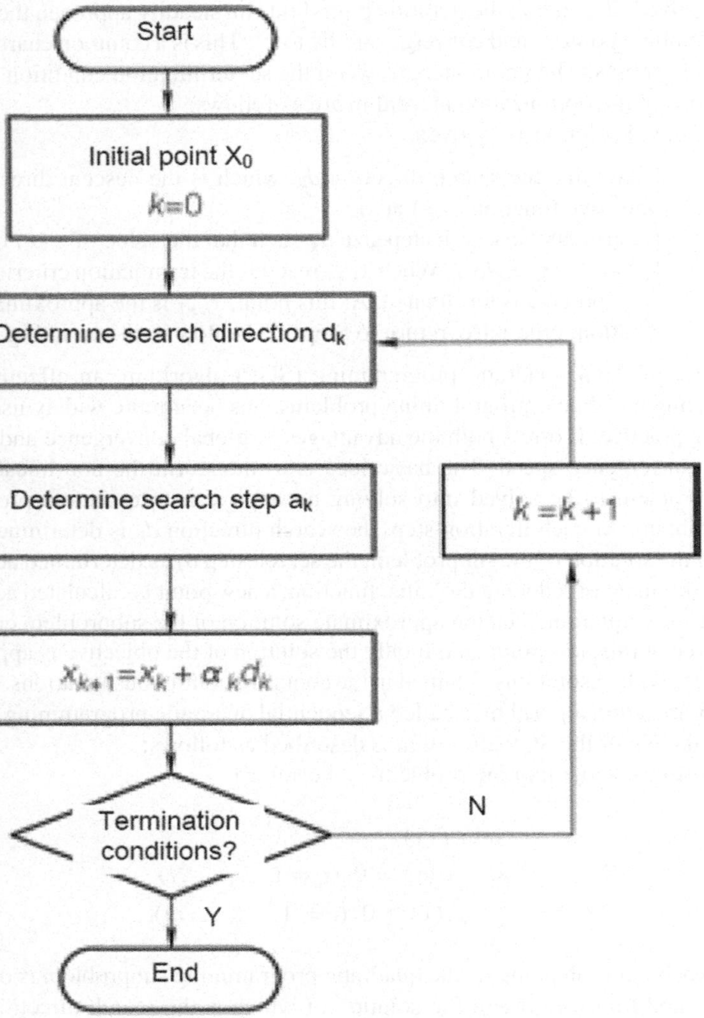

Fig. 7.26 Optimization algorithm flow

$$
\begin{aligned}
&\min\left[\frac{1}{2}(\boldsymbol{d}_k)^{\mathrm{T}}\boldsymbol{H}_k\boldsymbol{d}_k + \Delta\boldsymbol{f}(x_k)^{\mathrm{T}}\boldsymbol{d}_k\right]\\
&\text{s.t.}\ \ [\Delta\boldsymbol{h}_i(x_k)]^{\mathrm{T}}\boldsymbol{d}_k + \boldsymbol{h}_i(x_k) = 0\\
&\qquad [\Delta\boldsymbol{g}_i(x_k)]^{\mathrm{T}}\boldsymbol{d}_k + \boldsymbol{g}_i(x_k) \geq 0
\end{aligned}
\tag{7.39}
$$

Its Lagrangian function is as follows:

$$L(d, \lambda_i^{k+1}, \mu_i^{k+1}) = \frac{1}{2}(d^k)^{\mathrm{T}} H^k d^k + \Delta f(x_k)^{\mathrm{T}} d^k + \sum_{i=1}^{p} (\lambda_i^{k+1})^{\mathrm{T}} [\Delta h_i(x_k) d^k + h_i(x_k)] + \sum_{j=p+1}^{q} (\mu_i^{k+1})^{\mathrm{T}} [\Delta g_i(x_k) d^k + g_i(x_k)] \tag{7.40}$$

After the transformation into an unconstrained problem, the Lagrange multipliers λ_i^{k+1} and μ_i^{k+1} for the optimal solution dk of Eq. (7.39) can be obtained according to the unconstrained optimization method. This optimal solution d_k can then become the search direction for xk in the optimization problem to be solved.

(2) Determining the search step size

Having already determined the direction of descent, the search step size needs to be further determined. According to the iterative equation:

$$x_{k+1} = x_k + \alpha_k d_k \tag{7.41}$$

In view of the computational speed and volume requirements, the trial-and-error method can be used to solve α_k. In addition, α_k needs to be such that the value of the Lagrange function decreases after the conversion of the constrained problem to be solved to an unconstrained one. That's to say,

$$L(x^{k+1}, \lambda_i^{k+1}, \mu_i^{k+1}) \le L(x^k, \lambda_i^k, \mu_i^k) \tag{7.42}$$

First, substitute $\alpha_k = 1$ into Eq. (7.41). If the inequality holds, then $\alpha_k = 1$; if not, substitute $\alpha_k = 0.8, 0.6, 0.4, \ldots\ldots$ into Eq. (7.41) until the inequality holds.

From the above process, the new point x_{k+1} can be calculated and then substituted into the termination condition as follows:

$$\frac{|f(x_{k+1}) - f(x_k)|}{|f(x_k)|} \le \mathrm{ACC} \tag{7.43}$$

where, ACC is the convergence accuracy. If the above termination condition is met, i.e. x_{k+1} is the optimal solution, then the iteration is terminated and the optimal solution x_{k+1} is output; if not, then the following approximate Hessian matrix correction is continued.

(3) Hessian matrix correction

Hk is the second order derivative array of the Lagrange functional Eq. (7.40) of the optimization problem Eq. (7.37) to be solved, i.e. the Hessian matrix. Because of the complexity of the Hessian matrix, the SQP algorithm replaces the direct calculation of the Hessian matrix with a variable scale method. In other words, H^k is replaced by a symmetric positive definite matrix, A^k, which is approximated by a stepwise correction during the iterative calculation. The correction formula is as follows:

$$A^{k+1} = A^k + \Delta A^k \tag{7.44}$$

where, ΔA^k is the correction matrix.

In order for A^{k+1} to remain symmetrical and positive in the correction, A^k needs to be symmetrical and positive with ΔA^k. So first let $A^0 = I$, after which the correction matrix ΔA^k can be calculated according to the following equation:

$$\Delta A^k = \boldsymbol{a}\boldsymbol{a}^{\mathrm{T}} - \boldsymbol{e}\boldsymbol{e}^{\mathrm{T}} \tag{7.45}$$

where, $\boldsymbol{a} = \boldsymbol{q}'/\sqrt{\boldsymbol{\xi}}, \boldsymbol{e} = \boldsymbol{A}^k\boldsymbol{t}\sqrt{\boldsymbol{\lambda}}$.

Let

$$\begin{gathered}
\boldsymbol{q} = \Delta L(x^{k+1}, \lambda^{k+1}, \mu^{k+1}) - \Delta L(x^k, \lambda^k, \mu^k) \\
\boldsymbol{t} = x^{k+1} - x^k = \alpha^k d^k \\
\boldsymbol{\xi} = \boldsymbol{t}^{\mathrm{T}} q \\
\boldsymbol{\lambda} = \boldsymbol{t}^{\mathrm{T}} \boldsymbol{A}^k \boldsymbol{t}
\end{gathered} \tag{7.46}$$

From Eq. (7.45), ΔA^k satisfies the requirements of symmetry and positivity. Since $t^T q$ should also be positive definite, in order to keep $t^T q$ constant, q can be replaced by q'

$$\boldsymbol{q}' = \theta \boldsymbol{q} + (1 - \theta)\boldsymbol{A}^k \boldsymbol{t} \tag{7.47}$$

where, θ is defined as:

$$\begin{aligned}
\theta &= 1 \qquad\qquad\qquad (\xi \geq 0.2\gamma) \\
\theta &= 0.8\gamma/(\gamma - \xi) \ (\text{other})
\end{aligned} \tag{7.48}$$

When A^k is positive definite, the modified matrix A^{k+1} is symmetric and positive definite. Therefore, it has a deterministic solution. Then, optimal solution or near-optimal solution can be obtained by the optimized SQP algorithm.

7.6.3 Simulation Results Analysis of the Optimal Heating Control Strategy

1. Analysis of the effects of AC heating based on EIS measurement data

In order to investigate the variation law of the battery impedance, EIS tests and model simulations are carried out in the current frequency range of 100 Hz. As an example, the curves plotted at −20 °C, −15 °C, −10 °C and −5 °C are shown in the subplot in Fig. 7.27. As can be seen from the subplots in Fig. 7.27a and

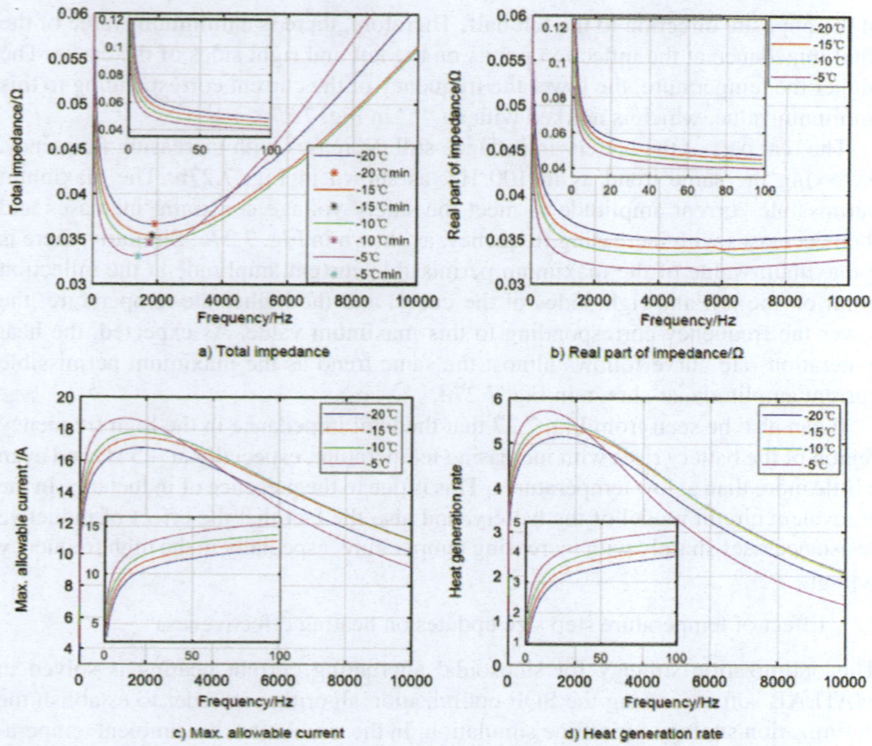

Fig. 7.27 Analysis of the effect of AC heating based on EIS data

b, the total impedance of the battery and the real part of the impedance decrease with increasing current frequency, and the lower the temperature, the greater the value of the impedance. The maximum safe current amplitude that can be applied to the battery under the condition of meeting the safe operating voltage constraint is obtained from the equation, as shown in the subplot in Fig. 7.27c. The maximum permissible safety current amplitude rises with increasing current frequency at all temperatures. The heat generation rate of the battery at the maximum permissible safety current amplitude applied can then be calculated from Eq. (7.19), as shown in the subplot in Fig. 7.27d. Similar to the pattern of variation of the maximum permissible safety current amplitude, the heat generation rate of the battery rises with increasing current frequency at all temperatures within 100 Hz.

However, when the current frequency exceeds 100 Hz, the trends of the individual curves differ significantly. The values within 10,000 Hz are obtained according to the above method. The curves are then plotted as shown in Fig. 7.27. As can be observed from Fig. 7.27a, the total impedance - frequency curve of the battery can be clearly divided into two parts. In the left half, the higher the frequency of the current, the lower the total impedance. In the right half, the total impedance varies with frequency in the opposite direction to the left half. Therefore, there is a minimum value of the total impedance at the inflection points on the left and right sides of the curve. The higher the temperature, the lower the frequency of the current corresponding to this minimum value, which is marked with an "*" in Fig. 7.27a.

The real part of the battery impedance still decreases with increasing frequency, following the same trend as in 100 Hz, as shown in Fig. 7.27b. The maximum permissible current amplitude to meet the safety voltage constraint increases and then decreases with increasing frequency, as shown in Fig. 7.27c. Similarly, there is a maximum value of the maximum permissible current amplitude at the inflection point on the left and right sides of the curve, and the higher the temperature, the lower the frequency corresponding to this maximum value. As expected, the heat generation rate curve follows almost the same trend as the maximum permissible current amplitude, as shown in Fig. 7.27d.

It can also be seen from Fig. 7.27 that the total impedance in the high frequency region of the battery rises with increasing temperature, especially at −5 °C and even a little more than at low temperatures. This is due to the presence of inductance in the equivalent circuit model of the battery, and also the fact that the effect of inductive resistance rises sharply with increasing temperature, especially in the high frequency region.

2. Effect of temperature step size updates on heating effectiveness

The optimization strategy for sinusoidal alternating current heating is solved in MATLAB software using the SQP optimization algorithm in order to establish the optimization strategy for offline simulation. In the simulation, the ambient temperature and the initial temperature of the battery are both set to −20 °C. The upper and lower voltage limits of the battery are 4.2 V and 2.8 V respectively.

In the case of a battery with SOC = 20%, the amplitude and frequency of the sinusoidal alternating current are optimized and updated at certain temperature step sizes, e.g. 0.5, 1 and 2 °C. The results of the optimized current amplitude versus frequency are shown in Fig. 7.28a and b. The corresponding temperature rise curve of the battery is shown in Fig. 7.28c. As can be seen from Fig. 7.28a, the maximum permissible current rises with increasing battery temperature and the slope of the current amplitude curve gradually increases after iteration and optimization. This indicates that the rate of increase of the optimized maximum permissible current magnitude is also gradually increasing. When the optimized and updated temperature step size is large, the maximum permissible current amplitude increases in a stepwise fashion. When the optimized and updated temperature step size is small enough, e.g. 0.5 °C, the maximum permissible current amplitude increases smoothly and continuously, and the maximum current amplitude increases as the updated temperature step size becomes shorter.

Similarly, it can be observed from Fig. 7.28b that as the battery temperature increases, the optimized current frequency decreases and the trend of the current frequency curve becomes more gentle. This indicates that the rate of reduction of the optimized current frequency is gradually becoming smaller. When the optimized and updated temperature step size is large, the current frequency decreases in a stepwise

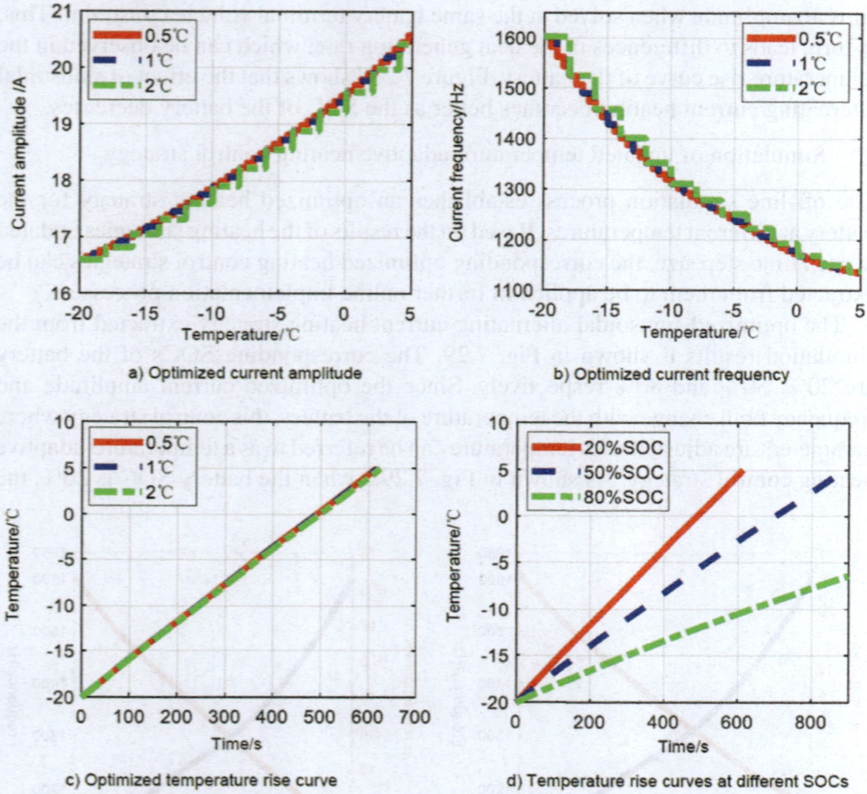

Fig. 7.28 Simulation curves for different temperature steps

fashion. When the optimized and updated temperature step size is small enough, e.g. 0.5 °C, the current frequency decreases smoothly and continuously, and the current frequency decreases as the updated temperature step size becomes shorter.

For the heat generation rate, the heating strategy corresponding to a temperature step size of 0.5 °C allows for a better heat generation rate at each optimization step, resulting in the fastest temperature rise rate, as shown in Fig. 7.28c. As the updated temperature step size becomes longer, the temperature rise rate of the battery gradually decreases.

In addition, as a comparison, the temperature rise of the battery at different SOCs is simulated, and the results of the battery temperature rise are shown in Fig. 7.28d. As can be seen from the figure, the steepest temperature rise curve, i.e. the greatest rate of temperature rise, is found at SOC = 20%, followed by the curve at SOC = 50% and the smoothest curve, i.e. the smallest rate of temperature rise, is found at SOC = 80%. This is due to the fact that the open circuit voltage of the battery varies considerably at different SOCs, which results in a large difference in the maximum permissible current amplitude when solved at the same battery terminal voltage constraint. This, in turn, leads to differences in the heat generation rate, which can be observed in the temperature rise curve of the battery. Figure 7.28d shows that the effect of sinusoidal alternating current heating becomes better as the SOC of the battery decreases.

3. Simulation of updated temperature-adaptive heating control strategy

The off-line simulation process establishes an optimized heating strategy for the battery at different temperatures. Based on the results of the heating strategies updated at each time step size, the corresponding optimized heating control strategies can be extracted from them to be applied in further online implementation processes.

The optimized sinusoidal alternating current heating strategy extracted from the simulation results is shown in Fig. 7.29. The corresponding SOCs of the battery are 20%, 50% and 80% respectively. Since the optimized current amplitude and frequency both change with the temperature of the battery, this control strategy where parameters are adjusted with temperature can be referred to as a temperature-adaptive heating control strategy. As shown in Fig. 7.29a, when the battery SOC is 20%, the

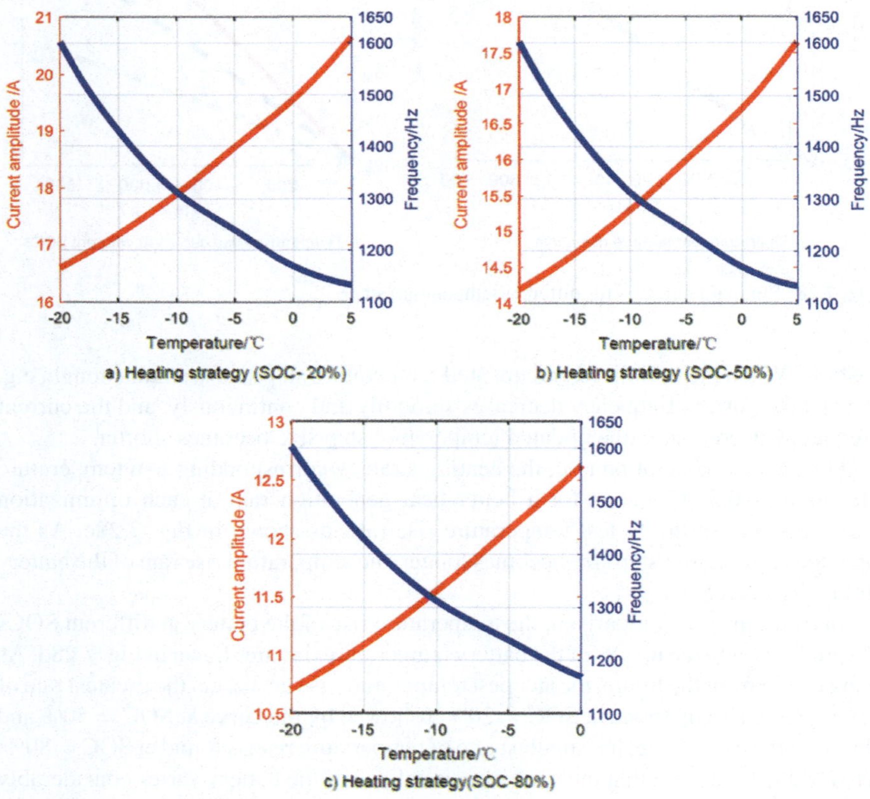

Fig. 7.29 Optimized temperature-adaptive heating strategy

optimized current amplitude gradually increases with increasing battery temperature from 16.6A at −20 °C to 20.7A at + 5 °C, showing a significant increase; in addition, the optimized current frequency decreases from 1602.5 Hz at -20 °C to 1133.4 Hz at + 5 °C with increasing battery temperature, showing a significant decrease.

As shown in Fig. 7.29b and c, when the battery SOC is 20%, the optimized current amplitude gradually increases from 14.2A at −20°C to 17.7A at + 5°C with the increase of the battery temperature, with a more obvious trend of increase; when the battery SOC is 80%, the optimized current amplitude gradually increases from 10.7A at −20°C to 12.6A at 0°C with the increase of the battery temperature, with a relatively slow increase. Furthermore, the trend of the frequency-temperature curve at SOC = 50% and SOC = 80% is not significantly different from that at SOC = 20%. This is due to the fact that the different SOCs of the battery affect the maximum permissible current amplitude meeting the safety voltage constraint, mainly by affecting the value of the open circuit voltage.

Overall, in this temperature-adaptive heating control strategy, the higher the battery temperature, the higher the optimized maximum current amplitude, the lower the corresponding optimized current frequency and the higher the generated heat generation rate. Therefore, during the application of sinusoidal alternating currents to batteries at low temperatures, it is necessary to gradually increase the current amplitude while gradually decreasing the current frequency in accordance with this optimization strategy in order to achieve the best possible heat generation effect.

During the online implementation of this optimization strategy, once the measured value of the battery temperature is given, the value of the optimized current amplitude and frequency corresponding to the current temperature can be obtained by checking the table. Furthermore, this optimized temperature-adaptive sinusoidal alternating current heating control strategy is feasible for the experimental equipment.

7.7 Simulation and Experiment of the Temperature Field of a Sinusoidal Alternating Current Heated Battery

In Sect. 7.2, in order to obtain the current amplitude and frequency of the AC heating, the battery is modeled as a whole for electro-thermal coupling, without taking into account the internal heat conduction of the battery. In contrast, the electrochemical - thermal coupling model developed in this chapter is intended to further analyze the internal temperature field distribution of the battery, which can be used to evaluate the heating effect and indirectly justify the simplification of the previous modeling. The modeling method of the 1D electrochemical model and the 3D thermal model is elaborated by establishing a coupled electrochemical-thermal model. Then, its correctness is verified with charge/discharge experiments and the internal temperature field distribution of the battery when heated by AC is simulated.

7.7.1 Modeling of Electrochemical-Thermal Coupling

The finite element simulation software is used to build an electrochemical - thermal coupling model, including a 1D electrochemical model and a 3D thermal model. The modeling principles are described in Chap. 3. The specific modeling principles are shown in Fig. 7.30 and the specific parameters used in this model are shown in Table 7.5.

Regarding the one-dimensional electrochemical model, a geometric model consisting of cathode and anode active materials, cathode and anode current collectors and diaphragms is developed according to the structural characteristics of a five-layer sandwich. Using the porous electrode correlation theory and the Butler-Volmer electrode kinetic equation, parameters such as the movement rate and concentration of lithium ions in the electrolyte and the cathode and anode active materials are obtained.

The heat generation of the battery when a sinusoidal alternating current is applied is introduced into a three-dimensional thermal model based on a three-dimensional differential equation of thermal conductivity to solve for the temperature distribution inside the battery. The 3D thermal model is a description of the heat transfer and temperature distribution of a battery with an internal heat source under Type III boundary conditions. Furthermore, the temperature in the thermal model is transferred to the electrochemical model, where the electrochemical heat generation and the sinusoidal alternating current heating are then solved for to achieve a coupled battery thermoelectric simulation.

Modeling also involves setting boundary conditions, dividing the grids and setting the relevant parameters of the solver.

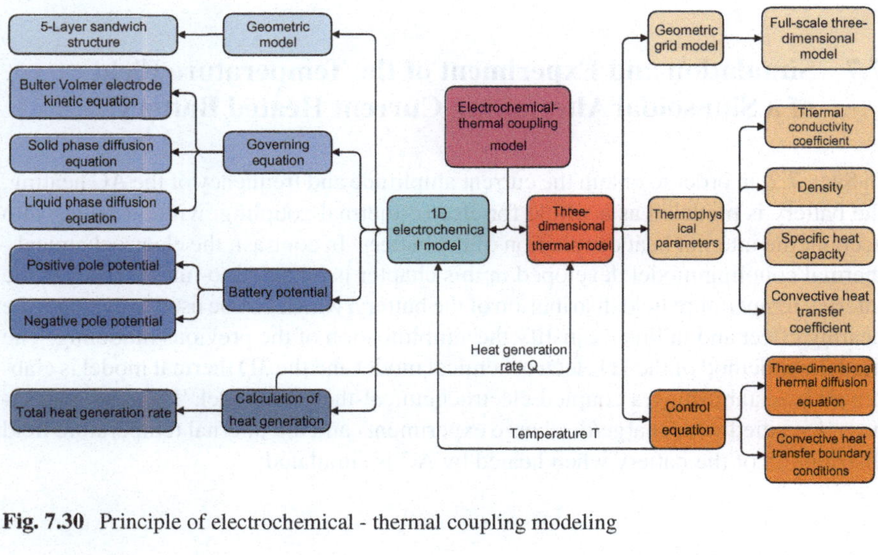

Fig. 7.30 Principle of electrochemical - thermal coupling modeling

Table 7.5 Parameters related to the electrochemical model

Symbol	Parameter name	Unit	Cathode	Anode
L	Length	m	L_{pos}	L_{neg}
σ_s^{eff}	Effective solid phase conductivity	S/m	$\sigma_s \varepsilon_s^{1.5}$	$\sigma_s \varepsilon_s^{1.5}$
σ_s	Solid phase conductivity	S/m	$100 \times g(\mathrm{T})$	$100 \times g(\mathrm{T})$
ε_s	Solid phase porosity	1	0.297	0.471
σ_e^{eff}	Effective liquid phase conductivity	S/m	$\sigma_e \varepsilon_e^{1.5}$	$\sigma_e \varepsilon_e^{1.5}$
σ_e	Liquid phase conductivity	S/m	$\sigma_e = f(C_e)$	$\sigma_e = f(C_e)$
ε_e	Liquid phase porosity	1	0.444	0.357
R_s	Particle radius	m	2×10^{-6}	4×10^{-6}
A_{cell}	Battery plate area	m^2	0.033	0.033
D_s	Solid phase diffusion coefficient	m^2/s	$5 \times 10^{-13} \times g(T)$	$5 \times 10^{-13} \times g(T)$
D_e^{eff}	Effective liquid phase diffusion coefficient	m^2/s	$\varepsilon_s^{1.5} D_e$	$\varepsilon_s^{1.5} D_e$
D_e	Liquid phase diffusion coefficient	m^2/s	7.5×10^{-11}	7.5×10^{-11}
t_0^+	Liquid phase transfer coefficient	1	1	1
R_{film}	SEI membrane resistor	$\Omega \cdot m^2$	0.001	0.001
α_n, α_p	Exchange current reaction rate coefficient	1	0.5	0.5
k	Exchange current reaction constant	m/s	2×10^{-11}	2×10^{-11}
$C_{s,max}$	Maximum lithium concentration in solid phase	mol/m^3	29,000	31,507

A 18,650 cell is made up of a number of "sandwich" units wound or stacked together. Each unit contains cathode and anode current collector, cathode and anode active material and diaphragm. Considering the speed of the model solution and the need to meet accuracy requirements, the battery is modeled as a whole and the differences between the layers are reflected by the setting of the thermal physical parameters.

After creating a geometric model based on the 3D shape parameters of the battery, the grid is divided and a triangular grid is chosen. Considering the variety of materials in the battery and the complexity of the heat generation in each layer, a more densely distributed and larger number of grids are used to solve them discretely. The 3D grid model of the battery is shown in Fig. 7.31.

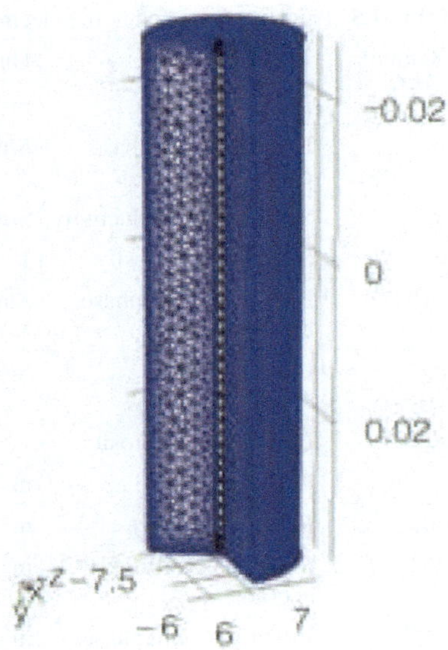

Fig. 7.31 3D grid model of a battery

The relevant parameters for the cathode and anode active materials, the cathode and anode current collectors and the diaphragm are shown in Table 7.6, from which the overall thermal parameters of the battery can be derived.

The cylindrical battery cell is made by the battery active materials wound in a spiral shape. Therefore, the thermal conductivity coefficient of the battery in the three-dimensional thermal model is therefore anisotropic, with the thermal conductivity coefficient along the length of the battery (axial) being somewhat greater than that along the radius of the battery (radial) (Chen et al. 2006).

The heat transfer coefficient h of the convective heat transfer surface between the battery surface and the surrounding environment is the result identified by genetic algorithm, with h taken as 9.93 $W/(m^2 \cdot K)$.

7.7.2 *Validation of an Electro-Thermal Coupling Model Based on Electrochemistry*

The correctness of the electrochemical-thermal coupling model is first verified experimentally using the battery 1C (2.15A) charge/discharge condition as an example. The voltage curve of a battery charged at 1C rate (constant current–constant voltage) is shown in Fig. 7.32a. The voltage curve for a battery discharged at 1C rate is shown in Fig. 7.32b. The temperature rise curve for a battery discharged at 1C rate is shown in Fig. 7.32c. The results of the finite element model accuracy evaluation based on the three curves in Fig. 7.32a–c are shown in Table 7.7.

Table 7.6 Thermal and physical parameters of materials used in batteries

Parameters	Material	Thickness/μm	Density/(kg/m3)	Specific heat capacity/[J/(kg·K)]	Thermal conductivity coefficient/[W/(m·K)]
Cathode material	$LiNi_{1/3}Mn_{1/3}Co_{1/3}O_2$	55	2500	1000	3.4
Anode material	Li_xC_6MCMB	55	1347	1437	1.04
Diaphragm	PVDF	30	940	1046	0.15
Cathode current collector	Cu	10	2770	875	170
Anode current collector	Al	7	8933	385	398

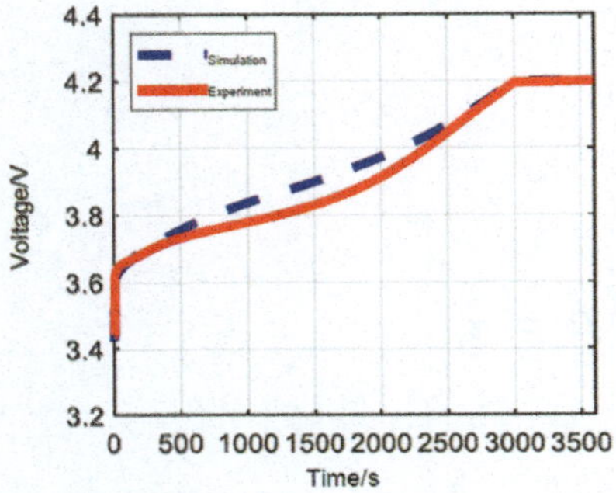

a) Voltage curve when the battery is charged at 1C rate

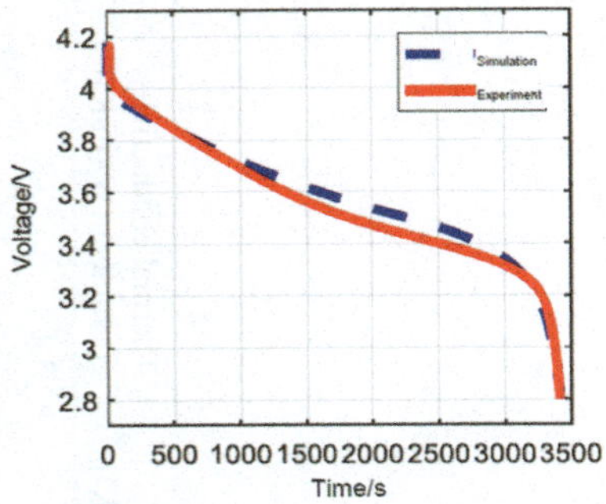

b) Voltage curve when the battery is discharged at 1.5C rate

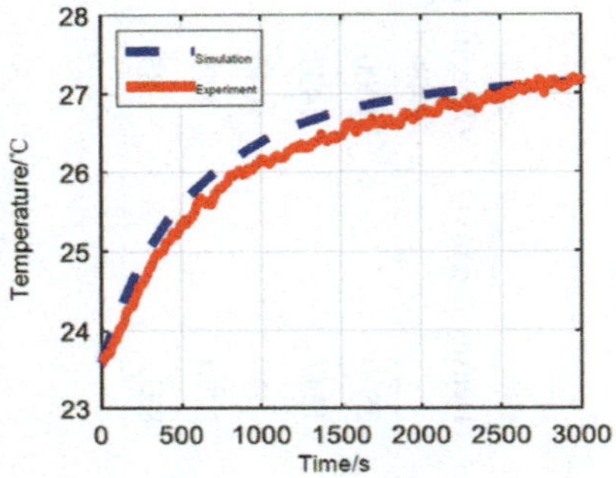

c) Temperature rise curve when the battery is discharged at 1C rate

Fig. 7.32 Battery voltage versus temperature curve at 1C rate charge/discharge

Table 7.7 Results of the accuracy evaluation of model

Parameters	1C rate charge voltage/V	1C rate discharge voltage/V	1C rate discharge temperature rise/°C
Maximum error	0.1607	0.1658	0.3672
Average error	0.0360	0.0400	0.2054
RMSE	0.0233	0.0225	0.0955

In particular, Fig. 7.32a and b allow for the validation of the one-dimensional electrochemical model. As can be observed from the figure, in the charging curve, the model simulation voltage curve and the experimentally measured voltage curve both show a vertical upward surge with a high degree of overlap in the initial phase of constant current charging. The slopes of both the voltage curves are initially small and then gradually increase during the phase when the charging voltage gradually rises. The slope of the simulation curve has a smaller rate of change than the experimental curve, but its voltage value is slightly larger than the experimental value. The time elapsed between the constant current charging phases of these two curves is essentially the same. In the constant voltage charging phase, the two curves are in good agreement. In the discharge curves, the plunging parts of these two voltage curves overlap well at the beginning and end of the discharge phase. The slope of both voltage curves is initially larger and then decreases during the gradual decrease of the discharge voltage. Again the slope of the simulation curve has a slightly smaller rate of change than the experimental curve, and its voltage value is slightly larger than the experimental value.

Overall, during the process of 1C rate charge and discharge, the model simulation curve of the battery terminal voltage is in general consistent with the experimental measurement curve. Moreover, the model prediction voltage is slightly higher than the experimental value, with a maximum error of 0.17 V, an average error of 0.04 V and an RMSE of less than 0.03, which is a satisfactory accuracy. Therefore, the developed one-dimensional electrochemical model better simulates the actual voltage change pattern of the battery during charge and discharge.

As shown in Fig. 7.32c, the temperature rise curve of the battery at 1C rate discharge allows the 3D thermal model to be validated. As can be seen from the figure, the model simulation curve overlaps well with the experimentally measured curve for the first half of the battery discharge time, and the two curves rise at a large temperature rise rate. The slopes of the two curves differ somewhat between 900 and 2000s. The experimental curve rises at a slightly smaller rate than the first half of the curve, while the simulation curve rises at a larger slope to a higher temperature and then at a slightly slower rate than the experimental curve. The slow-down of the temperature rise in both curves is mainly due to the fact that the resistance of the battery becomes lower as the temperature rises, which in turn makes the heat generation rate lower and the convective heat exchange with the environment higher.

Overall, the model simulation curve of the battery temperature during 1C rate discharge follows roughly the same pattern as the experimental measurement curve. Moreover, the calculated temperature of the model is slightly higher than the experimental value, with a maximum error of 0.37 °C, an average error of 0.21 °C and an RMSE of less than 0.1, which is within an acceptable range. As a result, the 3D thermal model developed more accurately describes the actual temperature rise of the battery.

In summary, Figs. 7.32 validate the one-dimensional electrochemical model and the three-dimensional thermal model that make up the electrochemical-thermal coupling model, respectively. The electrochemical-thermal coupling model provides an accurate picture of the variation of the battery terminal voltage and temperature, thus providing a basis for the subsequent simulations under sinusoidal alternating current heating.

7.7.3 *Simulation of the Temperature Field of a Sinusoidal Alternating Current Heated Battery*

1. Simulation of heating at constant amplitude and frequency

Firstly, a sinusoidal alternating current heating with constant amplitude and frequency is simulated. The ambient temperature is set to −20 °C and the initial temperature of the battery is also set to −20 °C. In the case of the 2C/1500 Hz sinusoidal alternating current heating in Sect. 7.3.4, the simulation time is 1083 s, the same as the experimental time.

The temperature rise obtained from the finite element simulation is shown in Fig. 7.33a. The MATLAB simulation results of Fig. 7.21b are also plotted against the experimental results in Fig. 7.33a for comparison purposes. It can be observed from the figure that the battery warms up from −20 °C to −9.22 °C in 1083 s. The temperature rise curve from the finite element simulation follows the same trend as the MATLAB simulation curve and the experimental curve with a good degree of overlap. This indicates that the electrochemical-thermal coupling model developed in this chapter is able to accurately describe the temperature change of the battery during the heating process of sinusoidal alternating current. The variation of the battery heat generation rate is shown in Fig. 7.33b. The heat generation rate gradually decreases from 8.4 × 104 W/m3 at the beginning of heating to 7.7 × 104 W/m3 at the end of heating, due to the fact that the current amplitude remains constant during the heating process, but the real part of the battery impedance gradually decreases as the temperature increases.

The distribution of the temperature field inside the battery during the heating process is shown in Fig. 7.34. Figure 7.34 shows the temperature field distribution for simulation times of 0 s, 200 s, 400 s, 600 s, 800 s and 1083 s, where the temperature units are K. As can be seen from the figure, during the sinusoidal alternating current

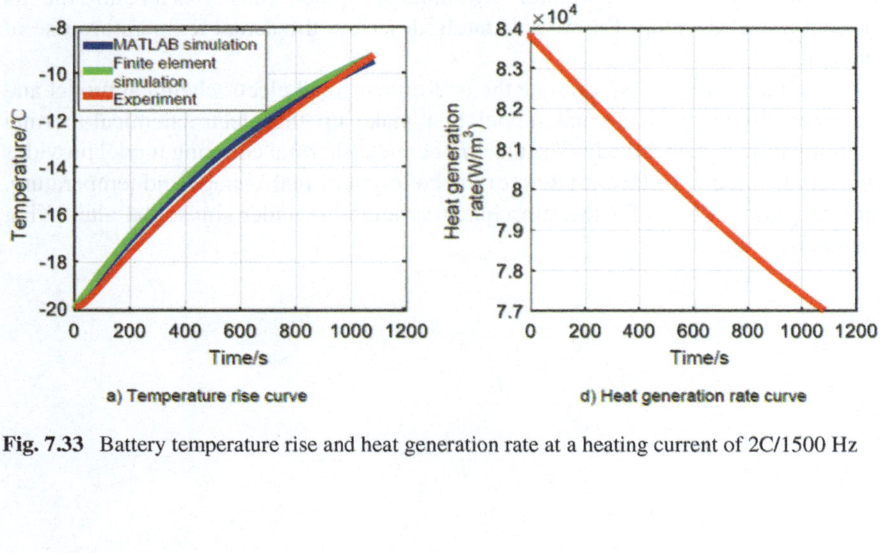

Fig. 7.33 Battery temperature rise and heat generation rate at a heating current of 2C/1500 Hz

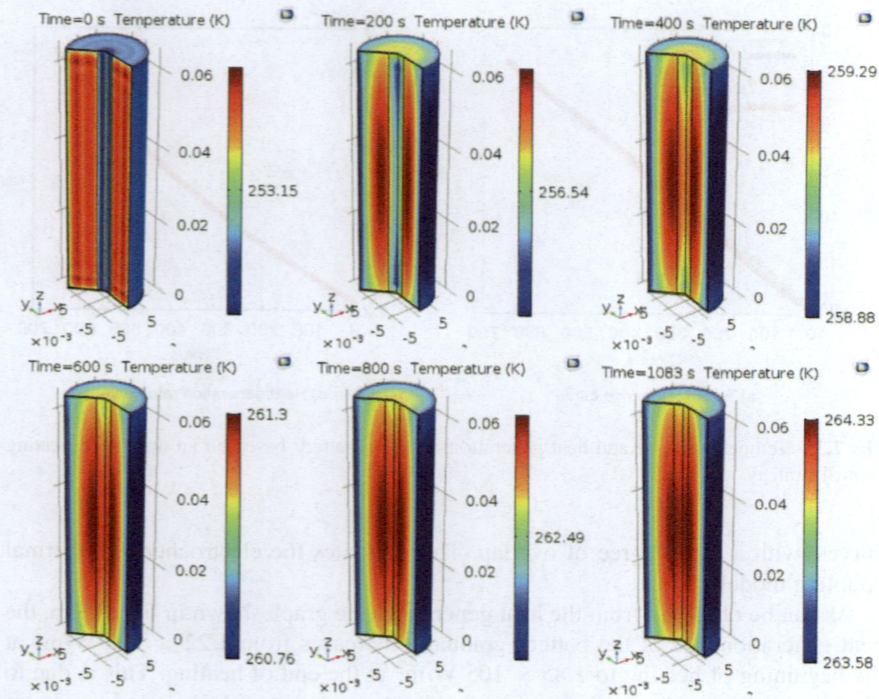

Fig. 7.34 Temperature field distribution of the battery at a heating current of 2C/1500 Hz

heating process, the temperature of the battery gradually increases from the inside to the outside; the temperature of the centre of the battery active material is the highest and the temperature of the surface case is the lowest, with a maximum temperature difference of less than 1 K. Moreover, the temperature field inside the battery is more evenly distributed, indicating that the sinusoidal alternating current heating method can achieve uniform heating of the battery.

2. Simulation based on an optimized heating control strategy

The low temperature heating of the battery is simulated according to the optimized temperature-adaptive heating control strategy proposed in Sect. 7.6.1. The other settings are the same as when an alternating current of constant amplitude and frequency is used. In the case of the battery SOC = 20%, the simulation time is 654 s, the same as the experimental time.

The temperature rise curve of the battery obtained from the simulation is shown in Fig. 7.35a. The MATLAB simulation results for Fig. 7.37a are also plotted in Fig. 7.35a along with the experimental results. As can be seen from the figure, the battery warms up from −20 °C to 6.75 °C in 654 s, and the temperature rise curve from the finite element simulation follows roughly the same pattern as the other two

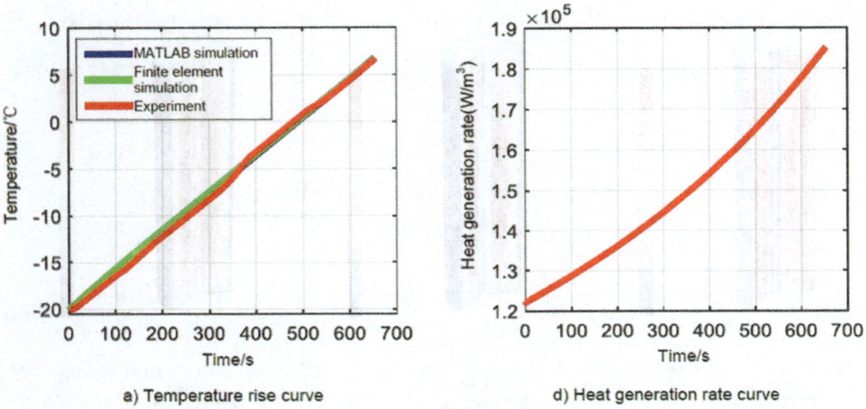

Fig. 7.35 Temperature rise and heat generation rate of a battery based on an optimized heating control strategy

curves, with a high degree of overlap. This validates the electrochemical-thermal coupling model.

As can be observed from the heat generation rate graph shown in Fig. 7.35b, the heat generation rate of the battery gradually increases from 1.22 × 105 W/m^3 at the beginning of heating to 1.85 × 105 W/m^3 at the end of heating. This is due to the fact that during the heating process, while the real part of the battery impedance gradually decreases with increasing temperature, the current amplitude gradually increases, and the current amplitude has a greater influence on the heat generation rate than the real part of the impedance. In comparison with Fig. 7.33b, this optimized temperature-adaptive heating control strategy keeps the heat generation rate of the battery at a large value throughout the heating process, and this rate increases as the battery temperature increases.

The distribution of the temperature field for simulation times of 100 s, 200 s, 300 s, 400 s, 500 s and 654 s is shown in Fig. 7.36. Likewise, during the heating process, the temperature of the battery gradually increases from the inside to the outside; the temperature of the centre of the battery active material is the highest and the temperature of the surface case is the lowest, with a maximum temperature difference of less than 1 K. Moreover, the temperature field inside the battery is evenly distributed. This shows that the sinusoidal alternating current heating method can achieve uniform and rapid heating of the battery.

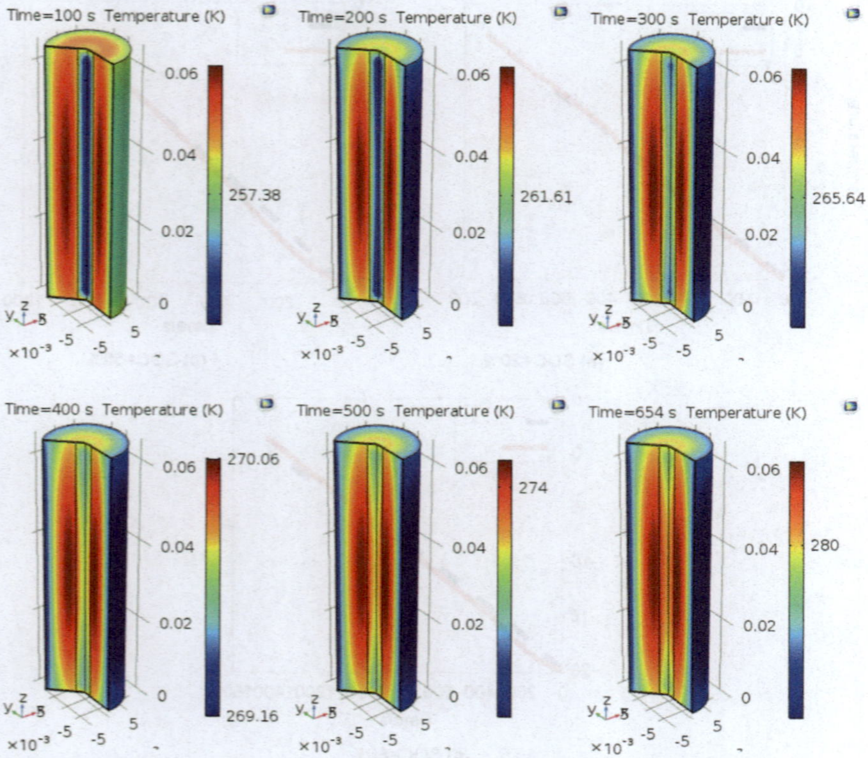

Fig. 7.36 Temperature field distribution of a battery based on an optimized heating control strategy

7.7.4 Experimental Verification Based on an Optimized Heating Control Strategy

After the optimized temperature-adaptive control strategy has been obtained from the simulation, the optimized strategy needs to be verified experimentally with sinusoidal alternating current heating. In the experiment, the temperature of the thermostat is set to −20 °C and the 3 battery modules are placed in the thermostat for 5 h so that the internal temperature of the battery is the same as the ambient temperature. The parameters of the bipolar power supply are set to sinusoidal alternating current output mode. After setting the initial amplitude and frequency of the sinusoidal alternating current, the battery module is heated by pressing the output key. Furthermore, the changes in surface temperature and voltage of the battery cells are observed and collected using a data acquisition instrument.

For module 1 with SOC = 20%, the initial amplitude of the sinusoidal alternating current is set to 16.6A (2.57C) and the initial frequency is set to 1602.5 Hz; for module 2 with SOC = 50%, the initial amplitude of the sinusoidal alternating current is set

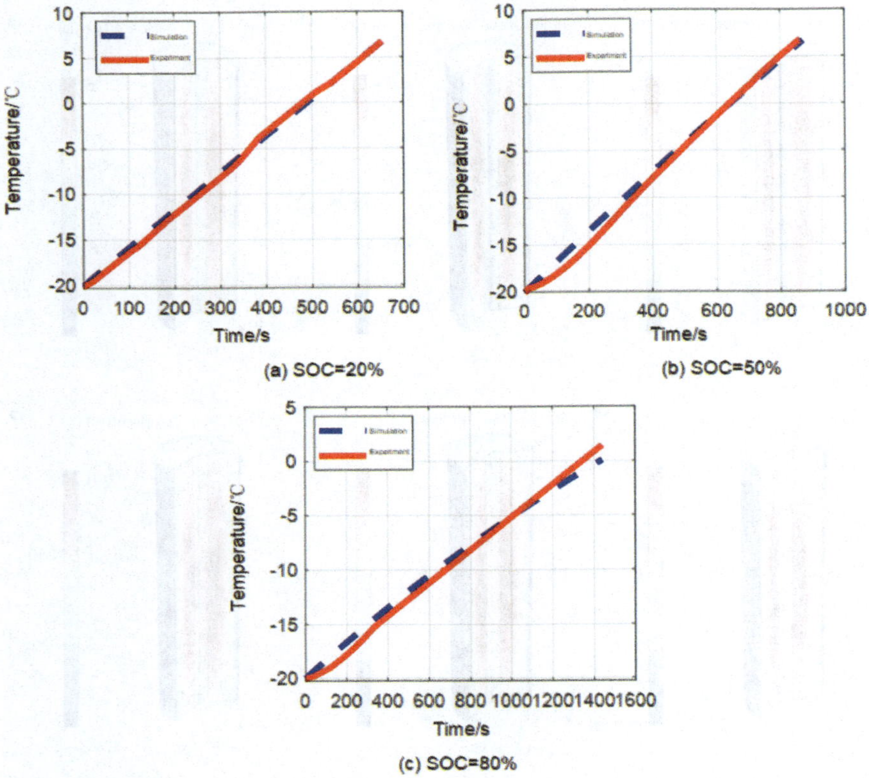

Fig. 7.37 Battery temperature rise curve based on temperature-adaptive heating strategy

to 14.2A (2.20C) and the initial frequency is set to 1602.5 Hz; for module 3 with SOC = 80%, the initial amplitude of the sinusoidal alternating current is set to 10.7A (1.66C) and the initial frequency is set to 1602.5 Hz. During the heating process, the temperature rise curve for each battery is obtained according to the amplitude and frequency of the sinusoidal alternating current output from the temperature adaptive heating control strategy shown in Fig. 7.29, as shown in Fig. 7.37.

In general, at the beginning of the heating period, the battery temperature rises relatively slowly. There are two reasons for this. On the one hand, the amplitude of the sinusoidal alternating current at this point is relatively small; on the other hand, due to the long resting time of the battery at −20 °C, the lithium ions inside the battery move slowly and the resistance to both mass transfer and diffusion movement is higher, leading to the slow electrode reaction. Then, at approximately −15 °C, although the battery impedance decreases with increasing temperature, the current amplitude has increased and the electrochemical activity within the cell has increased. This results in a significant increase in the temperature rise of the battery, which is reflected in a gradual increase in the slope of the experimental temperature rise curve.

Table 7.8 Evaluation results based on the prediction accuracy of the optimized model

Evaluation parameter	SOC = 20%	SOC = 50%	SOC = 80%
Maximum temperature difference/°C	0.6927	1.5933	1.3524
Average temperature difference/°C	0.3848	0.6654	0.6429
RMSE	0.1887	0.5071	0.3304

As shown in Fig. 7.37a, when the battery SOC = 20%, it takes only 480 s (8 min) to warm up the battery from −20.03 °C to 0 °C, with an average temperature rise rate of 2.50 °C/min. Furthermore, it takes 613 s (10.2 min) to warm up the battery to 5 °C. The results predicted by the model are almost identical to the experimentally measured temperature rise curve. The maximum temperature difference between the simulation and experimental results is 0.6927 °C, with the mean temperature difference of 0.3848 °C and the root mean square error (RMSE) of only 0.1887 (see Tables 7.8), which are satisfactory.

As shown in Fig. 7.37b, when the battery SOC = 50%, it takes only 640 s (10.7 min) to warm up the battery from -20.06°C to 0 °C, with an average temperature rise rate of 1.87°C/min; furthermore, it takes 800 s (13.3 min) to warm up the battery to 5 °C. The battery temperature curve predicted by the model largely overlaps with the actual surface temperature curve. The maximum temperature difference between the simulation and experimental result is 1.5933°C, with the mean temperature difference of 0.6654°C and the root mean square error (RMSE) of only 0.5071, which are satisfactory.

As shown in Fig. 7.37c, when the battery SOC = 80%, it takes 1343 s (22.38 min) to warm up the battery from −20.04 °C to 0 °C. The average temperature rise rate is 0.90 °C/min, which is much slower than that when SOC = 20% and SOC = 50%. The model simulation results are in good agreement with the experimental measurements. The maximum temperature difference between the simulation and experimental result is 1.3524°C, with the mean temperature difference of 0.6429 °C and the root mean square error (RMSE) of only 0.3304, which are within reasonable limits.

In summary, the experimental results fully demonstrate the correctness and feasibility of the proposed optimized temperature-adaptive heating control strategy. The error between the simulation and experimental results is mainly caused by the fitting error of the battery impedance in the model predictions.

This temperature-adaptive heating control strategy allows the heat generation rate of the battery to be maintained at a relatively large value throughout the sinusoidal alternating current heating process and to meet the safe operating voltage conditions of the battery at all times. As the battery temperature rises, the heat generation rate is further increased to achieve efficient and rapid heating. Furthermore, since the optimal frequency of this proposed optimized heating control strategy is distributed in the medium to high frequency region, as shown by the analysis in Sect. 7.5, no lithium ion deposition occurs and the optimized maximum permissible current amplitude is somewhat larger than that in the low frequency region. Therefore, a better heat generation effect can be obtained. The proposed optimized temperature-adaptive sinusoidal alternating current heating control strategy is therefore safe, efficient and feasible.

7.8 Implementation of a Sinusoidal Alternating Current Heating Scheme

7.8.1 Self-Heating System Schemes for Motor Vehicles

In the previous subsections, the sinusoidal alternating current heating method has been investigated in terms of modeling, experimentation and optimization of the control strategy. In this subsection, the application scheme of this heating method to electric vehicles will be investigated to demonstrate the practical value of this heating method.

When designing a sinusoidal alternating current heating system suitable for lithium-ion battery packs in electric vehicles, the source of the sinusoidal alternating current must be taken into account. In the experimental study in the previous section, the KIKUSUI PBZ20-40 bipolar power supply is used to provide sinusoidal alternating current to heat the battery module at low temperatures. Although it is possible to generate sinusoidal alternating current in an electric vehicle by configuring a bipolar power supply, this method requires an external add-on device, which is costly, complex and not conducive to widespread use.

This section presents a self-heating method for lithium-ion battery systems for electric vehicles. To be specific, the bridge arm in the motor controller and the motor inductor are used to generate sinusoidal alternating current for self-heating, without the need for an additional external power supply, which therefore helps to reduce the cost significantly. Furthermore, by adjusting the amplitude and frequency of the sinusoidal alternating current, fast and efficient heating of the lithium-ion battery pack at low temperatures can be achieved.

A model is used as an example to design a sinusoidal alternating current heating scheme for a complete vehicle. The parameters of the selected model are shown in Table 7.9.

The structure of the electric vehicle with self-heating system in this solution is shown in Fig. 7.38, which mainly consists of a lithium-ion battery system, an electric motor, a motor controller, a power transmission mechanism and a drive wheel. The motor controller is connected in parallel to the main positive bus MPL and the main negative bus MNL and is connected to the motor. The motor controller converts the drive power (DC power) provided by the lithium-ion battery system into AC power to drive the motor. On the other hand, the motor controller converts the regenerative power (AC power) generated in motor generation mode into DC power for delivery to the lithium-ion battery system. The inverter circuit in a motor controller is a three-phase bridge circuit consisting of power switching tubes, such as IGBTs,

Table 7.9 Basic powertrain parameters for a model

Parameters	Value
Rated voltage of battery	400 V
Rated capacity of battery	75 kW·h
Max. range	469 km
Total motor power	386 kW
Total motor torque	525 N·m
Charge time of battery	4.5 h for fast charging/10.5 h for slow charging
Acceleration time from 0 to 100 km/h	4.4 s
Max. speed	225 km/h

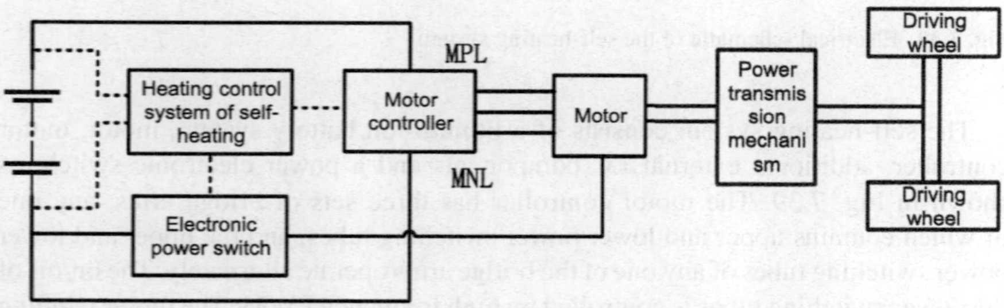

Fig. 7.38 Structure of electric vehicle with self-heating system

MOSFETs, etc. By controlling the conduction and switch-off of the individual power switch tubes, the DC power supplied by the lithium-ion battery system is converted into three-phase AC power to drive the motor. The drive from the electric motor is transmitted to the drive wheels via the power transmission mechanism to drive the vehicle.

The self-heating system uses a voltage sensor to collect the voltage signal of the lithium-ion battery system, a current sensor to collect the input and output current signal of the lithium-ion battery system and a temperature sensor to collect the temperature signal of the lithium-ion battery system. At the same time, the lithium-ion battery system can solve for the state of charge (SOC) of the battery system based on these collected voltage, current and temperature information, and then pass the resulting SOC results to the self-heating control system. The self-heating control system can be a battery management system. Based on these information collected, it can be determined whether sinusoidal alternating current self-heating is required. The self-heating circuit is switched on or off by controlling a power electronic switch. Then, the amplitude and frequency of the sinusoidal alternating current are determined in accordance with the framework of the heating strategy in this chapter.

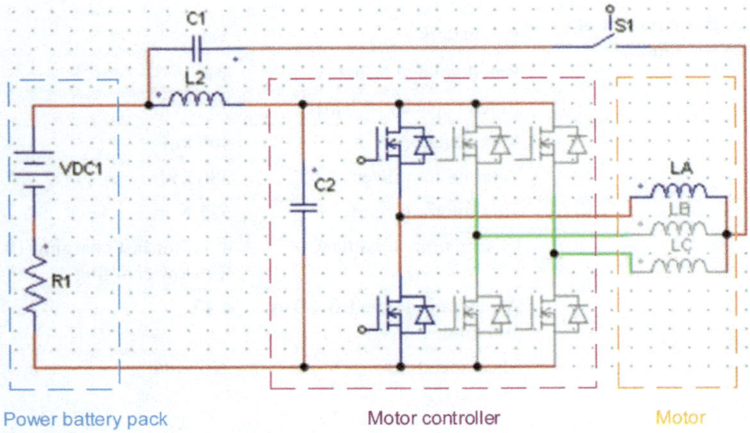

Fig. 7.39 Electrical schematic of the self-heating system

The self-heating system consists of a lithium-ion battery system, motor, motor controller, additional external LC components and a power electronic switch, as shown in Fig. 7.39. The motor controller has three sets of bridge arms, any one of which contains upper and lower power switching tubes, and the upper and lower power switching tubes of any one of the bridge arms operate alternately. The on/off of the power switching tubes is controlled by high frequency PWM. The motor winding has a three-phase resistive inductive load. The self-heating system allows the motor controller to select any one of the three sets of bridge arms and any one of the three phase windings of the motor to form the self-heating circuit. The self-heating control system only controls the power electronic switch to close when the cryogenic lithium-ion battery system needs to be self-heated in the shutdown condition.

7.8.2 Battery Pack Parameter Matching

As can be seen from Table 7.9, the 18,650 ternary lithium-ion battery pack used in this chapter has a DC voltage rating of 400 V and a capacity of 75kWh. The battery cell has a capacity of 2.15Ah and an operating voltage range of 2.8 to 4.2 V. The number of battery cells in series required to form the battery pack of a complete vehicle is:

$$N_S = \frac{400\text{V}}{4.2\text{V}} = 95.2 \approx 96 \tag{7.49}$$

The number of battery cells in parallel required to form the battery pack of a complete vehicle is:

$$N_P = \frac{75\text{kW} \cdot \text{h}}{400\text{V} \times 2.15\text{A} \cdot \text{h}} \approx 87.2 \approx 87 \tag{7.50}$$

The battery system is therefore arranged as 87P96S. In other words, every 87 cells are connected in parallel to form a group, 6 groups are connected in series to form a module, and then 16 modules are connected in series to form the battery pack for the whole vehicle, containing a total of 8352 cells.

To simplify the calculations, the consistency differences between battery cells are ignored, i.e. they are treated as if the battery cell parameters were identical. The preceding section uses a small battery module consisting of 3 battery cells connected in parallel. Therefore, the real part of the total impedance of the battery panel of the complete vehicle is calculated as follows:

$$Z_{\text{Res}} = Z_{\text{Re}} \frac{N_S}{N_P} = 0.037\Omega \times \frac{96}{87/3} = 0.122\Omega \tag{7.51}$$

The optimized current amplitude is:

$$I_S = \frac{\Delta U_S}{|Z_S|} = \frac{96\Delta U}{\frac{96}{87/3}|Z|} = 29I \tag{7.52}$$

For example, with SOC = 20% and a temperature of −17.3 °C, an optimized current amplitude of $I = 16.9$ A and a frequency of $f = 1500$ Hz can be achieved according to the temperature-adaptive heating control strategy. The optimized current amplitude for the complete vehicle system is therefore $I_S = 29I = 29 \times 16.9 = 490.1$, with a frequency of $f = 1500$ Hz.

7.8.3 Design and Simulation of a Self-Heating System Circuit

1. Determination of circuit parameters

The circuit simulation schematic of this self-heating system is shown in Fig. 7.40, where VDC1 is the lithium-ion battery pack voltage, which is 400 V; R1 is the total equivalent impedance of the lithium-ion battery pack, which is 0.122Ω; MOS1 and MOS2 are the two power switching tube MOSFETs of either set of bridge arms in the motor controller; C2 is the capacitor connected in parallel to the front of the MOSFET bridge arm in the motor controller, which is 600μF; L1 is the inductance of any one of the three phase windings of the motor, which is 3000μH; C1 is the external capacitor, which is 3.75μF; L2 is the external inductor, which is 1876.3μH. Ammeter I1 is used to measure the current applied to the terminals of the lithium-ion pack. Voltmeter V3 is used to measure the voltage across capacitor C2. The SPWM control technique is used to control the conduction and switch-off of the two power switching tube MOSFETs. VAC is a sinusoidal alternating current voltage source, which is used as the modulating wave signal and whose voltage value is measured

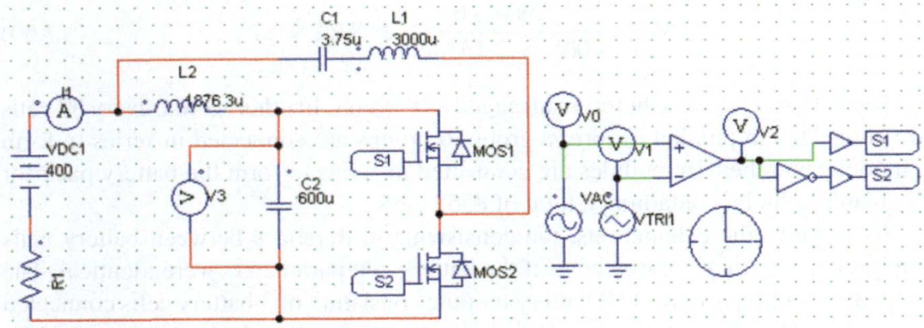

Fig. 7.40 Simulation circuit for a self-heating system

by voltmeter V0. VTRI1 is a triangular wave AC voltage source, which is used as the carrier signal and whose voltage value is measured by voltmeter V1. Voltmeter V2 is used to measure the voltage signal of the SPWM wave output from the modulating wave and the carrier wave via the comparator.

For the L_1 and C_1 branches, when $X_L = X_C$ or $2\pi f L = \frac{1}{2\pi f C}$, then

$$\varphi = \arctan \frac{X_L - X_C}{R} = 0 \tag{7.53}$$

This means that the voltage u is in phase with the current i. At this point the circuit resonates at a resonant frequency of:

$$f = f_0 = \frac{1}{2\pi\sqrt{LC}} \tag{7.54}$$

That is, resonance occurs when Eq. (7.55) is satisfied between the supply frequency f and the circuit parameters L and C.

The parameters of the external LC element can be solved for by using the series resonance to select the signal and suppress interference. Taking the case of SOC = 20% and a temperature of −17.3 °C as an example, the current frequency of the lithium-ion battery pack system is optimized to 1500 Hz and let $f_{01} = 1500$ Hz. As the motor inductance is a constant value of approximately 3000 μH, the external capacitor C1 = 3.75 μF can be derived from Eq. (7.55).

The resonant frequencies of the L2 and C2 branches should be much lower than those of the L1 and C1 branches in order to give them sufficient frequency selectivity. For example, if $f_{02} = 150$ Hz is taken, and since the capacitor C2, which is connected in parallel to the front of the MOSFET bridge arm in the motor controller, is a constant value of approximately 600μF, the external inductor L2 = 1876.3μH can be found according to Eq. (7.55).

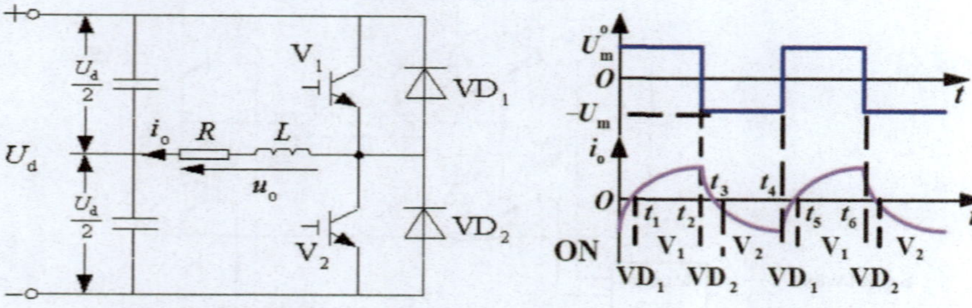

a) Inverter circuit b) Waveform

Fig. 7.41 Inverter circuit and waveform

2. Operating principle of self-heating system

The inverter circuit of this solution is shown in Fig. 7.41a, with both bridge arms containing power switching tubes and anti-parallel diodes. Two large capacitors are connected in parallel on the DC side, the connection point of which is the neutral point of the DC supply, with a resistive load connected between the connection points of the two bridge arms.

The corresponding operating waveforms for signals that are complementary to V_1 and V_2 and have semi-circle positive and negative bias respectively are shown in Fig. 7.41b. The output voltage uo is a rectangular wave with amplitude $U_m = U_d//2$ and the output current io waveform varies depending on the load. V_1 is set to be on and V_2 to be off at t_2. When the off and on signals are sent to V_1 and V_2 respectively at t_2, V_1 turns off. However, as the load is inductive, its current i_0 cannot change direction immediately, so VD_2 continues to conduct. At t_3, when i_0 decreases to 0, VD_2 cuts out, V_2 turns on and i_0 reverses. At t_4, V2 is switched off and V_1 is switched on, V_2 is switched off and VD_1 is switched on. At t_5, V1 is switched on. The conduction of the individual components during operation is shown in the lower part of Fig. 7.41b.

When V_1 or V_2 is switched on, energy is transferred from the DC side to the load. In this case, the current in the load is in the same direction as the voltage. When VD_1 or VD_2 is switched on, the inductive load feeds reactive energy to the DC side. At this moment, the current in the load is in the opposite direction to the voltage and this feedback energy is first stored in the DC side capacitor.

The PWM control is based on the principle of area equivalence. This paper uses the modulation method to generate SPWM waveforms. That is, the modulating signal is a sine wave and the carrier signal is a triangle wave, and the desired SPWM waveform is obtained by means of modulation.

The PWM inverter circuit is shown in Fig. 7.42. The waveform of the bipolar control is shown in Fig. 7.43. V_1 and V_2, V_3 and V_4 have complementary on–off states. The modulation wave ur is a sinusoidal signal. Within half a period of ur, both the triangular carrier wave uc and the PWM wave have positive and negative levels. Within one cycle of ur, the PWM waveform has ± Ud levels. The on/off of

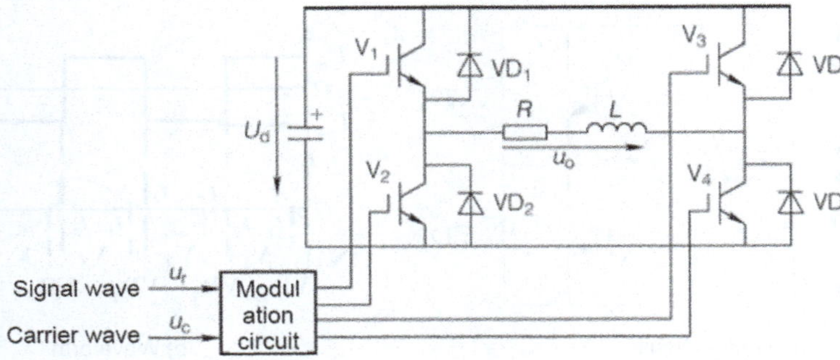

Fig. 7.42 Single-phase bridge PWM inverter circuit

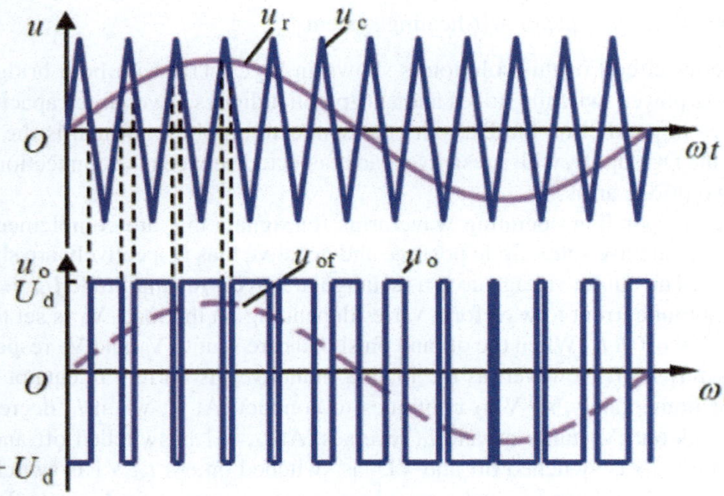

Fig. 7.43 Bipolar PWM control mode waveform

the switching tubes are controlled at the intersection of ur and uc. When ur > uc, V_1 and V_4 are on and V_2 and V_3 are off. If io > 0, V_1 and V_4 are on; if io < 0, VD_1 and VD_4 are on, and uo = Ud in the above two cases. When ur < uc, V_2 and V_3 are on and V_1 and V_4 are off. If *io* < 0, V_2 and V_3 are on; if io > 0, VD_2 and VD_3 are on, and uo = −Ud (Zhaoan and Jinjin 2009) in the above two cases. This allows the SPWM waveform uo to be output.

The equivalent circuit of this self-heating system is shown in Fig. 7.44. In the analysis of the AC circuit, the lithium-ion battery pack in the DC section can be equated to an impedance R1. In other words, L2 is connected in series with C2 and then in parallel with R1, and later in series with the branches of C1 and L1.

u_r u_c $\pm U_d$ $u_r > u_c$ $i_O > 0$ $i_O < 0$ $u_O{=}U_d$ $u_r < u_c$ $i_O < 0$ $i_O > 0$ $u_O{=}-U_d$ u_O

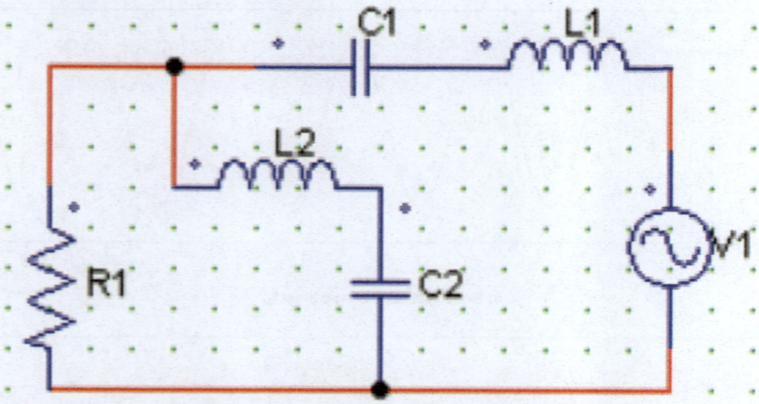

Fig. 7.44 Equivalent circuit of a self-heating system

The total impedance is as follows:

$$Z = \frac{R_1(\omega^2 L_2 C_2 + 1)}{R_1 \omega C_2 + \omega^2 L_2 C_2 + 1} + \omega L_1 + \frac{1}{\omega C_1} \tag{7.55}$$

3. Simulation analysis of self-heating system

The individual outputs from the simulation of the self-heating circuit shown in Fig. 7.40 using circuit simulation software are shown in Fig. 7.45.

The waveform of the sinusoidal alternating current I1 applied to both ends of the lithium-ion pack, which is obtained by the inverter circuit, is shown in Fig. 7.45a. As can be seen from the figure, the output current waveform is basically sinusoidal in shape, with a small degree of distortion within an acceptable range; moreover, the amplitude and frequency of the sinusoidal alternating current are within the optimized range, which indicates that the self-heating scheme is feasible and effective. As shown in Fig. 7.45b, the voltage U3 across capacitor C2, which is connected in parallel with the inverter bridge arm, remains at around 400 V with small fluctuations. The effect of this solution on the DC side voltage is therefore negligible. Figure 7.45c shows the sine waveform of the modulating waveform V0, the triangular waveform of the carrier V1 and the modulated SPWM waveform, which is consistent with the theory relating to SPWM waveform modulation described above.

Considering that the total impedance of the battery decreases with increasing temperature during heating, this has an effect on the output current amplitude. To investigate the extent to which the total impedance affects the current amplitude, a constant modulation ratio (ratio of modulating wave amplitude to carrier wave amplitude) is used throughout the heating process, i.e. an initial modulation ratio of i = 0.3215 at −20 °C. The current amplitude output from the model at different temperatures is simulated. The results are collated in the first three columns of Table 7.10. According to the temperature adaptive heating strategy proposed in Sect. 7.3.4,

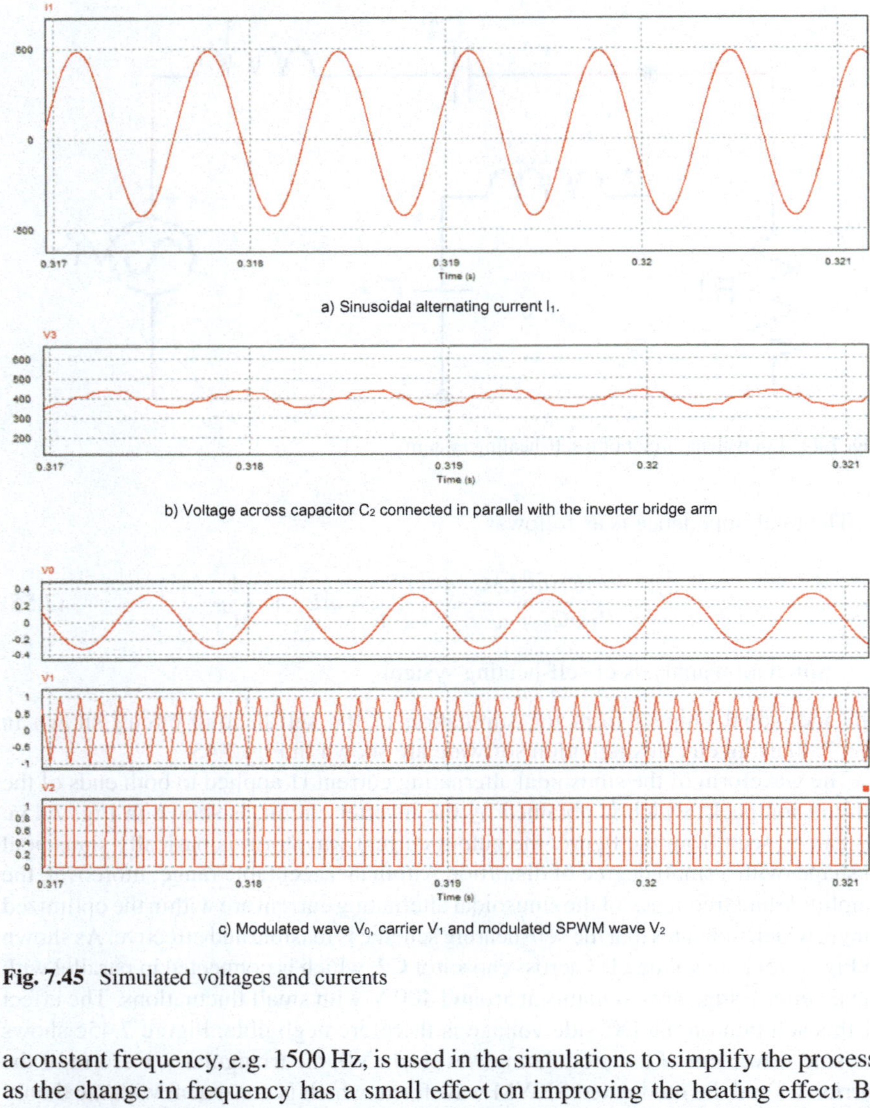

a) Sinusoidal alternating current I_1.

b) Voltage across capacitor C_2 connected in parallel with the inverter bridge arm

c) Modulated wave V_0, carrier V_1 and modulated SPWM wave V_2

Fig. 7.45 Simulated voltages and currents

a constant frequency, e.g. 1500 Hz, is used in the simulations to simplify the process, as the change in frequency has a small effect on improving the heating effect. By doing so, the optimized current amplitude is output by varying the modulation ratio and the results are shown in Table 7.10.

As can be seen from Table 7.10, when a constant initial modulation ratio is used, the total battery impedance decreases with increasing temperature, which results in a gradual increase in the output current amplitude with increasing temperature. However, the increase in current amplitude is less than that when the modulation ratio is updated. This suggests that increasing the current amplitude by relying solely on the change in total impedance with temperature helps to gradually increase the heat generation rate, but no optimum results can be achieved. Therefore, the model can be

Table 7.10 Effect of total impedance and updated modulation ratio on current amplitude at different temperatures

Temperature T/°C	Total impedance of battery Z_{Re} /Ω	Current of initialImagnitude I1/A	Updated modulation ratio i	Optimized current amplitude I2/A
−20	0.123	481.4	0.3215	481.4
−18	0.122	484.8	0.3230	487.2
−16	0.120	490.4	0.3242	495.9
−14	0.118	497.2	0.3244	501.7
−12	0.116	505.3	0.3246	510.4
−10	0.114	510.6	0.3249	519.1
−8	0.112	517.4	0.3257	527.8
−6	0.110	525.2	0.3286	536.5
−4	0.108	532.7	0.3320	547.1
−2	0.106	537.9	0.3330	556.8
0	0.104	544.8	0.3342	567.4

controlled according to the results in Table 7.10 when updating the modulation ratios, so that the optimum current amplitude can be output at each temperature to obtain the maximum permissible heat generation rate. The output currents at −20 °C, −12 °C and −6 °C are shown in Fig. 7.46. Furthermore, through comparing Fig. 7.45a with Fig. 7.46, it can be observed that the total impedance of the battery biases the output current to a certain extent, and the smaller the total impedance, the smaller the bias, due to the fact that the total impedance R1 is affected by the resonance of the LC branch in the dry circuit. Taking this effect into account, this paper uses the maximum value of the current amplitude at each temperature as the output current amplitude at that temperature to ensure that the battery voltage is within the range of safe operating voltages.

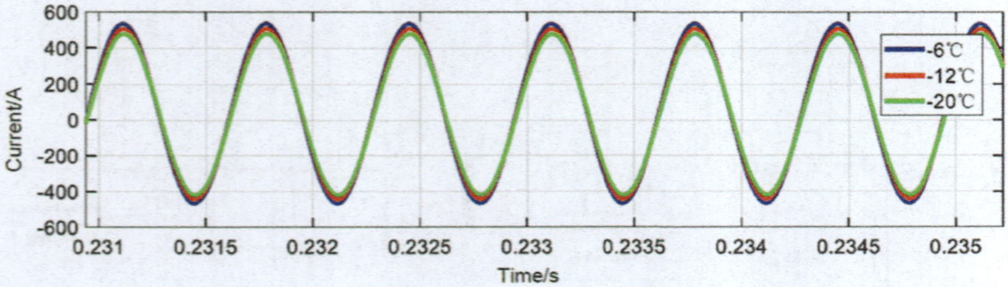

Fig. 7.46 Output currents at −20 °C, −12 °C and −6 °C

7.8.4 Battery Pack Performance Simulation Before and After Self-Heating

In order to investigate the effect of the sinusoidal alternating current heating method on enhancing the performance of lithium-ion battery packs, pulse charge and discharge capability simulations are carried out before (−20 °C) and after (0 °C) heating. Taking the above case of 20% SOC as an example, the battery pack is heated by sinusoidal alternating current at −20 °C for 480 s according to the optimized heating strategy described above, and the temperature of the battery pack rises to 0 °C after heating. In order to study the dynamic characteristics of lithium-ion battery packs in more detail, this paper improves the traditional hybrid pulse power characteristic (HPPC). Three sets of charge and discharge currents of different rates, i.e. 0.5C (93.525A), 1C (187.05A) and 1.5C (280.575A), are selected to form a compound pulse, as shown in Fig. 7.47a, to obtain the charge and discharge performance of the battery pack.

The voltage curve and power curve of the simulation output are shown in Fig. 7.47b and c. As can be seen in Fig. 7.47b, the battery pack can be discharged at a 0.5C

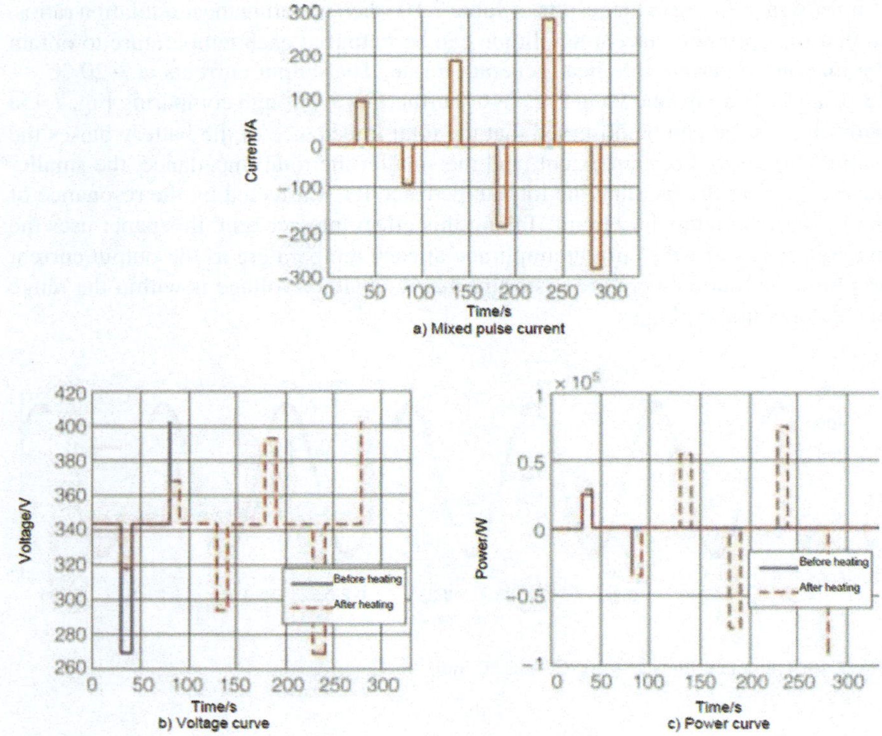

Fig. 7.47 Mixed pulse characteristic test curve at 20% SOC

rate before heating. However, when the battery is charged at 0.5C rate, the voltage of the battery pack reaches the safety limit and the charge and discharge process is terminated. After heating, the battery pack can be charged and discharged at 0.5C or 1C rate, or can be discharged at 1.5C rate. When the battery is charged at a 1.5C rate, the voltage of the battery pack reaches the safety limit and the charge and discharge process is terminated. The comparison shows that the heated battery pack can withstand significantly higher charge/discharge current rate, with significantly improved performance. As can be observed from Fig. 7.47c, before heating, the maximum discharge power that can be achieved by the battery pack is 2.5911 × 104 W and the maximum charge power is 3.3903 × 104 W; after heating, the maximum discharge power increases to 7.6025 × 104 W and the maximum charge power increases to 9.2679 × 104 W, which indicates that the sinusoidal alternating current heating method can significantly improve the performance of the battery pack.

Summary

In this chapter, an electrical-thermal coupling modeling method for the heating of a lithium-ion battery with sinusoidal alternating current (AC) is proposed from the perspective of internal heating of the battery and by using a cylindrical ternary lithium-ion battery as the research object. Furthermore, this chapter verifies the accuracy of the model through experiments and analyses the effect of AC amplitude and frequency on the heating effect. In this chapter, an optimized temperature-adaptive heating control strategy is further proposed and a safe operating voltage constraint for the battery is introduced. Furthermore, the heating strategy is optimally solved based on a sequential quadratic programming (SQP) algorithm, and its correctness is verified experimentally. Finally, a solution is designed for the application of sinusoidal alternating current heating in electric vehicles. The lithium-ion battery pack, motor, motor controller and additional LC components form the sinusoidal alternating current self-heating system, and the circuit simulation software is used to simulate the circuit of the solution and verify the effectiveness of the design. The main conclusions are as follows:

(1) The correctness of the model has been validated by the results of the battery temperature rise obtained when heating the battery with a sinusoidal alternating current at two constant amplitudes and frequencies, and the effect of current amplitude A and frequency f on the heating effect has been further explored. It was found that at the same frequency, the heating effect of the battery became better as the current amplitude increased; at the same amplitude, the heating effect of the battery became better as the current frequency decreased. Moreover, an increase in amplitude has a more pronounced effect on improving the heating effect than a decrease in frequency. However, an excessive amplitude and an extremely low frequency may cause the battery voltage to exceed the safety limit, and an extremely low frequency may cause lithium precipitation, which can damage the structure of the battery and even lead to safety accidents. Therefore, when selecting sinusoidal alternating current parameters, the limits of the battery safety voltage need to be taken into account, and medium to

high frequencies need to be chosen as far as possible to prevent lithium ion deposition.

(2) Based on the principle of electrochemical reactions in the process of sinusoidal alternating current heating of lithium ion batteries at low temperatures, an equivalent circuit model of the battery in the frequency domain is established. The resistance, capacitance and inductance in this model are functions of temperature and AC current frequency, reflecting the current–voltage response of the lithium-ion battery at low temperature. Based on the principles of energy conservation and convective heat transfer, a thermal model of the battery is developed to calculate the change in battery temperature during sinusoidal alternating current heating. The equivalent circuit model is then combined with the thermal model to develop a coupled electrical-thermal model to solve for the combined electrical-thermal performance of the battery during the AC heating process.

(3) In order to achieve fast and safe heating of the battery at low temperatures, the sinusoidal alternating current heating control strategy is studied and optimized, and a specific framework for offline simulation and online implementation of the optimized heating control strategy is established.

(4) The safe operating voltage constraint of the battery is introduced and the heating strategy is optimally solved based on the SQP algorithm. An optimized temperature-adaptive heating strategy is established through simulation. According to this strategy, as the battery temperature increases, the optimized current amplitude gradually increases, the optimized current frequency gradually decreases and the generated heat generation rate gradually increases.

(5) The correctness and feasibility of the proposed optimized heating strategy is verified by means of sinusoidal alternating current heating experiments at 20%, 50% and 80% of SOC respectively. Furthermore, when the battery SOC = 20%, the battery heats up from −20.03 to 0 °C in just 480 s (8 min), with an average temperature rise rate of 2.50 °C/min, achieving efficient, fast and safe heating.

(6) Based on the structure of an electric vehicle, a solution for applying the sinusoidal alternating current heating method to electric vehicles is proposed. In other words, the lithium-ion battery pack, motor, motor controller and additional LC components constitute a sinusoidal alternating current self-heating system. In addition, an optimized current frequency is selected based on LC resonance and the inverter circuit is controlled using SPWM control technology to generate sinusoidal alternating current, and the solution is simulated and analyzed using circuit simulation software.

References

Chen SC, Wang YY, Wan CC (2006) Thermal analysis of spirally wound lithium batteries. J Electrochem Soc 153(4):A637–A648

Fan W (2018) Study on the low temperature heating control strategy of electric vehicle power battery system based on sinusoidal alternating current. Beijing University of Technology, Beijing

Karimi G, Li X (2013) Thermal management of lithium-ion batteries for electric vehicles. Int J Energy Res 37(1):13–24

Li X (2017) SOC estimation of battery pack based on adaptive extended Kalman filter. Kunming University of Technology, Kunming

Li D, Tong X (2005) Numerical optimization algorithms and theory. Science Press, Beijing

Lai Y, He G (2007) Optimization methods and theory. Tsinghua University Press, Beijing

Li J, Barillas JK, Guenther C et al (2013) A comparative study of state of charge estimation algorithms for LiFePO4 batteries used in electric vehicles. J Power Sources 230:244–250

Somasundaram K, Birgersson E, Mujumdar AS (2012) Thermal–electrochemical model for passive thermal management of a spiral-wound lithium-ion battery. J Power Sources 203:84–96

Von Lüders C, Zinth V, Erhard SV et al (2017) Lithium plating in lithium-ion batteries investigated by voltage relaxation and in situ neutron diffraction. J Power Sources 342:17–23

Wang F, Zhang J, Wang L (2013) Design of air electric heating device for automotive power battery pack. Power Technol 7:37(7)

Zhang X (2011) Thermal analysis of a cylindrical lithium-ion battery. Electrochim Acta 56(3):1246–1255

Zhang SS, Xu K, Jow TR (2004) Electrochemical impedance study on the low temperature of Li-ion batteries. Electrochim Acta 49(7):1057–1061

Zhaoan W, Jinjin L (2009) Power electronics technology. Mechanical Industry Press, Beijing

Zhu J, Sun Z, Wei X et al (2016) An alternating current heating method for lithium-ion batteries from subzero temperatures. Int J Energy Res 40(13):1869–1883

Zinth V, von Lüders C, Hofmann M et al (2014) Lithium plating in lithium-ion batteries at sub-ambient temperatures investigated by in situ neutron diffraction. J Power Sources 271:152–159